Tobias Heinroth • Wolfgang Minker
Editors

Next Generation Intelligent Environments

Ambient Adaptive Systems

Springer

Editors
Tobias Heinroth
Institute of Information Technology
University of Ulm
Albert-Einstein-Allee 43
89081 Ulm
Germany
tobias.heinroth@uni-ulm.de

Wolfgang Minker
Institute for Information Technology
University of Ulm
Albert-Einstein-Allee 43
89081 Ulm
Germany
wolfgang.minker@uni-ulm.de

ISBN 978-1-4899-8857-7 ISBN 978-1-4614-1299-1 (eBook)
DOI 10.1007/978-1-4614-1299-1
Springer New York Dordrecht Heidelberg London

Printed on acid-free paper

Springer is part of Springer Science+Business Media (www.springer.com)

Foreword I

Recent advances in Intelligent Environments research give a glimpse into the future of our planet and reveal exciting visions of smart everything – smart cities, smart homes, smart workplaces, smart hotels, smart schools, and much more. Driven by technological evolution offering low power many-things and wireless almost-everything (e.g., IEEE 802.15.4 radio, wireless sensor networks, sensor platforms), we could, in only a decade, envision and prototype impressive cyber-physical systems and applications. In most of these systems, the goal has been clear and convincing, and the technology proved to be promising and exciting.

But prototyping is only a beginning, and much remains to be innovated and done before such Intelligent Environments (IE) become common places. Many research disciplines must collaborate among and within themselves, including domain experts (of the particular environment, e.g., gerontologist for assisted living spaces), behaviour scientists, engineers, computer scientists, to mention just a few. Collaboration within the computer science and engineering discipline is key to the success of IE. Systems support and middleware are essential foundation to building any systems – IE are no exception. Software engineering is urgently needed to understand and support the full life-cycle of IE. New programming models are also needed for developing safe and adaptive applications and services. New notions of trust must be formulated and supported to ensure symbiotic relationship between the users and their environments. Understanding human computer interaction is crucial and in some environments, affecting persuasion is of paramount importance. But this is not all. Without machine learning and computational intelligence techniques, the potential utility and "ceiling of goals" of IE would be severely limited.

"Next Generation Intelligent Environments" provides an excellent compendium of collaborative research efforts in support of *ambient ecologies* orchestrated within a mega project funded by the European Community – The Adaptive and TRusted Ambient eCOlogies (ATRACO).

The book goes beyond prototypes and addresses the fundamental generalities and necessary ecologies that can lead to better design, development, operation and adaptation of IE. The book covers many important areas of collaborative research within computer science and engineering, including system support and middleware, applied knowledge and ontology, user interactions, artificial intelligence, user experience and much more. The coverage is unique in that all chapters are interrelated and aligned, showing key dependencies and themes. One overarching theme is adaptation that appears in many chapters addressing the environment itself as well as the user and their interactions.

It is gratifying to finally see a book contributed to the IE research community that brings unprecedented depth in the treatment of such delicately interdependent topics that make up the core areas of IE. It is a timely and a much needed contribution that will help shape the curricula in emerging smart systems specialities and programs around the world.

Gainesville (USA), February 2011 *Sumi Helal*

Foreword II

Technological advances have been shaping our world for centuries. The believe that technology is beneficial to human kind has deep roots in our societies, since initial technological developments were closely linked to survival conditions like hunting and building. Nowadays technology is around us everywhere and the lives of most of humanity have become closely intertwined with more or less intensity.

Many western societies are at this point in history investing on technology that can pervade all levels of daily life to assist humans in their activities, whether this is at home, at the office, at school or shopping. The idea is that there is technology already available that can help identifying some situations where humans need help and also to deliver some of that assistance, in a more or less automated way. This has led to the idea that we can create "Intelligent Environments" which can actively pursue benefits for the humans who inhabits those environments.

This is no trivial enterprise. Technology can fail. Humans are complex beings. We live in a dynamic world and situations can change substantially in short periods of time. The engineering of such systems require the careful blending of cutting edge technology and expertise. Previous advances in Computer Science, Engineering, Architecture, Social Sciences and other areas are supporting the development of a new generation of technological developments which are aiming at helping you wherever you are. So far humans have to invest substantial effort to understand how to interact with a computer and benefit from it. This technology aims at bringing benefits to human without demaning a specific technological expertise from the human.

The advances provided by the ATRACO project are a good example of a well rounded solution which provides a dynamic middleware and adaptive networking architecture which can facilitates the creation of Intelligent Environments making available a dynamic range of services. Intelligent adaptation of the environment and flexible interaction with humans complete a technological solution which adapts to context and situations to be most useful. Another important contribution of this project is that these technological advances are designed with ethical principles in mind.

Surely this will not be the last word in the area. Humans seem to have developed an insatiable appetite for technological developments that can potentially make their lives easier. Whether you just become interested on doing serious work in this area or you want to keep yourself updated in the latest developments that help you to design the world of tomorrow this book is a good place to start to get engaged with this development which can shape our world.

Jordanstown (UK), February 2011 *Juan Carlos Augusto*

Preface

This book is based on the work that has been conducted within the ATRACO (Adaptive and TRusted Ambient eCOlogies) project[1] as part of the European Community's 7th Framework Programme (FP7/2007-2013) under grant agreement n° 216837. The aim of ATRACO project is to contribute to the realization of trusted ambient ecologies. Interactive appliances, collaborative devices, and context aware artefacts, as well as models, services, software components are parts of ambient ecologies. A context-aware artefact, appliance or device uses sensors to perceive its context of operation and applies an ontology to interpret this context. It also uses internal trust models and fuzzy decision making mechanisms to adapt its operation to changing context. Finally, it employs adaptive dialogue models to communicate its state and interact with people.

Ambient ecologies form the infrastructure that supports user activities. In ATRACO, each activity is modelled as a "bubble" using finite resources to achieve the goals of its owner and having clearly marked borders, which realize the privacy requirements. The user tasks that compose an activity are supported by an ad-hoc orchestration of ubiquitous computing services, which are manifested via an ecology of smart artefacts. The bubble adapts to different contexts by re-negotiating its borders, adopting suitable interaction modes and employing resource management models. In ATRACO, adaptation will be researched in terms of artefact operation, ecology composition, network election and man-machine interaction with respect to user context and behaviour.

The book edition consists of eight chapters each covering a detailed look on a specific scientific area within the field of Intelligent Environments. The first chapter describes a middleware architecture that has been developed for the ATRACO prototype. The second chapter deals with the networking aspects, which are crucial within the context of ambient intelligent systems. The third chapter provides a detailed insight into on the theoretical and the practical approaches to ontology-based knowledge management. Chapter 4 presents one of the most important adaptation mechanism used within ATRACO realized as advanced fuzzy mechanism. Chal-

[1] `http://www.uni-ulm.de/in/atraco`

lenges and novel approaches to adaptive user interaction between users and Intelligent Environments are discussed in Chapter 5. Planning and artificial intelligence is the main matter of the sixth chapter. Since privacy and trust issues are especially pertinent to ambient systems and computer-based systems that are used in our daily lives we have dedicated the Chapter 7 to this topic. The edition concludes with the evaluation methods and results of the social evaluation that has been conducted within the framework of the ATRACO project.

We are convinced that computer scientists, engineers, and others who work in the area of Ambient Environments, no matter if in academia or in industry, may find the edition interesting and useful to their own work. Graduate students and PhD students specialising in the area of Intelligent Environments more generally, or focusing on issues related to the specific chapters in particular, may also use this book to get a concrete idea of how far research is today in the area and of some of the major issues to consider when developing Intelligent Environments in practice. We would like to express our sincere gratitude to all those who helped us in preparing this book. Especially we would like to thank all reviewers who through their valuable comments and criticism helped improve the quality of the individual chapters as well as the entire book.

Ulm (Germany),
June 2011

Tobias Heinroth
Wolfgang Minker

Acknowledgements

The editors would like to thank Juan Carlos Augusto and Sumi Helal for their interesting forewords that allow the audience to quickly step into the book. We furthermore want to thank the reviewers for their valuable input and for all the objections and suggestions they had. A list of the reviewers can be found in the Appendix at the end of the book.

The research leading to these results has received funding from the European Community's 7th Framework Programme (FP7/2007-2013) under grant agreement n° 216837 and from the Transregional Collaborative Research Centre SFB/TRR 62 "Companion-Technology for Cognitive Technical Systems" funded by the German Research Foundation (DFG).

Contents

List of Contributors

Yacine Bellik
National Center for Scientific Research (LIMSI-CNRS) BP 133, 91403, Orsay cedex, France, e-mail: yacine.bellik@limsi.fr

Julien Bidot
Institute of Artificial Intelligence, Ulm University, Ulm, D-89069 Germany, e-mail: julien.bidot@uni-ulm.de

Susanne Biundo
Institute of Artificial Intelligence, Ulm University, Ulm, D-89069 Germany, e-mail: susanne.biundo@uni-ulm.de

Dimitris Economou
inAccess Networks S.A., Athens, Greece, e-mail: decon@inaccessnetworks.com

Christos Goumopoulos
The Hellenic Open University, Patras, Hellas, e-mail: goumop@cti.gr

Hani Hagras
University of Essex in Colchester CO4 3SQ, United Kingdom, e-mail: hani@essex.ac.uk

Tobias Heinroth
Institute of Information Technology, Ulm University, Ulm, 89081 Germany, e-mail: tobias.heinroth@uni-ulm.de

Achilles Kameas
The Hellenic Open University and DAISy research unit, Patras, Hellas, e-mail: kameas@cti.gr

Bastian Könings
Institute of Media Informatics, Ulm University, Ulm, 89081 Germany, e-mail: bastian.koenings@uni-ulm.de

Ioannis Liverezas
inAccess Networks S.A., Athens, Greece, e-mail: iliverez@inaccessnetworks.com

Apostolos Meliones
University of Piraeus, Department of Digital Systems, and inAccess Networks S.A., Athens, Greece, e-mail: meliones@unipi.gr

Wolfgang Minker
Institute of Information Technology, Ulm University, Ulm, 89081 Germany, e-mail: wolfgang.minker@uni-ulm.de

Gaëtan Pruvost
National Center for Scientific Research (LIMSI-CNRS) BP 133, 91403, Orsay cedex, France, e-mail: gaetan.pruvost@limsi.fr

Lambrini Seremeti
The Hellenic Open University and DAISy research unit, Patras, Hellas, e-mail: seremeti@cti.gr

Joy van Helvert
University of Essex in Colchester CO4 3SQ, United Kingdom, e-mail: jvanhe@essex.ac.uk

Christian Wagner
University of Essex in Colchester CO4 3SQ, United Kingdom, e-mail: chwagn@essex.ac.uk

Michael Weber
Institute of Media Informatics, Ulm University, Ulm, 89081 Germany, e-mail: michael.weber@uni-ulm.de

Björn Wiedersheim
Institute of Media Informatics, Ulm University, Ulm, 89081 Germany, e-mail: bjoern.wiedersheim@uni-ulm.de

Acronyms

AA	Artefact Adaptation
AE	Ambient Ecology
AS	Activity Sphere
FTA	Fuzzy Task Agent
IA	Interaction Agent
IE	Intelligent Environment
LO	Local Ontology
NA	Network Adaptation
OM	Ontology Manager
PA	Planning Agent
PM	Privacy Manager
SA	Sphere Adaptation
SM	Sphere Manager
SO	Sphere Ontology
TM	Trust Manager
UBA	User Behaviour Adaptation
UIA	User Interaction Adaptation
UPnP	Universal Plug and Play
UX	User Experience

Chapter 1
A Middleware Architecture for Ambient Adaptive Systems

C. Goumopoulos

Abstract Ambient adaptive systems have to use mechanisms to regulate themselves and change their structure in order to operate efficiently within dynamic ubiquitous computing environments. First of all we outline a survey on existing middleware solutions for building ambient adaptive systems. After, discussing the limitations of the existing approaches, we present our propositions for a middleware architecture to support dynamic adaptation within ambient environments. Our approach is based on the Service-Oriented Architecture (SOA) paradigm which can be considered as an evolution of the component-based design paradigm. The aim is to use component interfaces for the identification and automated connection of components acting as service providers/consumers. The proposed middleware provides a solution that supports the adaptation of applications at the structural level, where the structure of the application can change through dynamic service composition. We call this adaptation 'polymorphism' in analogy with the synonymous term found in the object-oriented programming paradigm. Besides SOA, we use a set of intelligent agents to support adaptive workflow management and task realization based on a dynamically composed ontology of the properties, services and state of the environment resources. An experimental prototype is provided in order to test the middleware developed.

1.1 Introduction

Intelligent environments (IE), like smart homes, offices and public spaces, are featured with a large number of devices and services that help users in performing efficiently various kinds of tasks. Combining existing services in pervasive computing environments to create new distributed applications can be facilitated by middleware architectures, but this should accomodate special design considerations, including

Christos Goumopoulos
Research Academic Computer Technology Institute, Patras, Greece, e-mail: goumop@cti.gr

context awareness, adaptation management, device heterogeneity, and user empowerment [6].

Traditional middleware, such as Remote Procedure Calls [4], OMG CORBA [11], Java Remote Method Invocation (RMI) [44] and Microsoft Distributed Component Object Model (DCOM) [20] facilitate the development of distributed applications and help to resolve problems such as tackling the complexity of programming inter-process communication and the need to support services across heterogeneous platforms. However, traditional middleware is limited in its ability to support adaptation.

Ambient adaptive systems which are a special category of distributed systems operate in a dynamic environment. The dynamicity of the environment may relate with evolving user requirements and varying execution context due to the diversity of available devices, user preferences and services. Consequently there is a need for both applications and infrastructure to be designed for change. The evolution of user requirements calls for system evolution. The dynamic execution environment calls for dynamic adaptation. In order to allow evolution, the internal structure of the system must be made open in order to support proactive and reactive system reconfiguration.

In this work, we present firstly a survey of the state-of-the-art on existing middleware solutions for building adaptive ambient systems. After, discussing the limitations of the existing approaches, we present our propositions for middleware architecture to support dynamic adaptation within ambient environments. Our approach uses the service-oriented architecture paradigm coupled with agents and ontologies. The aim is to use component interfaces for the identification and automated connection of components acting as service providers/consumers. The proposed middleware provides a solution that supports the adaptation of applications at the structural level, where the structure of the application can change through dynamic service binding. Behavioural adaptation, not examined here, is also possible when the application logic is changed as a result of learning. An experimental prototype is provided in order to test the middleware developed.

1.2 Related Work

Three key paradigms that can be used to build adaptive systems are computational reflection, Aspect-Oriented Programming (AOP) and service oriented architectures. Researchers have also explored the possibility to combine different paradigms such as AOP and reflection in middleware systems to increase support for the development of dynamic distributed systems [19]. In the following we examine how each one of these paradigms can support the development of adaptive systems.

1.2.1 Reflective Middleware

In principle *computational reflection* allows a program to observe and modify its own structure and behavior at runtime, by providing a self-representation that can be accessed and changed by the program [31]. More importantly, these changes must be causally reflected to the actual computations performed by the program. In particular, the architecture of reflective systems follows a kind of "white-box" approach that provides comprehensive access in the internal details of a system allowing dealing with highly dynamic environments, for which run time adaptation is required. This is conceptually contrary to the encapsulation principle in objectoriented programming followed by traditional middleware that adopt the remote object model. The reflection technique was used initially in the domain of programming languages as a means to help designing more open and extensible languages. The reflection is applied also in other domains including operating systems and distributed systems. Recently, reflection has been also applied in middleware, which needs to adapt its behavior to changing requirements when operating in a dynamic environment [27]. The dynamic modification of the middleware implementation allows for the adaptation of the behavior of distributed applications that are based on this middleware. Typically, reflective middleware provides adaptation of behavior of distributed applications in terms of non-functional requirements such as QoS, security, performance and fault tolerance.

A reflective system is organized into two levels called *base-level* and *meta-level*. The former represents the basic functionality of the system. The latter models the structural and computational aspects of the base level in order to observe or modify the behavior of the objects that exist in the base level, assuming an object-oriented system. The reflective approach supports the inspection and the adaptation of the underlying implementation (base-level) in run-time. A reflective system provides a *meta-object protocol* (MOP) in order to determine the services that are available in meta-level and their relationship to the base-level objects[26]. The meta-level can be accessed via a process called reification. Reification is the disclosure of certain hidden aspects of the internal representation of the system in terms of programming entities that can be manipulated at runtime. The "opening of" implementation offers a simple mechanism in order to interpose some behavior (e.g., add a method in an object, save the state of the object, check security issues, etc.) with a view to watch or alter the internal behavior of the system.

Reflection enables an application to adjust its behaviour based on a reflective middleware that allows inspecting and adapting its own behavior according to application's needs. Figure 1.1 shows schematically a simple example of this situation. A meta-object (`mObj`) has been defined at the meta-level and is associated through MOP to a base-level object (`Obj`) belonging to some middleware implementation. An application calls a method of the middleware (`Obj.Method()`) which is reified through MOP and a defined object with a specified reference (`ref`). This object is passed to the associated meta-object that executes a method (`mObj.mMethod(ref)`). This method executes a logic specified by the MOP interface and then passes control to the original method through a reflection method (`base_Method(ref)`). The meta-object receives the results, performs any post-processing specified by the MOP and returns to the calling application.

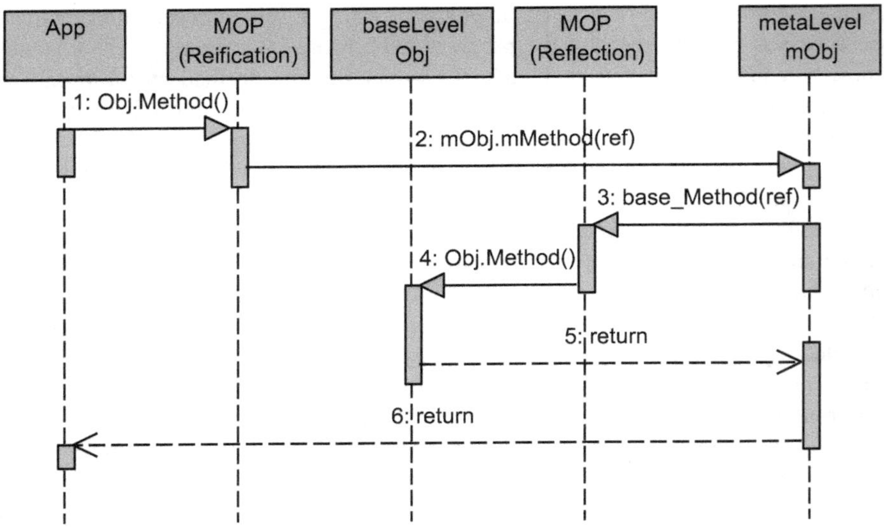

Fig. 1.1 An application calling a method in a reflective middleware.

A category of reflective systems support a higher level reflection in the sense that they can add or remove methods from the objects and classes dynamically and even change the class of an object in run-time. The practical result is to be able to restrict the size of middleware with a minimal total of operations that can run in devices with limited resources. On the contrary, other systems are focused in simpler reflective forms in order to achieve a better performance. Their reflective mechanisms are not part of the normal flow of control and are only called when needed.

Middleware systems that integrate reflection in their architecture have been developed as research prototypes. In the following we cite a few examples of reflective middleware. A number of early systems such as FlexiNet [22], OpenCorba [30], dynamicTAO [28] and OpenORB [5] were based on CORBA and targeted flexibility and dynamic reconfigurability of Object Request Broker (ORB). However, these systems suffered from the heavy computational load imposed by CORBA. Capra et al in [7] discuss CARISMA, which uses reflection to support dynamic adaptation of middleware behaviour to changes in context (e.g., adapting a streaming encoder binding in variable QoS conditions) and ReMMoC, which uses reflection to handle heterogeneity requirements imposed by both applications and underlying device platforms. Both approaches target a minimal reflective middleware for mobile devices where pluggable components can be used by developers to specialize the middleware to suit different devices and environments, thus solving heterogeneity issues. QuA middleware explores the principle of mirror-based reflection to design a reflective API according to the programming abstractions defined by a language

[14]. In the QuA middleware approach a mirror can be defined to reflect a service, in terms of middleware abstractions like type, interface, service and binding, without being dependent the running instances.

Even though reflection is a powerful mechanism to construct adaptive systems there are still issues that need to be understood and solved. The performance of reflective middleware is a matter that is open for further research. The majority of reflective systems impose a rather heavy workload that would cause significant performance deterioration in devices with limited resources and there is always a trade-off issue between performance and scope of adaptability. Another issue that needs to be addressed is dynamically tuning the scope of changes when reconfiguring the system based on adaptation semantic information.

1.2.2 Aspect Oriented Programming

Aspect-Oriented Programming (AOP) [24] is a software development paradigm that emphasizes decomposition of complex programs in terms of intervened cross-cutting aspects, such as QoS, security, persistence, fault-tolerance, logging and resource utilization. This is different from other programming paradigms which emphasize functional decomposition breaking a problem into units like procedures, objects and modules. For instance, object-oriented programming uses inheritance hierarchies to abstract commonalities among classes, however global aspects (affect many classes) are implemented in an ad-hoc manner and become tightly intermixed with classes, which makes changes to the program difficult and error prone. On the other hand, AOP supports the concept of separation of concerns to counter this problem. AOP defines the methods and tools to separate cross-cutting aspects during development time. The source program consists of modules that deal with the different aspects that are described independently. All these modules are integrated during compile (statically) or run time (dynamically) to form a global application with new behaviour using a composition tool called *aspect weaver*.

AOP combines principles of object-oriented programming and computational reflection discussed in the previous section. AOP languages have functionality similar to, but more restricted than meta-object protocols and use a few key concepts: *join points*, *point cuts*, and *advices*. A *join point* is a place, in the source code of the program, where aspect-related code can be inserted. A join point needs to be addressable and understandable by an ordinary programmer to be useful. It should also be stable across typical program changes in order for an aspect to be stable across such changes. Aspect weaving relies on the concept of *point cut*, i.e. the specification of a set of join points according to a given criterion, and *advice*, i.e. the definition of the interaction of the inserted code with the base program. An advice specifies whether the inserted code should be executed before, after, or in replacement for the operations located at the point cuts. Two Java based composition tools that implement the AOP paradigm are AspectJ [25] and JAC [34].

AOP benefits outlined above are important to adaptive middleware. Such approach enables the separation of middleware cross-cutting concerns (e.g., security, logging) at development time and later at compile or run time, where these concerns can be selectively woven into application code. Using AOP, tailored versions of middleware can be generated for application-specific domains. In the following we cite a few examples of adaptive middleware developed based on AOP principles. Yang et al in [45] proposes a systematic approach for preparing an existing program for adaptation and defining dynamic adaptations. The approach uses a static AOP weaver at compile time and reflection during run time. This basic scheme has been followed by others researchers also. Frei et al in [13] present an architecture supporting dynamic AOP that establishes an event infrastructure to extend existing application's behavior at runtime. When the application extension is activated, the dynamic AOP platform inserts an AOP aspect into the AOP platform which intercepts the application's execution and monitors its progress. Whenever the application reaches selected points in the execution, the AOP platform redirects the execution to the appropriate application extension. The memory footprint of the platform is however quite heavy (1MB) to run on resource-constrained devices. Similarly, Maciel da Costa et al in [9] discuss an adaptive middleware architecture, based on aspects, which can be used to develop adaptive mobile applications. A mail server prototype was implemented based on Web Services, Java and AspectJ technologies to evaluate the architecture regarding operation adaptation depending on resource utilization (e.g., power consumption).

AOP has advantages, such as separation of cross-cutting concerns, but presents also difficulties. Programming in terms of aspects requires much more than just identifying the different aspects of concern. It requires being able to express those aspects of concern in a way that is precise and that makes the relations among the aspects of concern precise. This is what enables the aspect weaver to work, and is also what makes possible reasoning about the code or debugging the code. Another important problem related to AOP is the composition of aspects. For instance, if different pieces of aspect-related code are inserted at the same join point, the order of insertion may be relevant if the corresponding aspects are not independent. Such issues cannot usually be settled by the weaver and call for additional specification. Finally, most AOP approaches do not support adequate point cut descriptions to capture join points based on context data and business-level semantics.

1.2.3 Service-Oriented Architecture

The Service Oriented Architecture (SOA) paradigm has been envisioned as an evolution of the component-based engineering paradigm centered on the concept of service [12]. This can be applied in the design of distributed applications that are seen as a composition of services. In addition, the service concept can be applied recursively, since a system component can provide a service, but simultaneously it can encapsulate a composition of services from its service requestors.

In an SOA environment, resources on a network are made available as independent services that can be accessed without knowledge of their underlying platform implementation. A provided service is usually embodied in a set of interfaces, each of which represents an aspect of the service. In general this set contains the operations that a service supports, and some information on how to access these operations. Service interfaces can be published in registries, which also provide services themselves (publish and discovery services), allowing the potential service requestors to discover and access these services (Figure 1.2).

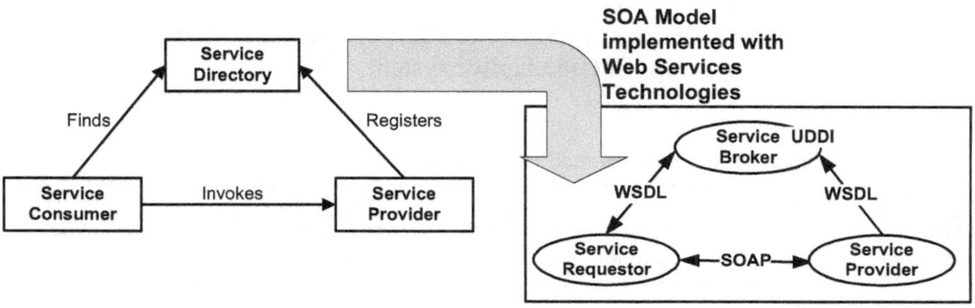

Fig. 1.2 SOA conceptual model.

The independent deployment of services enables late binding which is essential for adaptive systems. *Late binding* provides the opportunity for dynamic composition of services or for swapping two compatible services at run time through a well defined interface. In the SOA paradigm, we can view two abstraction levels of the service concept. *Elementary services* are basic functionalities, usually provided by resources (e.g., devices) in an AmI environment. *Composite services* assemble a set of functionalities in relation to user tasks, and thus are closer to user actual goals. Service composition has widely been addressed in the Web Service field. Existing composition frameworks [8] enable expressing and enacting complex service compositions. However, they rely on explicitly named services, which are not discovered dynamically. On the contrary, the Semantic Web Services (SWS) approach [32] is a step toward dynamic service discovery and composition [40], [10], where intelligent systems try to build composite services from abstract user requirements with or without manual selection of services. SWS leverage knowledge representation techniques, with ontologies describing a domain in a formal manner, and AI planning methods to make composition systems more autonomous.

Although, Web Services are a key implementation technology of the SOA paradigm, the main standards defined to implement the SOA paradigm (i.e., WSDL, UDDI, SOAP) emphasize interoperability rather than the capability to accommodate seamless changes at runtime. Frameworks based on ontologies, such as METEOR-S [42], also lack flexible mechanisms for the distribution of information about services as they require the adoption of shared ontologies that impose the distribution policy. Regarding composition, Business Process Execution Language (BPEL) is the de-facto standard [1]. It takes a workflow-oriented approach to the coordination of cooperating services and provides a good solution for the design-time composition of heterogeneous components wrapped as WSDL services. However, runtime identification of partner services is not addressed and thus the degree of dynamism and flexibility is limited.

1.2.4 Overview

Based on the above discussion we give in the following table an overview of the relative advantages/shortcomings of the three middleware design paradigms regarding their support to the development of adaptive systems.

Table 1.1 Pros and cons of the three middleware design paradigms.

Paradigm	Pros	Cons
Reflective Middleware	• a system can modify its structure and behavior at runtime • achieves variable system size to suit different devices and environments • changes made to the self-representation are immediately mirrored in the underlying system's actual state and behavior (causally connected)	• conceptually contrary to the encapsulation principle • usually heavy computational load and low performance • dynamically tuning the scope of changes based on adaptation semantic information is still an open issue
Aspect Oriented Programming	• supports the concept of separation of concerns • separates cross-cutting aspects during development time (e.g., security, logging) • combines principles of object-oriented programming and computational reflection	• may give large memory footprint • programming in terms of aspects is not easy • the order of insertion may be relevant if the corresponding aspects are not independent (composition of aspects) • does not support adequate point cut descriptions to capture join points based on context data and business-level semantics
Service-Oriented Architecture	• modular design appropriate for adaptation and reconfiguration • late binding of services • composite services can be defined from simple ones • main standards defined to implement SOA provide support for interoperability	• automatic service composition is not trivial • main standards defined to implement SOA provide limited capability to accommodate seamless changes at runtime • limited degree of dynamism and flexibility

In this work, we describe an approach based on the SOA paradigm. Besides SOA a novel mechanism is proposed to achieve different kinds of adaptation centered upon the management of knowledge, which is encoded in multi-layered ontologies, which are used by intelligent agents.

1.3 ATRACO Architecture

Ambient Intelligence (AmI) is a paradigm that puts forward the criteria for the design of the next generation of UbiComp environments [37]. In this context we have introduced the *Ambient Ecology* (AE) metaphor to conceptualize a space populated by connected devices and services that are interrelated with each other, the environment and the people, supporting the users' everyday activities in a meaningful way [16].

In the context of the EU funded R&D project ATRACO [18] we aim to extend the AE concept by developing a conceptual framework and a system architecture that will support the realization of adaptive and trusted AEs which are assembled to support user goals in the form of *Activity Spheres* (ASs). Our approach is based on a number of well established engineering principles, such as the distribution of control and the separation of service interfaces from the service implementation, adopting a SOA model combined with intelligent agents and ontologies. Agents support adaptive planning, task realization and enhanced human-machine interaction while ontologies provide knowledge representation, management of heterogeneity, semantically rich resource discovery and adaptation. ATRACO ASs are dynamic compositions of distributed, loosely-coupled and highly cohesive components that operate in dynamic environments.

Therefore the architecture and the system we propose operate in an AmI environment, which is populated with people and an AE of devices and services. Our basic assumption is that the AE components are all autonomous, in the sense that (a) they have internal models of their properties, capabilities, goals and functions, and (b) these models are proprietary and "closed", that is, (i) they are not expressed in some standard format and (ii) they can only be changed by the owner components. However, each component can be queried and will respond using a standardized protocol.

1.3.1 ATRACO World Model

The concepts discussed below constitute a critical subset of the ATRACO conceptual framework defined for building AmI applications.

The basic terms and concepts of the ATRACO world model are encoded in the ATRACO Upper Level Ontology (ULO). In general, ontology is used as the means to share information among heterogeneous parties in a way that is commonly understood [21]. An ontology is a network of concepts and entities, which can be associated with different types of relations (the most common being the hierarchical association, or is-a relation). More concrete (or domain) ontologies contain also instances of these entities with specific properties and values. More powerful ontologies contain constraints and rules that cause inferences for the entities. Figure 1.3 illustrates in UML representation the AS domain model which is also encoded as ontology in the ATRACO ULO.

Table 1.2 ATRACO main concepts and corresponding descriptions.

Concept	Description
Ambient Ecology (AE)	The set of heterogeneous artefacts with different capabilities and provided services that reside within an Intelligent Environment (IE).
Activity Sphere (AS)	It is formed to support an actor' specific goal. An AS represents both the model and the realization of the set of information, knowledge, services and other resources required to achieve an individual goal within an IE. The concept of AS is a "digitization" of the concept of "bubble" used by the psychologist Robert Sommer [39] to describe a temporary defined space that can limit the information coming into and leaving it.
Intelligent Environment (IE)	A territory that has both physical properties and offers digital services. It is the container of AE. ASs are instantiated in an IE using the resources provided by its AE.
Artefact	A tangible object which bears digitally expressed properties; usually it is an object or device augmented with sensors, actuators, processing, networking unit etc. or a computational device that already has embedded some of the required hardware components.
Actor	Any member of AE capable of setting and attaining goals by realizing activities. Within the AE actors are users or agents.
Goal	Each actor may have its own set of goals and plans to achieve them. A goal is described as a set of abstract tasks, which is described with a task model.
Task Model	It may be *abstract* or *concrete*. An abstract task model describes what should be done, without details of how it should be done or by the use of what kind of modality; these are described in the corresponding concrete model. The abstract task model may also contain several decomposition rules modelled as a set of subtasks.
Local Ontology (LO)	Each member of the AE stores locally descriptions of its properties, services and capabilities. It is a sub-class of the class Ontology.
Sphere Ontology (SO)	The SO results from the LO of those AE members that are required to achieve the AS's goal based on the resolution of its task model. Apart from device and service ontologies, it may contain user profiles, agent rule bases and policies. It is another sub-class of the class Ontology.
Agent	A software module (is a kind of actor) capable of pursuing and realizing plans in order to achieve specific goals based on tasks. It includes three types of agents: *Task Agent* (e.g., Fuzzy systems based Task Agent or FTA), who manipulates sensors and actuators in order to realize specific tasks; *Planning Agent* (PA), who resolves an abstract task hierarchy into concrete tasks using the resources of the AE; and *Interaction Agent* (IA), who manages user-system interaction using a mixed-initiative dialogue model.
User	The actor that uses the available services and devices in order to perform a task. When a user performs a task, this can be subdivided into different activities. Users use devices, which provide them with services. Devices run these services in a physical environment (context). Users use these services according to personal conditions (user profile) and within a physical context.
Aim	It is attributed to a user; it is decomposed into a set of interrelated goals, which are distributed to the components of the AS.
Policy	Actors specify high-level rules for granting and revoking the access rights to and from different services. Examples of policy ontologies are privacy policy ontology, interaction ontology and conflict resolution policy ontology.
Service	The entity which describes the service offered by a device.
Device	The entity that has physical/digital properties and offers a specific service.
Resource	A resource can be the *space*, an *entity*, or a *component*, such as managers (e.g. Ontology Manager, Sphere Manager) or other basic components.

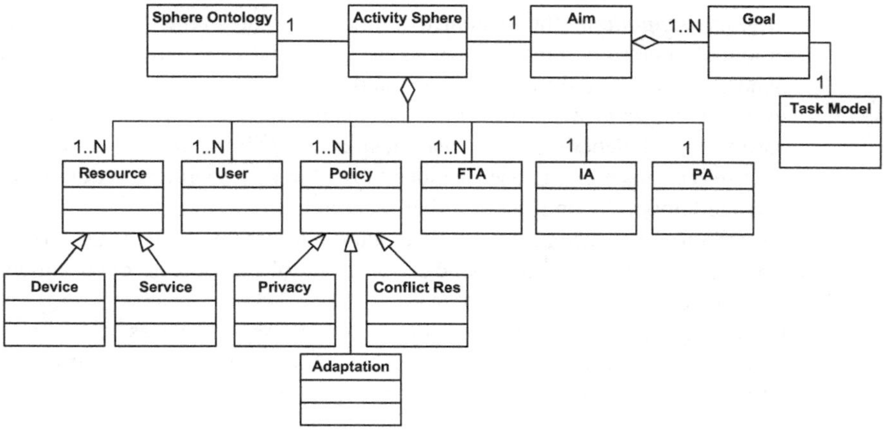

Fig. 1.3 Activity Sphere domain model (part of ATRACO ULO).

1.3.2 System Requirements

For the requirement analysis and design of the ATRACO architecture we followed a process where initially application scenarios were defined and application requirements from a user perspective were identified. In addition as a separate process we defined high-level requirements on system perspective. Then initial requirements were used as input for a process of abstraction that allowed identifying a set of challenges that the architecture has to address, in order to frame further design [18]. These challenges are organized in the following categories:

Challenge 1: Assemble/Dissolve The first challenge has naturally to do with the formation and the dissolution of ATRACO applications (ASs). ASs encapsulate the Ambient Ecology resources that are necessary to serve the goal for which the AS has been created. ATRACO supports adaptation and trust requirements of ASs by integrating into the AS services of the system components that develops.

Challenge 2: Adaptability Adaptability implies that an AS should attempt to continuously provide expected behaviour by adapting to unexpected conditions such as changes to the resources constituting the system or changes in the behavior of the user.

To this effect, the ATRACO system components are defined that adapt task-based usage of the sphere to the changing user behaviour, environment conditions and context. ATRACO implements mechanisms that support adaptability in several forms:

- *activity sphere adaptation*, in terms of *structural adaptation*: the persistent achievement of the goal when changes on the type or cardinality of the available resources occur
- *behavioural adaptation*, where the application logic is changed as a result of changes in the user and/or device behavior. This category is specialized as

  - *artefact adaptation* (the system examines how an artefact can adapt its model of operation in reaction to changes in the device characteristics e.g., handling a partial failure of a heater) and
  - *user behaviour model adaptation* (an agent will learn and adapt its rule base to face the changes in the user desires and preferences by monitoring the user actions e.g., the user decides to read in bed, therefore requires her bedside lamp to be on instead of her reading light).
- *user interaction adaptation* specifies adaptation interacting with the user using different devices/modalities depending on available resources, environment characteristics, tasks and user profile.
- *network adaptation* to allow the uniform and transparent access to devices and services present in the networked environment supporting the realization of activity spheres across a mixture of heterogeneous networks.

Challenge 3: Semantic heterogeneity A basic assumption is that an AmI space is available to host an ambient ecology and devices and services are inherently heterogeneous and contain heterogeneous descriptions of their capabilities and services in the form of local ontologies. Thus, in order to achieve collaboration among them, firstly one has to deal with these forms of heterogeneity. However, the issue raised by the heterogeneity of ontologies and how to achieve semantic interoperability between systems using different ontologies is a challenge. In ATRACO, the approach that is followed in order to address this challenge is to research, develop and test theories of ontology alignment to achieve task-based semantic integration of heterogeneous devices and services.
To this effect, a Sphere Ontology is defined and an Ontology Manager administrates its use by performing ontology updating, ontology querying and ontology matching services.

Challenge 4: Trustworthiness The interactions in the activity sphere should be trustworthy. The ambient ecology will behave in a dependable manner and will not adversely affect information, other components of the system or people.
To this effect, policies and rules are defined in the ontology and mechanisms are defined for the management of the identity of service requestors and service providers as well as access control on services and context information.

Summarizing the above discussion, ATRACO architecture should provide:

- support for realizing user goals (activity spheres), by resolving abstract tasks to a workflow of concrete tasks;
- support for executing workflows by applying service composition and control policies in the form of rules (obligation policies);
- support for establishment and management of associations between service clients and service providers (as described in task workflows);
- support for maintaining the sphere ontology which contains the contextual knowledge necessary to realize the concrete tasks;
- support for ontology alignment and lookup;
- support for adaptation of the given tasks according to the user desires and behaviour (personalization and learning over-time);

- support for use of heterogeneous network capabilities for communication (network adaptation);
- support for discovery of services, devices, networks and resources;
- support for usage of services offered within ATRACO infrastructure or by third parties (e.g., external Web Services);
- support for privacy enforcement and access control through policies;
- support for the possibility of adapting the user interaction depending on available interactive devices and objects;
- support for management of user profiles and preferences;
- support for gathering, processing and distribution of context information;

In the next section we outline the ATRACO system design that accommodate the system requirements and then we discuss in more detail the service composition framework for deploying adaptive workflows in IEs to achieve structural adaptation of ATRACO applications, which is the focus of this presentation.

1.3.3 System Design

In ATRACO, we propose a combination of the SOA model with agents and ontologies (Figure 1.4). We adopt SOA both at the resource level to integrate resources, such as devices, sensors and context in applications and at the system level to combine ATRACO services that provide adaptation and trust features into applications. ATRACO aims to empower users with the ability to interact in environments with many resources such as devices (UPnP devices), web-services, content (music/video file, contacts) and applications (e.g., media player) using adaptive user interfaces. The functionality in these environments is exposed as semantically rich services which an actor (either a user or an agent) can discover and then compose to form ATRACO Activity Spheres.

Each service is associated with at least one semantic description which shields the actor from the complexity of the Resource Layer realization and makes it easy for the actor to employ these services in accomplishing interesting and useful tasks. Figure 1.4 shows a conceptual layered view of the ATRACO architecture. The ATRACO infrastructure consists of SOA services. On the one hand, these context-aware services are built on "Core distributed middleware" and rely on "'Network and resources" layer. On the other hand, the ATRACO infrastructure supports basic services such as context management and reasoning, communication management, user profiling and service (discovery), as well as adaptation and privacy services that form the basis for ATRACO systems (i.e., ASs).

The ATRACO architecture consists of ontologies, active entities, passive entities, and the user who as the occupant of the IE is at the centre of each AS. Active entities are agents and managers. The role of the ATRACO agents is to provide task planning (Planning Agent or PA), adaptive task realization (Fuzzy systems based Task Agent or FTA) and adaptive human-machine interaction (Interaction Agent or IA).

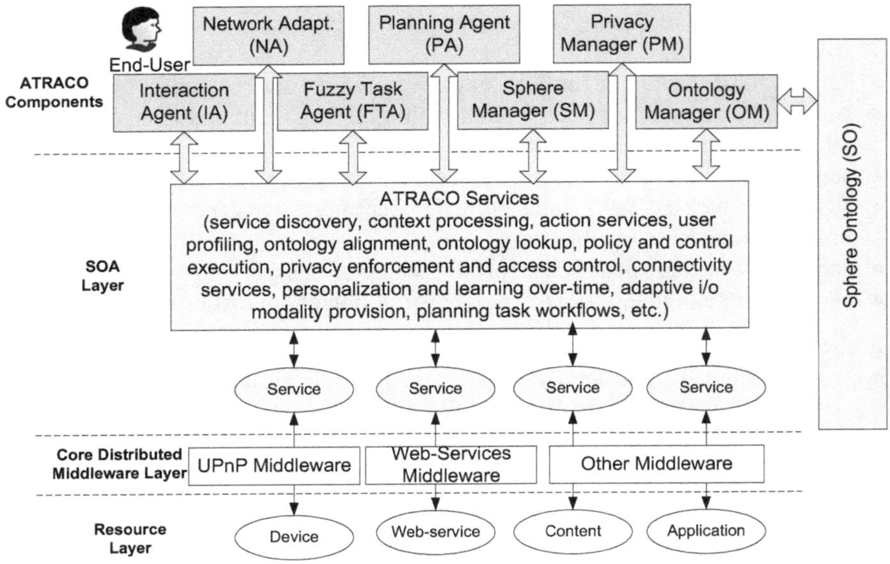

Fig. 1.4 ATRACO architecture.

The PA encapsulates a search engine that exploits hierarchical planning and partial-order causal-link planning to select atomic services that form a composite service (workflow) [3]. One or more FTAs oversee the realization of given tasks within a given IE. These agents are able to learn the user behavior and model it by monitoring the user actions. The agents then create fuzzy based linguistic models which could be evolved and adapted online in a life learning mode [43]. The IA provides a multimodal front end to the user. Depending on a local ontology it optimizes task-related dialogue for the specific situation and user [35]. The IA may be triggered both by the FTA and the PA to retrieve further context information needed to realize and plan tasks by interacting with the user. On the other hand, ontologies complement agents regarding adaptation by tackling the semantic heterogeneity that arises in IEs by using ontology alignment mechanisms to generate the so-called, Sphere Ontology (SO). There are two main kinds of ontologies: local ontologies, which are provided by both active and passive entities and encode their state, properties, capabilities, and services and the SO, which serves as the core of an AS by representing the combined knowledge of all entities [38].

The Sphere Manager (SM) and Ontology Manager (OM) components are responsible for the formation, adaptation and evolution of the user applications (modeled in ATRACO as ASs) and will be further examined in this paper. In the current version of the system there is also a Privacy Manager (PM) that provides a set of privacy enhancing techniques in order to support privacy in an adaptive and individualized way. Finaly, devices in the IE that may come from heterogeneous networks (e.g., LonWorks, ZigBee, Z-Wave, etc.) and services (e.g., Network Time, VoIP, Real Time Streaming, etc.) are accessed transparently through a service representation layer exporting them to the ATRACO clients as UPnP services. This layer is implemented in the Network Adaptation (NA) component [33].

1.4 Adaptive Workflows and Structural Adaptation

In many respects, a composite service can be modeled as a workflow [36]. The definition of a composite service includes a set of atomic services together with the control and data flow among the services. Similarly, a workflow is the automation of a business process, in whole or part, during which documents, information, or tasks are passed from one participant to another for action, according to a set of procedural rules [23]. Workflows have been used to model repeatable tasks or operations in a number of different industries including manufacturing and software. In recent years, workflows have increasingly used distributed resources and Web services through resource models such as grid and cloud computing. In this section, we argue that workflows can be used to model how various services should interact with one another as well as with the user in IEs depending on available resources, environment characteristics, user tasks and profile.

In this section we describe how SOA can support AS adaptation. The structural adaptation (a form of polymorphism) is possible because the workflow model represents abstract services and binding to real devices can be accomplished at runtime. ATRACO-BPEL, a streamlined version of BPEL, has been defined as the specification language to describe workflows of abstract services.

1.4.1 Scenarios

In order to test our framework and to illustrate how workflows can be used to fit user interaction with an IE, as well as the structural adaptation mechanism of ASs, we use two simple scenarios. The first example corresponds to an AS that supports the realization of goal named *"Feel comfortable upon arrival at home"*.

Martha arrives at the door of her smart apartment. The system recognizes her, through an RFID card, and opens the door. On entering the space the system greets Martha by saying "Welcome home" and then when she has entered the living space the lights and A/C are switched on and brightness and temperature are automatically adjusted according to her profile, season, and time of day, to make her feel comfortable. Martha then sits at the sofa to relax and after a while, the system asks "Would you also like some music?" Martha responds positively and the music plays (according to predetermined preferences). Following this, the system asks "Would you like to view yesterday's party photos?" Martha responds positively and a rolling slide show appears in a picture frame in front of her. After a while, Martha gets up, walks towards the window and opens it. Fresh air pours into the room. Temperature level drops. Brightness level increases. Some of the lights are automatically switched off, in an attempt to maintain the previous level of brightness in the room. After a while, the A/C is switched off because of the open window. Suddenly, the picture frame goes off! The system finds a proper replacement and as a result, photos are displayed in the TV set, while Martha is informed on the event.

The second example corresponds to an AS that supports the realization of goal named *"Studying AS"*.

Suppose that John is using a number of objects to support the studying activity at his writing desk, according to his profile (his preferable level of light, temperature, etc.). In this case, John has set as a goal to study. This goal can be decomposed in a hierarchy of abstract tasks that constitute a task model for the goal: sit on a chair, move the chair in proximity to the desk; take the book; place the book on top of the desk; turn on the light. In the AmI environment an AS is formed to support the specific goal, by using four artefacts, a lamp, a chair, a desk and a book. The application logic can be stated as follows: when the chair is occupied and it is near the desk and the book is open on the desk, the lamp is turned on (reading activity has been inferred). The implementation of such a task specification can be represented as a graph of connected services provided by the artefacts.

Furthermore, John can move in the room and change his reading spot at the sofa. This causes an adaptation in the configuration of the Studying AS since a new artefact (sofa) is added and one is removed (desk). Another implication of this mobility is that the light service will adapt to the new reading spot. While reading at the desk the desk light is used, and when he moves to the sofa the lamp near the sofa is used. This implies that device selection for instantiating/adapting an AS depends on user location.

Since workflows are essentially graphs of activities, it is useful to express those using UML activity diagrams. 1.13 describes the sequence of activities for the example scenario. Note that the tasks "AdjustLights", "AdjustAC", "ShowPhotos", and "PlayMusic" can run in parallel and therefore they have been enclosed in a fork-join block. Note also that the exception events are not part of the workflow description but they are handled by the corresponding ATRACO active entities.

1.4.2 Late Binding

We have developed a service composition mechanism which includes 3 phases: *task workflow planning*, *dynamic service binding* and *execution management and control* as illustrated in Figure 1.5.

The planning problem can be stated as "discover an execution path of services (tasks) given some state of the world to achieve a goal". In ATRACO, we use a library of abstract plans which model specific user goals. An abstract plan contains a sequence of abstract services which are actually ontological descriptions of service operations that cannot be directly invoked, but will be resolved by the SM during runtime. Having an abstract service workflow description, which is given in a BPELlike language, the Dynamic Service Binding module of the SM applies a

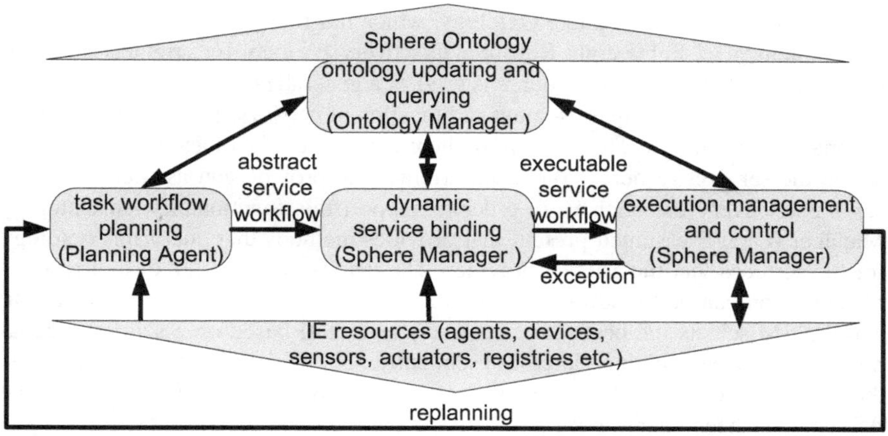

Fig. 1.5 Late service composition process in ATRACO.

semanticbased discovery mechanism and uses information about available services and context to discover suitable services or devices in registries able to perform each abstract service. The output of this process is an executable service workflow. In the execution management and control phase the SM executes and continuously monitors the deployed services and the termination condition of the workflow.

This adaptation has been inspired by the subtype polymorphism found in the object-oriented programming paradigm [2]. The concept is that we can adapt the instantiation of the AS to different environments provided that a late binding mechanism is in place that determines the exact resources that will be used in the AS (i.e. the specific artefacts, e.g. "the lamp in the corner"). The different resources that may be involved only need to present a compatible interface to the clients (i.e., in our case, a UPnP interface). Figure 1.6 gives a conceptual view of the dynamic service binding process. A workflow is mapped into a number of tasks and a workflow task is mapped into one or more abstract services. In addition, each service would also require certain physical resources for its implementation. Mapping of the task to the services can be specified at design time by the PA as per users' functional requirements. However, mapping of the service to the actual human and physical resources is done at runtime, in keeping with service orientation. This dynamic binding is therefore dependent on the context in which the binding occurs.

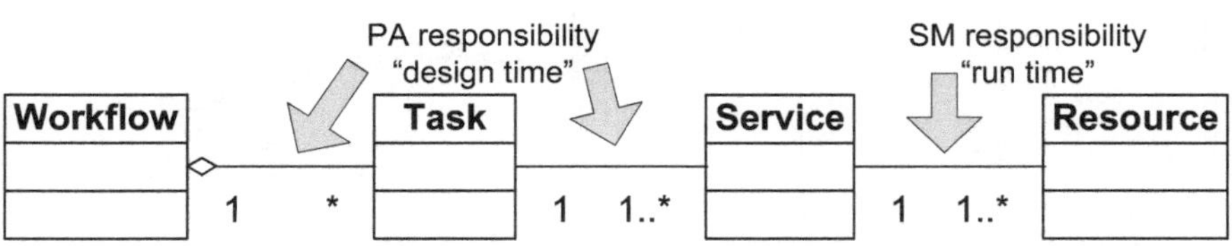

Fig. 1.6 Conceptual model for dynamic service binding.

In the absence of the Sphere Ontology, which has not been yet instantiated, the SM implements a lightweight Resource Discovery Protocol for artefacts or eEntities (eRDP) where the term resource is used as a generalization of the term service. eRDP is a protocol for advertisement and location of network/device resources with a semantic description. The assumption here is that there is a local ontology to describe the services/resources that each artefact can provide and as such assist the service discovery mechanism. In order to support this functionality, an Ontology Manager (OM) is assumed present that provides methods that query this ontology for the services that the artefact provides. The details of the eRDP design and implementation can be found in [17]. The matching resources are returned by eRDP and the SM selects the best set of device(s)/service(s) based on a scoring mechanism that will be explained later. Subsequently, the SM invokes the OM to create the Sphere Ontology (SO) which will include links to all the relative devices to the AS that have been discovered.

After service binding the SM starts any interaction task in conjunction with the IA and also any FTA task and executes the workflow preserving the precedence constraints or the conditions that are specified in the workflow. At runtime a Workflow object aggregates a number of Task objects where each object represents a task in the workflow. The services that this task requires for running are divided into input and output services and are connected with the appropriate resources. The resources that are bound to the Task object can be either devices that the Task directly controls (i.e., input sensors and actuation devices) or agents, such as the IA or the FTA. In either case the Task object is informed on the status of the resource and operates according to the pattern specified by its type. The sequence diagram in Figure 1.7 shows the basic interaction of the software components during the instantiation of the "Feel Comfortable" AS, which employs the dynamic service binding process mentioned earlier. In the diagram, this process is implemented by the methods used inside the two loops.

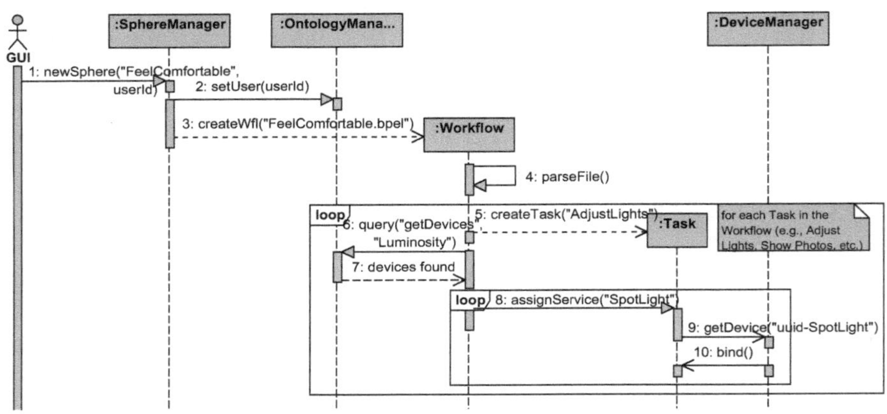

Fig. 1.7 Example AS instantiation and binding of devices to services.

The Task object "AdjustLights" is assigned to the FTA component to generate adaptive models for the individual devices/artefacts and for the user behaviours. Figure 1.8 illustrates the initialization of the FTA component to control the room lights in the example AS. The FTA is initialized by passing the input/output, light level related devices as well as light controls which are in turn retrieved from the Sphere Ontology which has been populated with the required ontologies during the AS instantiation. In addition, if the user profile stores initial light preferences (for example from previous executions of the FTA) these can be passed to the FTA in the form of a rule base.

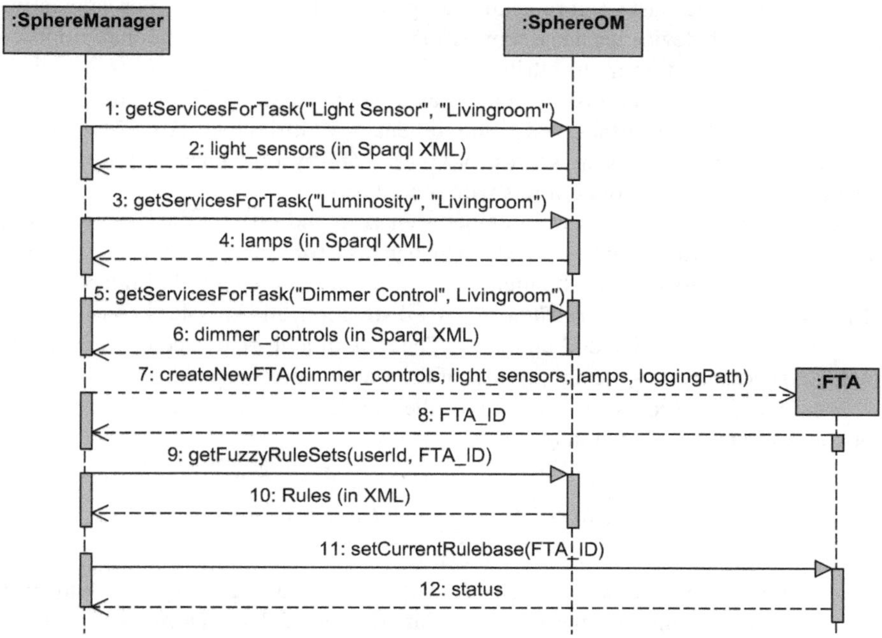

Fig. 1.8 FTA initialization for the AdjustLights task of the "Feeling Comfortable" AS.

In addition, the SM handles exception events that affect the configuration of the AS. For example, exceptions during the execution of the workflow, such as disconnection or failure of devices trigger an adaptation of the workflow by rebinding services to alternative devices. Context changes during the execution of the workflow may invalidate preconditions that were valid during the workflow instantiation. For example, if the user changes location and a follow-me property has been defined for a display service, then the execution state needs to be updated and a new display service instance to be scheduled. In order to achieve workflow adaptation, replanning capabilities may be required by the PA. Replanning comes into play when the dynamic binding fails during workflow execution or update. When replanning is requested a new planning problem is defined with the services that are actually available, and the PA solves the problem and delivers a new workflow.

During AS instantiation in IEs there could be multiple devices or services providing similar functionality from which the system will have to choose. Thus, the ATRACO system must provide mechanisms for selection between similar devices or services and decide which of them is the most suitable to participate in the AS. Device selection is based on criteria such as: task suitability, efficiency (as device's proximity to the user, quality of the service or device and stability), user distraction (the inconvenience a user experiences when the system selects different groups of devices than those that the user prefers or is used to use for a specific task) and conflicts with other tasks (more details are given in the Appendix). For calculating the rank for each device we use a scoring mechanism that is similar to that proposed in [29] and is based on multi-attribute utility theory (MAUT). The overall rank of a device given a specific task is defined as a weighted sum of its evaluation with respect to its relevant orthogonal value dimensions (attributes). For ATRACO the relevant value dimensions are scores for task suitability, efficiency, negative of user distraction and negative of conflicts with other tasks.

A ranking policy defines weights between zero and one for each of the above metrics. The scoring policies are defined per task (or task category) by the user and give priority to some of the metrics. E.g., if a task is urgent, the suitability and efficiency ranks must have priority over user distraction, and inter-task conflicts. The weights are normalized to add up to one. The rank of a given device D according to policy P is computed as the dot product of the vector weights specified by the policy with the vector of scores for each one of the metrics. Applying MAUT, the device rank is computed as shown in (1).

$$DR(D,TP) = \sum_{i=1}^{4} w_i(TP) * D(m_i) \qquad (1.1)$$

where DR is the overall rank of device D according to ranking policy TP for the task T, $w_i(TP)$ is the weight of metric i according to policy TP and $D(m_i)$ is the rank of device D for the metric m_i.

For the task suitability and efficiency we have $D(m) = DS(m)$, while for user distraction and inter-task conflict that have a negative meaning we have $D(m) = 1 - DS(m)$, where $DS(m)$ is the device's score for the metric normalized from 0 to 1.

1.4.3 Ontology Manager

The Ontology Manager (OM) component provides an interface to the SM to access AS related data, including personal and contextual information, represented in ontologies. The OM provides methods for querying and modifying User Profile Ontologies, Device Ontologies, the Privacy and Policy Ontology as well as the eventual Sphere Ontology (SO) that emerges from the alignment of all the previous ontologies. The ontology alignment process can be described as: given two ontologies, each describing a set of discrete entities (which can be classes, properties, rules, predicates, or even formulas), find the correspondences, e.g., equivalences or subsumptions, holding between these entities. Thus, under the request of the SM, the OM produces ontology alignments, responds to queries regarding the state or properties of sphere resources, and creates inferences in order to enrich the SO as specified in [38].

The OM has been developed as a wrapper around the Jena Framework (`http://jena.sourceforge.net/`). The OM interface provides comprehensive and simple methods for creating an RDF/OWL based ontology, importing and removing other RDF/OWL based ontologies, updating the ontology at run time, querying of the ontology using SPARQL, and saving the modifications in OWL files. Ontology alignment has been applied by using the Java Alignment API (`http://alignapi.gforge.inria.fr/align.html`). After the alignment, inference and querying is performed on a grid of imported ontologies, given the alignment points that have been produced using OWL class and individual equivalence assertions.

Figure 1.9 illustrates a small sample of the OM interface that is used, for example, to query an ontology using SPARQL syntax, and methods related to the User Profile Ontology e.g., for importing and exporting rules from the FTA.

```
public String queryForSparqlXML(String query, boolean autoPrefix,
String queryType)
```

Performs a query to the ontology. autoPrefix determines if OM will try to resolve known prefixes and the queryType can be ASK or Select. Returns the results in SparqlXML format.

```
public String[] getFuzzyRuleSets(String userId, String FTA_ID) throws
Exception
```

Returns in XML format the stored fuzzy rulesets that match the given userId and FTA_ID. Used by SM to retrieve stored fuzzy rule sets during initialization of the corresponding FTA.

```
public String getServicesForTask(String serviceDescription, String
location) throws Exception
```

Returns a Sparql XML string containing technical parameters for each device and service that matches the serviceDescription description tag and is within the location specified by the second parameter. Used by SM for resolving tasks to specific devices, services, actions, variables and values.

Fig. 1.9 A sample of the OM interface.

A number of ontologies have been developed for and used in the prototype for the representation of AS high level concepts (Figure 1.3), devices and their services, and users and their profile information.

The User Profile ontology holds personal information about the user. It consists of a local, private OWL ontology file that contains the actual user information in the form of individuals and assertions and a publicly accessible (via HTTP) ontology that contains the generic classes, properties and restrictions that describe a user profile. Currently User Profile Ontology contains assertions about the Social Profile (name, nickname, email, address etc.), the location, the activities (Goals and Plans) and the preferences of the user (in the form of stored fuzzy rule sets).

Figure 1.10 illustrates part of an instance of a device ontology for one of the spot lamps used in the prototype. The Service concept represents an abstract service that the device can provide enriched with descriptive tags e.g., a lamp can provide lighting service. In general, a device may offer more than one service and thus more Service instances may be defined. The `StateVariable` concept represents the abstract states of the corresponding Service. It encapsulates the linguistic variable and labels that are required by the FTA for the creation of adaptive device models. The device ontology includes technical characteristics and information about communication with the device in the context of a UPnP environment. Finally, the name, the owner, the location and physical properties of the device are included.

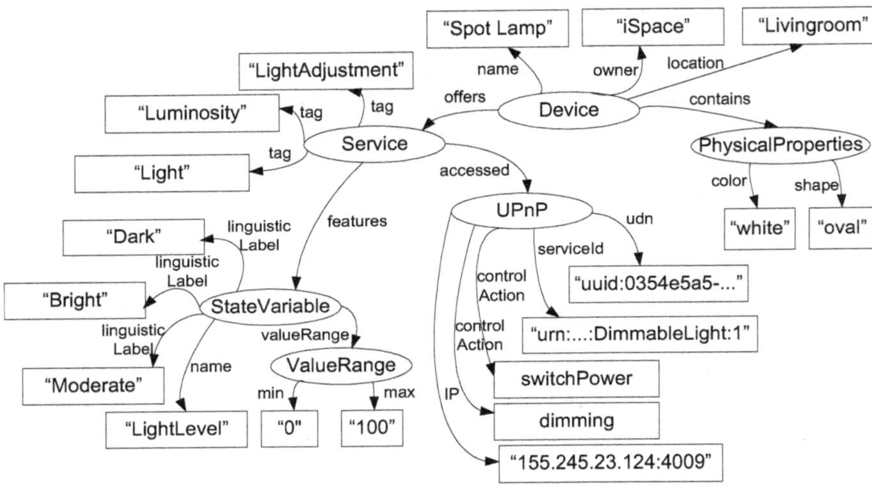

Fig. 1.10 Part of an instance of a device ontology for a spot lamp.

1.4.4 ATRACO-BPEL Workflow Specification

BPEL defines a model and a grammar for describing the behavior of a business process based on interactions between the process and its partners. It allows for creating complex processes by creating and wiring together different activities that can, for example, perform Web services invocations (<invoke>), waiting to be invoked by someone externally (<receive>), generate a response (<reply>), manipulate data (<assign>, throw faults (<throw>), or terminate a process (<exit>). In our case, the business process represents the process model of an AS and the partners can take the form, either of a service of a simple device, or the service of an ATRACO agent.

While BPEL is a suitable language for describing workflows, an ATRACO workflow description presents requirements that cannot be completely covered by BPEL. This is due to the following:

i. BPEL partners (partnerLinks) are bound statically to specific Web services. In the context of ATRACO, however, services are not bound at design time but dynamically during the execution of the workflow. Thus, there is a need to describe services in the workflow by their semantics which mainly define ontological related searching terms (for example, "Luminosity" for a light service).
ii. The limitation of the one-to-one mapping of services between communicating partners, supported by BPEL. On the other side, ATRACO tasks may need to handle two or more services that provide input or output to the task.
iii. BPEL supports a single coordinator that executes the orchestration logic. ATRACO workflows normally are centrally handled by the Sphere Manager which implements the workflow execution engine; however a more distributed scheme can also be followed by sharing parts of the workflow with collaborating agents (e.g., IA and FTA). This collaboration sets some special requirements in the description of the workflow.

Given the above requirements a variant of BPEL, called ATRACO-BPEL, was defined in order to provide those ATRACO specific features needed in order specify workflows. In the following we explain how using the ATRACO-BPEL formalism an example task is bound with the appropriate service(s). The task *AdjustLights* is associated with the parnterLink *AdjustLightsPL* as part of the orchestration logic section:

```
<bpel:invoke
  name="AdjustLights" partnerLink="AdjustLightsPL">
</bpel:invoke>
```

The partnerLink *AdjustLightsPL* has an input role called *ATRACO:lightStatus* and an output role (partnerRole) called *ATRACO:triggerLight*. The *Continues* type denotes that the execution of the activity is to be treated as a task that is running continuously, i.e., the workflow does not wait its termination.

```
<bpel:partnerLink
  name="AdjustLightsPL"
  partnerLinkType="ATRACO:Continuous"
  myRole="ATRACO:lightStatus"
  partnerRole="ATRACO:triggerLight">
</bpel:partnerLink>
```

The input role *ATRACO:lightStatus* denotes the appropriate abstract service that must be bound to fulfill the role (Luminosity) along with any other application specific details that are needed for its operation e.g., the task will be monitored by an ATRACO agent for learning user behavior with respect to light adjustments and all found light devices are to be used.

```
<ATRACO:role
  name="lightStatus" type="input" Agent="yes" IAmode ="none">
```

```
    <ATRACO:service semantics="Luminosity" trigger="Low" reset
        ="none" quantity="all" rules="">
    </ATRACO:service>
</ATRACO:role>
```

The corresponding definition for the output role will be:

```
<ATRACO:role
    name="triggerLight" type="output" Agent="yes" IAmode="
        withAgent">
    <ATRACO:service semantics="Actuate Light" trigger="On"
        reset="Off" quantity="all" rules="">
    </ATRACO:service>
</ATRACO:role>
```

In ATRACO-BPEL each partnerLink role is specialized as an ATRACO:role which is a new definition in ATRACO-BPEL. In each ATRACO:role the attributes listed in Table 1.3 are defined.

Table 1.3 ATRACO:role semantics in ATRACO-BPEL.

Attribute	Semantics
name	The name of the role.
type	Denotes the type of the role. Accepted values are **input/output**.
Agent	This attribute defines whether the task is monitored by an ATRACO agent or not. Accepted values are **yes/no**.
IAmode	Specifies the interaction mode with the ATRACO Interaction Agent. Accepted values are: **none** no interaction is needed; **pure** this value is used to indicate that a single interaction with the user through a dialog interface (spoken, tangible or software) needs to be provided either to provide a message or to receive an input for the system from the user in a form of question; **direct** this value is used when the IA needs to create an interface for an output device; **withAgent** this value is used to indicate that there is a need to find proper user inputs for the Agent monitored tasks.

Each ATRACO:role envelopes a set of services that are bound to it. Each role can have more than one abstract service. If the role type is input then the activity waits for all the services to deliver their result before proceeding. If the role type is output then, upon activity completion, all the services enveloped in this role are triggered. For each abstract service specific attributes are defined, providing the necessary support for device discovery and service operation. Table 1.4 summarizes the service-specific attributes in ATRACO-BPEL.

Table 1.4 Service-specific attributes in ATRACO-BPEL.

Attribute	Semantics
semantics	The semantics of the service as a set of keywords – these are used to find the specific device that can be bound to this abstract service.
trigger	**input role**: denotes a linguistic value that triggers the service. **output role**: denotes a linguistic value passed to the service.
reset	The reset state (linguistic value) that the service should apply in the case that the activity cannot be performed.
quantity	A number that defines how many devices providing this service are needed for the specific activity. If the value is "**all**" then all found devices are used.
rules	Any special constraints need to be met for binding the corresponding device(s).
IAdlg	This attribute is associated with the direct or pure interaction modes with IA in order to give it the proper interaction dialog type. Examples of accepted values are: GreetingMessage, LightInstructions, GrantGuestAccess, MusicQuestion, MusicControl, PhotoFrameQuestion, SlideshowControl.

1.5 Deployment

In order to test the AS adaptation mechanisms we have implemented an experimental prototype in the AmInOffice testbed. The AmInOffice is a testbed developed in the premises of Dynamic Ambient Intelligent Systems Research Unit at RACTI (`daisy.cti.gr`) and consists of a variety of sensors deployed in the office environment, a set of smart objects that support office tasks and the appropriate network infrastructure. In order to implement the above scenario we have set up AmInOffice with the following devices:

- An RFID reader near the door of the office to read RFID tags
- Two light sensors each one reading light level in a different spot in the office
- Two ceiling lamps controlling the ambient light
- Two lamps one placed at the desk and one near the sofa
- Speakers connected to the main PC for playing music and producing vocal messages
- A smart chair (eChair) able to sense if someone is sitting on it
- A smart sofa (eSofa) that can sense if someone is sitting on it and at which spot (left or right)
- Smart books (eBooks). Apart from smart readers (eReaders) this includes a typical book instrumented with bending sensors that can sense if the book is opened or closed.
- A smart desk (eDesk) that can sense objects on it and near it.

Figure 1.11 illustrates how the devices have been placed in the AmInOffice.

The ATRACO components that implement the necessary functionality in order to support AS formation and adaptation based on mechanisms discussed in this work are the Sphere Manager (SM) and Ontology Manager (OM). Interaction with other ATRACO components such as Planning Agent and Privacy Manager is assumed

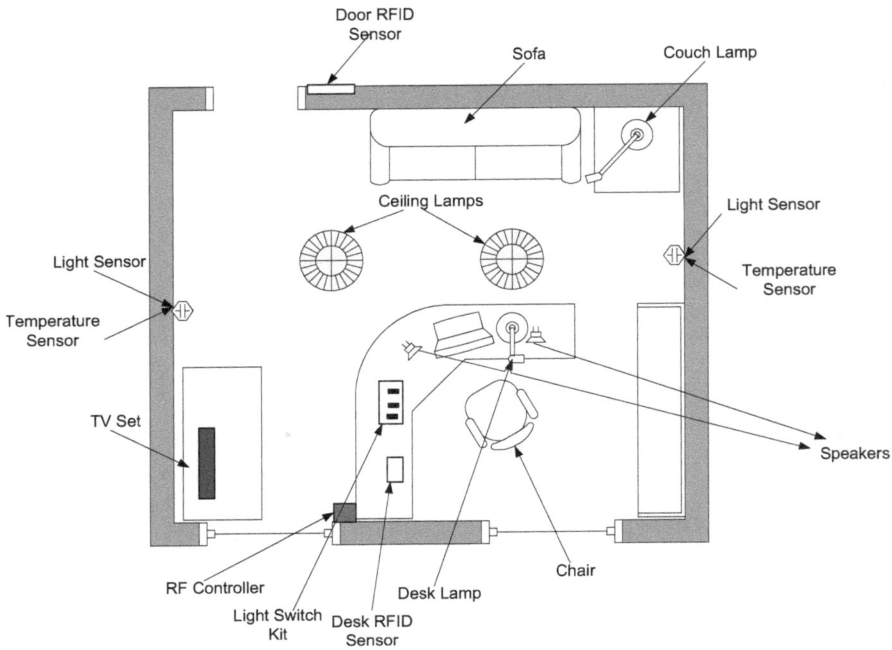

Fig. 1.11 AmInOffice setup for the experimental prototype.

and requires the interfaces specified in [15]. Third-party tools have been also used for performing alignment. The ontologies for all the artefacts used have been developed and a semantically rich UPnP device ontology was developed to support workflow-driven inclusion of UPnP compatible devices in a sphere. In the experimental prototype we have tested the following functionality:

- *Sphere initialization*: Initiate an AS through an ATRACO-BPEL file. Test workflow creation and execution.
- *Late binding of the devices*: Bind abstract services needed for each task to specific devices that exist in AmInOffice in collaboration with the OM at runtime.
- *Runtime application behavior*: Validate that the running tasks correspond to the scenario of the experiment.
- *Handling of adaptation events*: Test system response to adaptation events that affect the configuration of the AS categorized in the following types:
  - *User location change*: test system reaction when the location of the user associated with the AS changes.
  - *Resource not available*: test system reaction when a device bound to a task fails. Check if the system can find an appropriate replacement.
  - *New resource (service/device)*: test system reaction when a new device relevant to the task that is running is available.

– *New person*: test system reaction (in terms of security and privacy) when a new user is recognized by the system

Figure 1.12 illustrates in the form of an activity diagram the main tasks to be executed by the Sphere Manager component in order to handle each one of the above adaptation events. The starting point for running an AS is the generation of the cor-

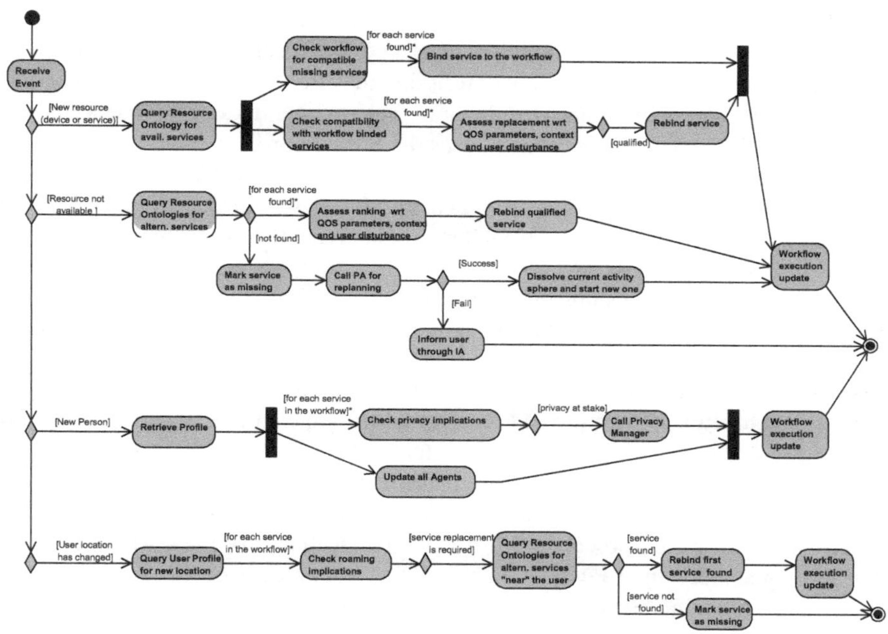

Fig. 1.12 Adaptation events handling.

responding workflow. Workflows are described in ATRACO-BPEL, but they can be represented in a more user friendly way with activity diagrams. The diagram in Fig. 1.13 illustrates the workflow for the "Feel Comfortable" AS of the example scenario. The diagram is annotated with labels from the source file in an attempt to close the gap between the high-level view of the diagram and the low-level view of the file. For example, the annotation in each box shows the activity type in the main sequence and the task name, the ontological searching term, as well as which ATRACO component, besides SM, has responsibility for running parts of this task.

The technical requirements for the deployment and testing of the ATRACO system include: the runtime versions of the ATRACO components with the specified service interfaces; the devices serving the scenarios, wrapped as UPnP devices; the domain and resource ontologies; the workflows specifying the tasks in each AS; and various third-party run-time libraries. The deployment of the system has been done in two IE testbeds using scenarios similar to the one discussed in this paper.

The implementation technologies and tools used are based on open frameworks and are compatible with the SOA paradigm. Java is the main programming language and UPnP enhanced with semantic descriptions [41] is used as the communication middleware for the integration of devices and services, instead of Web services. OWL has been used for the development of the ontologies as it provides a strong logical reasoning framework for the expression and enforcement of ATRACO policies and rules.

Although there are available (open source) execution engines for BPEL "programs" in ATRACO we need to build a layer upon such engines as a proxy in order to process the parts of the workflow description that are ATRACO-specific. In addition, most engines do not allow for dynamic binding and discovery of services. To address this limitation, the framework uses the SM as a proxy to communicate with service registries to obtain operational descriptions (e.g., UPnP or WSDL files) and instantiate services. This is achieved by encapsulating service search parameters in ATRACO-BPEL (see Table 1.4) as an input to the dynamic service binding process.

When the user changes position in the room the ATRACO system is notified for that change. While this location context can be provided either by using motion detection devices, or specific services such as Ubisense (used in iSpace), for our experiment we emulated such a device by using a WoZ interface and selecting the appropriate location. When the SM receives a location change event, it queries OM for the new location. Then for each service that is bound to the active task it checks if there are any requirements for device replacement. This is done by querying the OM with the new user location context. If the device that OM returns is not equal to the currently bound then it proceeds with service replacement for the appropriate task. The sequence diagram in Figure 1.14 shows the exact messages that are exchanged for the task "Reading" when the user changes location to the sofa.

1.6 Discussion

The SOA approach appears to be a convenient architectural style towards meeting one of the key objectives of the ATRACO project that is the need for adaptable and reconfigurable systems. Analyzing contemporary software technologies complying with the SOA architectural paradigm, such as OSGi, UPnP, and the Web services architecture appears that current software technologies do not meet the adaptability and interoperability requirements for the ATRACO project.

In the first case SOA provides little support on how adaptive services can be used to allow people to interact with an AmI environment in a seamless and unobtrusive manner. In other words, research into service composition has mainly focused on the composition mechanism rather than on guiding composition to enable the user to perform activities in the way they wish to do. A challenge here is how to automate the service composition process, so that the service offered to users appears to be

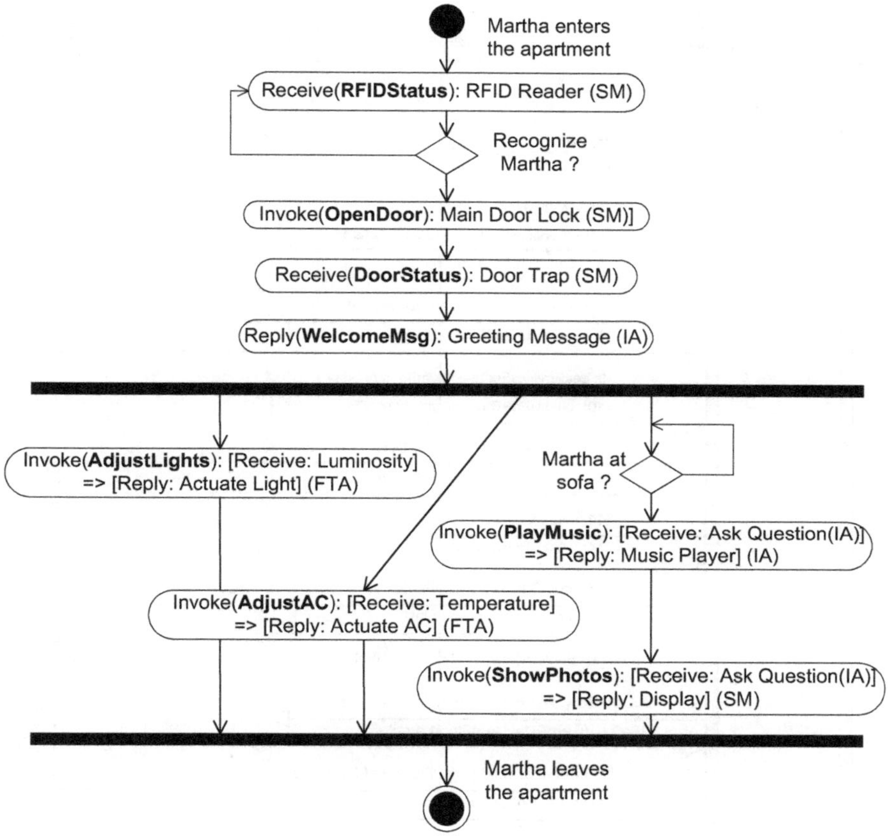

Fig. 1.13 Annotated activity diagram for the "Feel Comfortable" AS.

adaptive, in the sense that the service provided changes dynamically according to the task the user wishes to perform and the context in which they wish to perform it.

In the second case, current solutions provide little support for semantic-based interoperability, hence dealing with interaction between services based on syntactic description for which common understanding is hardly achievable in an open environment. The latter issue may be addressed using semantic modeling through ontologies. Ontologies can provide an extensible and flexible way of expressing the basic terms and their relations in a domain, task or service. However, the issue that can be raised by the heterogeneity of ontologies and how to achieve semantic interoperability between systems using different ontologies remains a challenge. In ATRACO the approach that is followed in order to address this challenge is to research, develop and test theories of ontology alignment to achieve task based semantic integration of heterogeneous devices and services. This issue is examined thoroughly in a separate chapter.

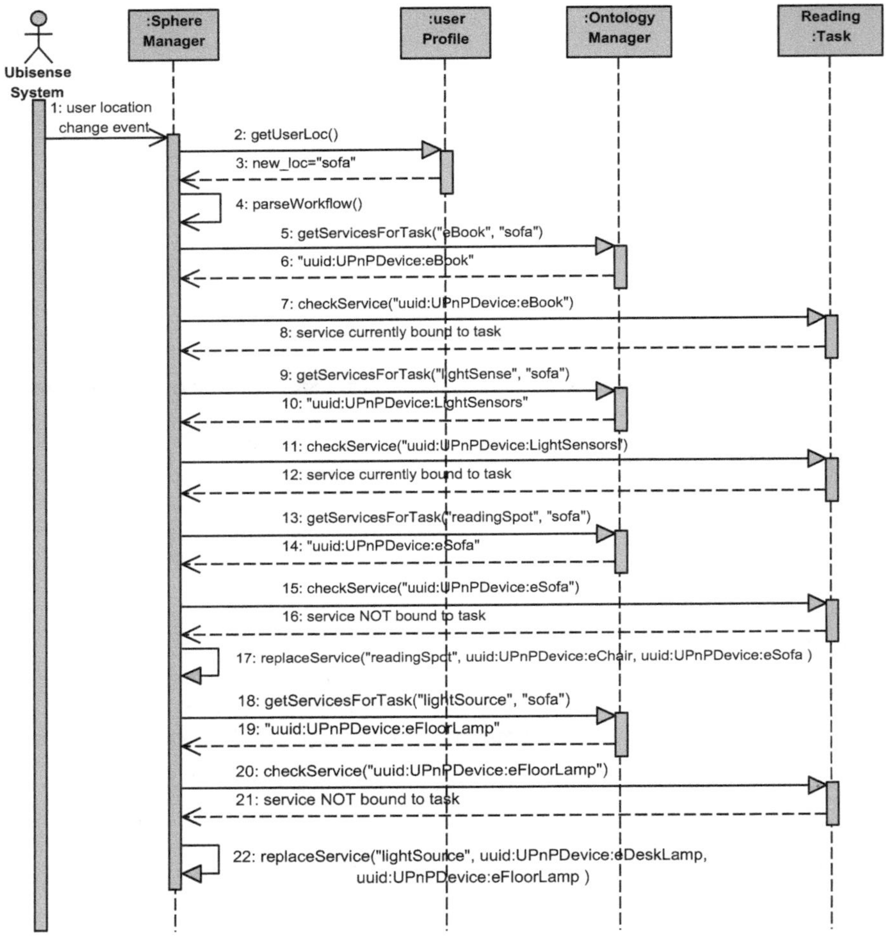

Fig. 1.14 Replacing the light providing device as a result of user mobility.

ATRACO concrete plans are described as mentioned previously by workflow specifications using an extension of BPEL (ATRACO-BPEL). Normally, workflow management systems have not been used for dynamic environments requiring adaptive behaviour. Typically an intranet-based workflow system executes, by using a collection of services that are owned and managed by the same organization. In this setting, service interruptions are rare and typically they are scheduled during system maintenance. On the contrary, in ATRACO we require adaptive workflows which need to react to varying environmental conditions. This transition from the static to dynamic and adaptive nature of workflows increases the runtime complexity of the management system, since the coordination mechanism must become more fault tolerant. On the other side our approach is viewed as collaborative problem solving approach where a set of autonomous agents work together to achieve a common goal. Our general idea then is that since a workflow describes the relationship between services and if an agent is represented by such a service, then the relationship between the agents would be possible to specify. Following such a combined agent-based and SOA approach means that a workflow could be used to establish the initial relationships of the ATRACO components. Applications can be specified then first with a workflow description using ATRACO-BPEL that defines the most common scenario and fault conditions. Once the basic system has been deployed, the agents could be working proactively so they can adapt to unforeseen circumstances and automatically handle the extension of the workflow description.

1.7 Conclusion

ATRACO supports the deployment and execution of applications that need to be adapted and reconfigured in dynamic environments. The need for adaptation and reconfiguration calls for a modular design approach, which the SOA paradigm tends to provide. Following this architectural style, each device provides services through which other components can obtain information or control its behaviour. When an application has to be adapted, either during application transition to a new environment or when a device running a service fails, a description of the structure of the application, which is modelled as a workflow of abstract services, is used by an adaptation module which makes use of ontologies, context information and defined policies to generate a new structure for the adapted application. The agent approach complements the SOA modular and flexible infrastructure by providing high level adaptation to user's tasks, as an intelligent control layer above SOA.

Appendix

Table 1.5 Metrics for device selection.

Metric	Description
Task suitability	This quantifies, as a percentage, how well a specific device or service is suitable for a specific task. Its value is calculated based on the semantic relevance between the task's description as it is presented in an abstract plan, and the device's or service's description provided by the device's ontology. E.g., for a service of providing light, both a lamp and a computer monitor can be candidates. The fact that the lamp's ontology states as primary purpose of the device the supply of light while the monitor's ontology state the light emission as a secondary attribute gives the lamp a higher score for this metric.
Efficiency	This metric measures the efficiency of a device or service for a certain goal. It expresses how well or to what degree the device is able to contribute in achieving a goal. Its value is calculated over a combination of other measures such as the device's proximity to the user (based on location info from User Profile Ontology (UPO) and device ontology), the quality of the service or device (as inferred from the specifications of the device that are encoded in its ontology) and the stability that quantifies how well a device will be able to perform a task to completion. The exact measures that participate in the calculation of efficiency are depended on the nature of the task and are derived from the policies encoded in the ATRACO ontologies. E.g., if the goal is to provide enough light for the user to read a book for an hour, a lamp located closer to the user has a strong potential to be selected. But if its light is weaker than the minimum needed for reading and another lamp exists a little more far, but in the range of the user, and can provide the desirable light level, the later should get a higher efficiency score. In a similar way if a device runs out of battery and will not last to achieve fully the goal its efficiency score should be discounted.
User distraction	User distraction expresses the inconvenience a user experiences when the system selects different groups of devices than those that the user prefers or is used to use for a specific task. User's preferences and habits are expected to be stated at the UPO or inferred from it. E.g., if a lamp with strong light is available near the user as he reads his book, but the user has expressed (directly or indirectly) its preference to use a specific one with weaker light when reading at this part of the room, the system should penalize the first lamp by increasing its user distraction score.
Conflicts with other tasks	This quantifies the number of other running tasks that will be blocked by selecting a device. E.g., if a monitor is currently used for watching a film, its conflicting score for selecting it for the task of displaying incoming emails should be greater than zero because it will obstruct the film watching task.

References

1. Alves, A., Arkin, A., Askary, S., Barreto, C., Ben, Curbera, F., Ford, M., Goland, Y., Guízar, A., Kartha, N., Liu, C.K., Khalaf, R., König, D., Marin, M., Mehta, V., Thatte, S., van der Rijn, D., Yendluri, P., Yiu, A.: Web Services Business Process Execution Language Version 2.0. Tech. rep., OASIS Web Services Business Process Execution Language (WSBPEL) TC (2007). URL http://docs.oasis-open.org/wsbpel/2.0/OS/wsbpel-v2.0-OS.html
2. Armstrong, D.J.: The quarks of object-oriented development. Commun. ACM **49**, 123–128 (2006). DOI http://doi.acm.org/10.1145/1113034.1113040. URL http://doi.acm.org/10.1145/1113034.1113040
3. Bidot, J., Schattenberg, B., Biundo, S.: Intelligent planner. Tech. rep., University of Ulm (2010)
4. Birrell, A.D., Nelson, B.J.: Implementing remote procedure calls. ACM Trans. Comput. Syst. **2**, 39–59 (1984). DOI http://doi.acm.org/10.1145/2080.357392. URL http://doi.acm.org/10.1145/2080.357392
5. Blair, G.S., Coulson, G., Andersen, A., Blair, L., Clarke, M., Costa, F., Duran-Limon, H., Fitzpatrick, T., Johnston, L., Moreira, R., Parlavantzas, N., Saikoski, K.: The design and implementation of open orb 2. IEEE Distributed Systems Online **2**, – (2001). URL http://portal.acm.org/citation.cfm?id=1435643.1436507
6. Brønsted, J., Hansen, K.M., Ingstrup, M.: Service composition issues in pervasive computing. IEEE Pervasive Computing **9**, 62–70 (2010). DOI http://dx.doi.org/10.1109/MPRV.2010.11. URL http://dx.doi.org/10.1109/MPRV.2010.11
7. Capra, L., Blair, G.S., Mascolo, C., Emmerich, W., Grace, P.: Exploiting reflection in mobile computing middleware. SIGMOBILE Mob. Comput. Commun. Rev. **6**, 34–44 (2002). DOI http://doi.acm.org/10.1145/643550.643553. URL http://doi.acm.org/10.1145/643550.643553
8. Chakraborty, D., Joshi, A.: Dynamic service composition: State-of-the-Art and research directions. Tech. rep., University of Maryland, Department of Computer Science and Electrical Engineering (2001)
9. Maciel da Costa, C., da Silva Strzykalski, M., Bernard, G.: An aspect oriented middleware architecture for adaptive mobile computing applications. In: Proceedings of the 31st Annual International Computer Software and Applications Conference - Volume 02, COMPSAC '07, pp. 81–86. IEEE Computer Society, Washington, DC, USA (2007). DOI http://dx.doi.org/10.1109/COMPSAC.2007.59. URL http://dx.doi.org/10.1109/COMPSAC.2007.59
10. Dustdar, S., Schreiner, W.: A survey on web services composition. Int. J. Web Grid Serv. **1**, 1–30 (2005). DOI 10.1504/IJWGS.2005.007545. URL http://portal.acm.org/citation.cfm?id=1358537.1358538
11. Emmerich, W.: OMG/CORBA: An Object-Oriented Middleware. In: J.J. Marciniak (ed.) Encyclopedia of Software Engineering, pp. 902–907. John Wiley & Sons (2002). URL http://www.cs.ucl.ac.uk/staff/w.emmerich/publications/Encyclopedia
12. Erl, T.: Service-Oriented Architecture: Concepts, Technology, and Design. Prentice Hall PTR, Upper Saddle River, NJ, USA (2005)
13. Frei, A., Popovici, A., Alonso, G.: Eventizing applications in an adaptive middleware platform. IEEE Distributed Systems Online **6**, 1– (2005). DOI 10.1109/MDSO.2005.20. URL http://portal.acm.org/citation.cfm?id=1069591.1069686
14. Gjørven, E., Eliassen, F., Lund, K., Eide, V.S.W., Staehli, R.: Self-adaptive systems: A middleware managed approach. In: SelfMan, pp. 15–27 (2006)
15. Goumopoulos, C., Calemis, I., Togias, K., Kameas, A., Pruvost, G., Wagner, C., Meliones, A., Wiedersheim, B., Bidot, J.: Integrated component platform for prototype testing and updated specification and design report. Tech. rep., Computer Technology Institute, ATRACO ICT 1.8.2 216837 D7 (2010)

16. Goumopoulos, C., Kameas, A.: Ambient ecologies in smart homes. Comput. J. **52**, 922–937 (2009). DOI http://dx.doi.org/10.1093/comjnl/bxn042. URL http://dx.doi.org/10.1093/comjnl/bxn042
17. Goumopoulos, C., Kameas, A.: Smart objects as components of ubicomp applications. International Journal of Multimedia and Ubiquitous Engineering, Special Issue on Smart Object Systems **4**(3), 1–20 (2009). URL http://www.sersc.org/journals/IJMUE/vol4_no3_2009/1.pdf. SERSC Press
18. Goumopoulos, C., Kameas, A., Hagras, H., Callaghan, V., Gardner, M., Minker, W., Weber, M., Bellik, Y., Meliones, A.: Atraco: Adaptive and trusted ambient ecologies. In: Proceedings of the 2008 Second IEEE International Conference on Self-Adaptive and Self-Organizing Systems Workshops, pp. 96–101. IEEE Computer Society, Washington, DC, USA (2008). DOI 10.1109/SASOW.2008.13. URL http://portal.acm.org/citation.cfm?id=1524875.1525041
19. Grace, P., Truyen, E., Lagaisse, B., Joosen, W.: The case for aspect-oriented reflective middleware. In: Proceedings of the 6th international workshop on Adaptive and reflective middleware: held at the ACM/IFIP/USENIX International Middleware Conference, ARM '07, pp. 2:1–2:6. ACM, New York, NY, USA (2007). DOI http://doi.acm.org/10.1145/1376780.1376782. URL http://doi.acm.org/10.1145/1376780.1376782
20. Grimes, R.: Professional Dcom Programming. Wrox Press Ltd., Birmingham, UK, UK (1997)
21. Gruber, T.R.: Toward principles for the design of ontologies used for knowledge sharing. Int. J. Hum.-Comput. Stud. **43**, 907–928 (1995). DOI 10.1006/ijhc.1995.1081. URL http://portal.acm.org/citation.cfm?id=219666.219701
22. Hayton, R.: Flexinet open orb framework. Tech. rep., APM Ltd., Poseidon House, Castle Park, Cambridge, UK (1997)
23. Hollingsworth, D.: Workflow management coalition - the workflow reference model. Tech. rep., Workflow Management Coalition (1995)
24. Kiczales, G.: Aspect-oriented programming. ACM Comput. Surv. **28** (1996). DOI http://doi.acm.org/10.1145/242224.242420. URL http://doi.acm.org/10.1145/242224.242420
25. Kiczales, G., Hilsdale, E., Hugunin, J., Kersten, M., Palm, J., Griswold, W.G.: An overview of aspectj. In: Proceedings of the 15th European Conference on Object-Oriented Programming, pp. 327–353. Springer-Verlag, London, UK (2001). URL http://portal.acm.org/citation.cfm?id=646158.680006
26. Kiczales, G., Rivieres, J.D.: The Art of the Metaobject Protocol. MIT Press, Cambridge, MA, USA (1991)
27. Kon, F., Costa, F., Blair, G., Campbell, R.H.: The case for reflective middleware. Commun. ACM **45**, 33–38 (2002). DOI http://doi.acm.org/10.1145/508448.508470. URL http://doi.acm.org/10.1145/508448.508470
28. Kon, F., Román, M., Liu, P., Mao, J., Yamane, T., Magalhã, C., Campbell, R.H.: Monitoring, security, and dynamic configuration with the dynamictao reflective orb. In: IFIP/ACM International Conference on Distributed systems platforms, Middleware '00, pp. 121–143. Springer-Verlag New York, Inc., Secaucus, NJ, USA (2000). URL http://portal.acm.org/citation.cfm?id=338283.338355
29. Kumar, R., Poladian, V., Greenberg, I., Messer, A., Milojicic, D.: Selecting devices for aggregation. In: WMCSA, pp. 150–159. IEEE Computer Society, Los Alamitos, CA, USA (2003). DOI http://doi.ieeecomputersociety.org/10.1109/MCSA.2003.1240776
30. Ledoux, T.: Opencorba: A reflective open broker. In: Proceedings of the Second International Conference on Meta-Level Architectures and Reflection, Reflection '99, pp. 197–214. Springer-Verlag, London, UK (1999). URL http://portal.acm.org/citation.cfm?id=646930.710404
31. Maes, P.: Concepts and experiments in computational reflection. SIGPLAN Not. **22**, 147–155 (1987). DOI http://doi.acm.org/10.1145/38807.38821. URL http://doi.acm.org/10.1145/38807.38821

32. McIlraith, S.A., Son, T.C., Zeng, H.: Semantic web services. IEEE Intelligent Systems **16**, 46–53 (2001). DOI http://dx.doi.org/10.1109/5254.920599. URL http://dx.doi.org/10.1109/5254.920599
33. Papadopoulos, N., Meliones, A., Economou, D., Karras, I., Liverezas, I.: A connected home platform and development framework for smart home control applications. In: Proceedings of the 7th IEEE International Conference on Industrial Informatics (INDIN09) (2009)
34. Pawlak, R., Seinturier, L., Duchien, L., Florin, G.: Jac: A flexible solution for aspect-oriented programming in java. In: Proceedings of the Third International Conference on Metalevel Architectures and Separation of Crosscutting Concerns, pp. 1–24. Springer-Verlag, London, UK (2001). URL http://portal.acm.org/citation.cfm?id=646931.710426
35. Pruvost, G., Kameas, A., Heinroth, T., Seremeti, L., Minker, W.: Combining agents and ontologies to support Task-Centred interoperability in ambient intelligent environments. In: Proceedings of the 2009 Ninth International Conference on Intelligent Systems Design and Applications, ISDA '09, pp. 55–60. IEEE Computer Society, Washington, DC, USA (2009). DOI http://dx.doi.org/10.1109/ISDA.2009.195. URL http://dx.doi.org/10.1109/ISDA.2009.195
36. Rao, J., Su, X.: A survey of automated web service composition methods. In: J. Cardoso, A. Sheth (eds.) Semantic Web Services and Web Process Composition, *Lecture Notes in Computer Science*, vol. 3387, pp. 43–54. Springer Berlin / Heidelberg (2005). URL http://dx.doi.org/10.1007/978-3-540-30581-1_5
37. Remagnino P Foresti, G.L.: Ambient intelligence: A new multidisciplinary paradigm. IEEE Transactions on Systems, Man and Cybernetics, Part A **35**(1), 1–6 (2005)
38. Seremeti, L., Goumopoulos, C., Kameas, A.: Ontology-based modeling of dynamic ubiquitous computing applications as evolving activity spheres. Pervasive Mob. Comput. **5**, 574–591 (2009). DOI 10.1016/j.pmcj.2009.05.002. URL http://portal.acm.org/citation.cfm?id=1630161.1630223
39. Sommer, R.: Personal Space: The Behavioral Basis of Design. Prentice Hall Trade, Englewood Cliffs, NJ, USA (1969)
40. Sycara, K., Paolucci, M., Ankolekar, A., Srinivasan, N.: Automated discovery, interaction and composition of semantic web services. Journal of Web Semantics **1**(1), 27–46 (2003)
41. Togias, K., Goumopoulos, C., Kameas, A.: Ontology-Based representation of upnp devices and services for dynamic Context-Aware ubiquitous computing applications. In: International Conference on Communication Theory, Reliability, and Quality of Service, pp. 220–225. IEEE Computer Society, Los Alamitos, CA, USA (2010). DOI http://doi.ieeecomputersociety.org/10.1109/CTRQ.2010.44
42. Verma, K., Sivashanmugam, K., Sheth, A., Patil, A., Oundhakar, S., Miller, J.: Meteor-s wsdi: A scalable p2p infrastructure of registries for semantic publication and discovery of web services. Inf. Technol. and Management **6**, 17–39 (2005). DOI 10.1007/s10799-004-7773-4. URL http://portal.acm.org/citation.cfm?id=1047575.1047628
43. Wagner, C., Hagras, H.: Toward general type-2 fuzzy logic systems based on zslices. Trans. Fuz Sys. **18**, 637–660 (2010). DOI http://dx.doi.org/10.1109/TFUZZ.2010.2045386. URL http://dx.doi.org/10.1109/TFUZZ.2010.2045386
44. Wollrath, A., Riggs, R., Waldo, J.: A distributed object model for the java system. Computing Systems **9**(4), 265–290 (1996)
45. Yang, Z., Cheng, B.H.C., Stirewalt, R.E.K., Sowell, J., Sadjadi, S.M., McKinley, P.K.: An aspect-oriented approach to dynamic adaptation. In: Proceedings of the first workshop on Self-healing systems, WOSS '02, pp. 85–92. ACM, New York, NY, USA (2002). DOI http://doi.acm.org/10.1145/582128.582144. URL http://doi.acm.org/10.1145/582128.582144

Chapter 2
Adaptive Networking

A. Meliones, I. Liverezas, D. Economou

Abstract Intelligent Environments have been commercially available already over ten years without becoming such a mass product as expected. The expectations of potential users of mentioned solutions have not been fulfilled yet due to missing globally accepted standards causing interoperability problems of different hardware and software components, complexity of configuration and use, lack of universal service consideration, and insufficient ROI for private residence owners. Clearly, a stronger emphasis is needed on device adaptation, usability and scalability, which can seamlessly accommodate new IE services. Although several research efforts have addressed the development of IEs through networking existing devices and resolving interoperability issues with the help of middleware, there has been little work on specifying at a high level of abstraction how such abstraction services would work together at the application level taking into account in a combined dynamic way the heterogeneous networks and services. This chapter presents a framework for network adaptation in IEs using OSGi and UPnP technology allowing the uniform and transparent access to devices and services present in the networked environment and supporting the realization of activity spheres across a mixture of heterogeneous networks. Our work specifically focuses on the issue of heterogeneity at the network level and defines adaptation as the systemic mechanism in order to deal with this issue. The proposed network adaptation framework has been implemented within the ATRACO project and supports an ambient ecology trial hosted in the iSpace in Essex demonstrating a number of futuristic user activity spheres.

Apostolos Meliones
University of Piraeus, Department of Digital Systems, and inAccess Networks, 12 Sorou Str., 15125 Maroussi, Athens, Greece, e-mail: `meliones@unipi.gr`

Ioannis Liverezas
inAccess Networks, 12 Sorou Str., 15125 Maroussi, Athens, Greece, e-mail: `iliverez@inaccessnetworks.com`

Dimitris Economou
inAccess Networks, 12 Sorou Str., 15125 Maroussi, Athens, Greece, e-mail: `decon@inaccessnetworks.com`

2.1 Introduction

Intelligent environments have been commercially available already over ten years without becoming such a mass product as expected. The expectations of potential users of mentioned solutions have not fulfilled yet due to missing globally accepted standards causing interoperability problems of different hardware and software components, complexity of configuration and use, lack of universal service consideration, and insufficient ROI for private residence owners. The main reason is that smart environment networks consists of a large variety of content sources (e.g. sensors), multiple information carriers (wired and wireless media) and communication standards which lead to problems of interoperability, administration and reducing the ease of use. At the same time, there is significant interest in home networking today, which stems from the availability of low-cost communication technology and readily available access to content and services from various sources and suppliers, extending to accommodate a number of smart applications, including management, control and security, in the home environment. Therefore a stronger emphasis needs to be set on device adaptation, usability and scalability, which can seamlessly accommodate new smart environment services.

Several research efforts have addressed the development of an Ambient Intelligent Environment (AIE) through networking existing devices and resolving interoperability issues with the help of middleware. The NGN@Home ETSI initiative [6] facilitates interoperability between the various home network end devices and various home hub technologies and will provide a standardized approach to Next Generation Networks at home and in home intelligent device technologies. TEAHA [8] addressed networked home control applications and their complementarity to audiovisual networked applications via an interoperable middleware having a hardware centric view for creation of universal solutions. EPERSPACE [3] concentrated on creating a Home Platform to link different devices at home to an interoperable network, to provide these devices information about what you or your friends need and make the system respond accordingly. AMIGO [1] aimed to develop ontology based middleware and interoperability of devices, artefact and services. SENSE [7] concentrated on developing methods, tools and a test platform for the design, implementation and operation of smart adaptive wireless networks of sensing components. HOME2015 [4] is a multidisciplinary research programme aiming to create future systems and technologies for future homes. MUSIC [5] develops a software development framework that facilitates the development of self-adapting, reconfigurable software that seamlessly adapts to dynamic user and execution context in ubiquitous computing environments. In [20] a home network is described as a set of interconnections between consumer electronic products and systems, enabling remote access to and control of those products and systems, and any available content such as music, video or data. The elements of the above definition are connections, access, and control. One point, implicit in the above definition, that must be emphasized is ease of use. For those involved in the consumer electronic industry, it is understood that the *home networking* implies fundamental simplicity.

Other efforts present solutions based on web services technology to solve the interoperability problem in smart home environments [16], permit users or agents to aggregate and compose networked devices and services for particular tasks [13], discuss how recent technological progress in the areas of visual programming languages, component software and connection-based programming can be applied to programming the smart home [12], introduce supporting infrastructures moving the burden of connectivity away from the end devices to the system core [10], enhance the OSGi standard to integrate many existing home protocols and networks with emphasis on realizing an effective converged service gateway architecture for smart homes [21], [14], as well as propose hardware architectures for ubiquitous computing in smart home environments where the devices may autonomously act and collaborate with each other using agent-based service personalization software architectures [22]. Other approaches to modelling and programming devices for intelligent environments model devices as collections of objects [16], as Web services [15], and as agents [17]. The lack of a de-facto standard middleware for distributed sensor-actuator environments has been identified by many researchers as one of the key issues limiting research on intelligent environment and the proliferation of intelligent environments from research environments to their deployment in our everyday lives, as for instance in [19] presenting a robotic middleware as glue between sensors, actuators and services for complex ubiquitous computing environments. Early attempts to adapt heterogeneous environments consisting of dissimilar networks and computing devices primarily address primitive system architectural issues and reconfiguration principles [23], but broader usability of the developed approaches is limited for AIE applications. Moreover, there has been little work on specifying at a high level of abstraction how such abstraction services would work together at the application level taking into account in a combined and dynamic way the heterogeneous networks and services.

One of the main objectives of the ATRACO project [2] is to research and design an architecture and provide a system specification that will lay the foundations for the development of adaptive and trusted ambient ecologies. The proposed system operates in an AIE, which is populated with people and an ambient ecology of devices and services. People have goals, which they attempt to attain by realizing a hierarchy of interrelated tasks. For each user goal and the corresponding task model, an Activity Sphere (AS) is initialized, which consists of all software modules, services and other resources necessary to support the user in achieving the goal. The ATRACO system architecture enables meaningful integration of relevant services and information during run-time and accomplishes that in a privacy-sensitive manner. ATRACO adopts a service-oriented architecture, combined with (a) intelligent agents that support adaptive planning, task realization and enhanced human-machine interaction and (b) ontologies that provide knowledge representation, management of heterogeneity, semantically rich resource discovery and adaptation using ontology alignment mechanisms. An ATRACO system supports adaptation at different levels, such as the changing configuration of the Ambient Ecology (AE), the realization of the same AS in different AIEs, the realization of tasks in different contexts, and the dynamic interaction between the system and the user.

In ATRACO, Network Adaptation (NA) is one of the key project dimensions allowing the uniform and transparent access to devices and services present in the networked environment and supporting the realization of activity spheres across a mixture of heterogeneous networks. Unlike other past and recent research efforts briefly presented above, our proposed framework specifies how network abstraction services should work together at the application level taking into account in a combined and dynamic way the heterogeneous networks and services present in the AIE. Our work additionally focuses on the issue of heterogeneity at the network level, concerning both devices and services, and defines adaptation as the systemic mechanism in order to deal with this issue.

NA in ATRACO offers a wealth of new exciting AIE experience on top of existing broadband service bundles. By exploiting the capability of continuous connectivity of an AIE to the Internet, it allows the provisioning of a set of services focused on the AIE offering at the same time a new experience for AIE management and control. Furthermore, it gives the operator the opportunity to expand its business by offering new customer services, such as security, safety, surveillance, energy management and monitoring, comfort and way of living management, and many other value added services. NA can also interface a wide variety of sensors and actuators by multiple suppliers according to the user needs. NA is fully scalable and new services can be seamlessly accommodated using a sophisticated service development framework. Other benefits include the generation of alerts and user notification to always be aware of the AIE condition, the time-scheduling of certain functions for carrying out time-dependent tasks (e.g. switch on/off electrical devices), the creation and management of scenes and scenarios to easily adapt the AIE according to the situation (e.g. scene creation for watching movies by applying the desired lighting), as well as the recording and presentation of historical data and statistics to provide the user with useful information that can be further exploited for his own benefit (e.g. energy saving, user behavior adaptation etc.). The following sections present in detail the network adaptation framework.

2.2 Intelligent Environment Network

An intelligent environment platform usually consists of different software components running in different network elements such as GUI clients for mobile phones and PC, portal server for remote login and user authentication and gateway software for accessing various home devices like automation equipment and network cameras. The user is able to have control over his environment either from inside using a PC or remotely using a PC or a mobile phone. Remote access is usually provided through a dedicated portal server after the user has authenticated himself. The connections on the portal server are encrypted while the connection between the home gateway and the portal is initiated from the gateway and is encrypted as well. This mechanism on one hand provides the user with advanced protection from harmful internet attacks and on the other hand offers simple and easy installation since it does not need any network configuration on the modem/router. Figure 2.1 depicts the intelligent environment network architecture, addressing a smart home, contributed to the ATRACO project.

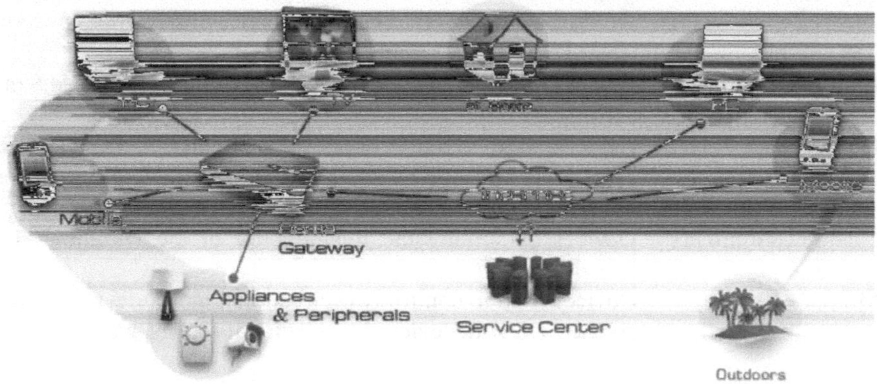

Fig. 2.1 Intelligent Environment Network Architecture.

The smart home platform contributed to the ATRACO project provides a set of services focused on the home and offers a new experience for home automation and management, including security, safety, surveillance, energy management and monitoring, comfort and way of living management, and other exciting smart living services. The networking technology supported by the smart home platform can be easily installed without the need for extra wiring. A wide variety of COTS sensors and actuators can be seamlessly integrated in the home network according to the user needs. Main platform benefits include the generation of alerts and user notification through email, phone or SMS to always be aware of home condition, the time-scheduling of certain functions for carrying out time-dependent tasks (e.g. switch on/off electrical devices), the creation and management of scenarios to adapt the home environment according to the situation (e.g. scene creation for watching movies by applying the desired lighting), as well as the processing of persistent data in order to provide the user with useful information (e.g. towards energy saving).

The Intelligent Environment Network is a network integrating wired and wireless peripherals from multiple technologies. The user may select from a long list of wired and wireless automation peripherals to be integrated with the IE, giving unlimited capabilities in connectivity with other devices, user access and service provision. Devices have to be based on one of the automation technologies supported by the smart home platform, e.g. Sienna powerline, Lonworks PL/TP, or RF Z-Wave. Each type of device is supported using a driver per different technology adapting the functionality to an upper network independent layer. Automation peripherals are grouped depending on the functionality type:

Sensors	Devices that can sense a medium and provide information about its state (binary). This group encapsulates devices like wall mounted switches, remote controls, motion sensors and door traps.
On/off	Devices used to open or close a circuit on demand. Such devices may be used to control appliances like coffee makers, water heaters, floor lights etc.
Combined on/off	Devices that combine the functionality of the on/off and Sensor types.
Dimmers	Devices used to open or close a circuit on demand and can also control the electrical current flow. Mostly used for dimmable lights.
Combined dimmers	Devices combining the functionality of the Dimmers and Sensor types.
Motors	Devices used to control two-directional motor devices. They can be used with devices like electrical shades, garage doors, valves etc.
Combined motors	Devices combining the functionality of the Motors and Sensors types.
Thermostats	Devices giving feedback of a room temperature and control the central heating.

The Intelligent Environment Platform contributed to the ATRACO project (see Fig. 2.2) integrates a set of Java tools and components that enable an OSGi-literate engineer to quickly design, develop and deploy new intelligent environment services utilising the provided OSGi service platform and widely adopted automation technologies. The Home Controller is used to integrate connectivity with home devices of various home control technologies. The Service Platform embeds the use of OSGi technology in the home controller, ensuring interoperability with home devices and an easy way to integrate new home services. The base OSGi platform is extended by a set of OSGi network subsystems integrating various automation technologies supported by the home controller. The different network subsystems are interfaced in a common way through a Network Adaptation Layer. This layer adapts any network subsystem supported by the service platform, providing unified interfaces for devices from different networks, adding any specific support required for each type of device. The main intelligent environment functionality is actually offered and built by using the Network Adaptation API. Following the Network Adaptation API specification, a developer may build various applications, such as presentation layer applications (e.g. a web based UI), monitoring applications that collect data and send them to a backbone server and other control applications in the IE.

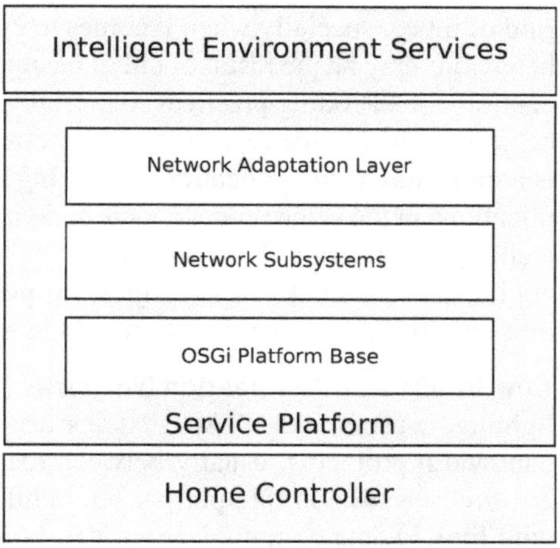

Fig. 2.2 Intelligent Environment Development Framework

2.3 Intelligent Environment Service Design Considerations

The design and development of an intelligent environment service should take into account a number of considerations, such as embedded systems performance, other embedded system constraints, service responsiveness, low bandwidth automation networks, battery operated peripherals etc.

Performance Smart home controllers are embedded systems having enough processing power to load featured operating systems and OSGi applications, however they are more than far the capabilities of a desktop system. That is why it is really important to consider performance when developing intelligent environment applications. There are generally three rules to consider when developing applications for systems with resource constraints: (i) do not do what you don't really need to do, (ii) do not allocate memory if you can avoid it, and (iii) search for a faster implementation with the same result.

Storage Constraints Smart home controllers use flash memory components with limited write operations comparing to traditional storage components. Taking into account that the controllers' flash memory components are not modular and replaceable, they should never be used for frequent storage operations. In smart home controllers, flash memories are used to store the operating system filesystem, except the /tmp filesystem which is stored in RAM. So, any file created not under /tmp, is stored in controller's flash memory. On the other hand, any file stored in /tmp filesystem does not affect the lifetime of flash memory, however it is lost in case of an unexpected shutdown of the controller. In case of critical information or useful log information, other resources should be considered: USB memory stick, backbone logging servers etc.

Responsiveness Any intelligent environment service providing user feedback and interaction should be responsive, that is not hanging or freezing for significant period of time, especially when it comes to every day operations. Such non responsive behaviour may be the result of either design choices not taking into account user perception or bad coding practices. Especially regarding OSGi, synchronization should be used with extreme care, especially between cooperating bundles: OSGi platform is a multi-threaded application, managing the execution of other multi-threaded applications in the same process context. Synchronisation should thus be using custom techniques, ensuring that two bundles will not interoperate in a way that produces application deadlocks or long hanging periods. Especially regarding OSGi, see a presentation regarding OSGi best coding practices from OSGi alliance [11].

Low Bandwidth Automation Networks Home automation networks dealing with lighting, sensors and HVAC systems usually utilise physical mediums using low bandwidth protocols, usually between the range of 5kbps to 80kbps with variable error/retransmission rate per packet. While 5-40 kbps seems enough for protocols with limited data communication, we should always consider how scalable the behaviour would be when the numbers of network nodes is increased. Basic rule is to avoid sending data to a device when this is not really needed: polling a thermostat for the temperature value, while knowing that the thermostat will send an update when the temperature changes, is not needed. In case the thermostat does not provide such an automatic update message, we should carefully consider the polling period required for a satisfying user perception or for other applications purposes: since the temperature usually will not fall significantly in five minutes, unless a door opens (and the outer temperature is below zero), the polling period could be set to five minutes. In case we would like a more frequent update, try for a one or two minute periods, however more frequent periods would likely be unreasonable: in such a case, it is likely that the device application is not intended for the use you desire and they should be replaced with a different device.

Battery operated peripherals Battery operated devices, e.g. motion sensors, are used in some automation networks. The battery lifetime is limited, usually specified by the manufacturer between one to five years. However, such specifications refer to wise use of the battery cycles. The three parameters that affect more the battery life are:

Sleeping period	Usually a device sleeps for some time and then wakes up to check for data available; keep the sleeping period of a device as long as possible.
Off delay	The period used until a device checks again for a change of its state, for example a motion sensor to declare that it no longer detects motion.
Polling operations	In case each time the device wakes up it receives a number of polling commands, then the battery is exhausted sooner.

2.4 Network Adaptation Definition

In the ATRACO project we address the design of AIE systems using a service-oriented approach, in which resources in the environment provide independent, heterogeneous, loosely-coupled, primitive services. The ATRACO infrastructure consists of SOA services. It supports basic services such as context management and reasoning, communication management, user profiling and service discovery, as well as adaptation and privacy services that form the basis for ATRACO ambient ecologies. NA is a set of functions and protocols allowing an AIE system to interact with network resources. The main goals of the NA layer are to allow devices and services to be used seamlessly by an AIE system and to simplify the access to networks in the AIE (usually for control, data sharing, communications, and entertainment). NA will achieve this by defining the appropriate functions and designing and implementing the needed abstraction layers.

Network resources in an AIE can be either devices or services. In general, devices include wired or wireless automation devices and entertainment devices. All these devices, intelligent or not, can use different networking technologies, thus could be part of different networks in a specific environment. For example, in a household the lights could be controlled using Lonworks technology while the motion sensors could be wireless Z-Wave for flexible and easy installation. The access from a single control point on both lights and motion sensors demands the use of two different networking protocol stacks: LonTalk and Z-Wave. This complicates the implementation of a control application since its design should cope with different protocol architectures and software structuring. The problem escalates when the control networks increase in numbers and the complexity of the control application gets maximized. The term intelligent device refers to a networked device with sufficient processing power to run network protocols. An intelligent Lonworks lamp or motor has built in the Lonworks technology so it can be controlled over the Lonworks network from another Lonworks device. Non intelligent devices can be converted to intelligent ones by connecting them on other intelligent nodes which are called network adapters.

Figure 2.3 illustrates the internal layering of the ATRACO NA component. To cope with the complexity of accessing diverse control networks, a common device representation layer is introduced. This representation layer should be architected in such a way in order to handle different networks (LonWorks, KNX, ZigBee, Z-Wave, X10 etc.) while it hides the complexity and details of each one of them. The application which should be placed on top of this representation layer, could not tell the difference between a Z-Wave lamp and a LonWorks lamp. The device representation layer creates a unique representation of each different device type across networks, which simplifies the evolution of applications and services. Beside devices there are various services which in most of the cases are services provided over IP networks. Examples of such services are the NTP (Network Time Protocol) to synchronize time to a reference time source, VoIP (Voice over IP) to implement internet Telephony, RTSP (Real Time Streaming Protocol) to deliver voice and/or video stream with real time properties etc. Such types of services are of interest to an AIE so they must be shaped as AIE services and provided through the NA layer.

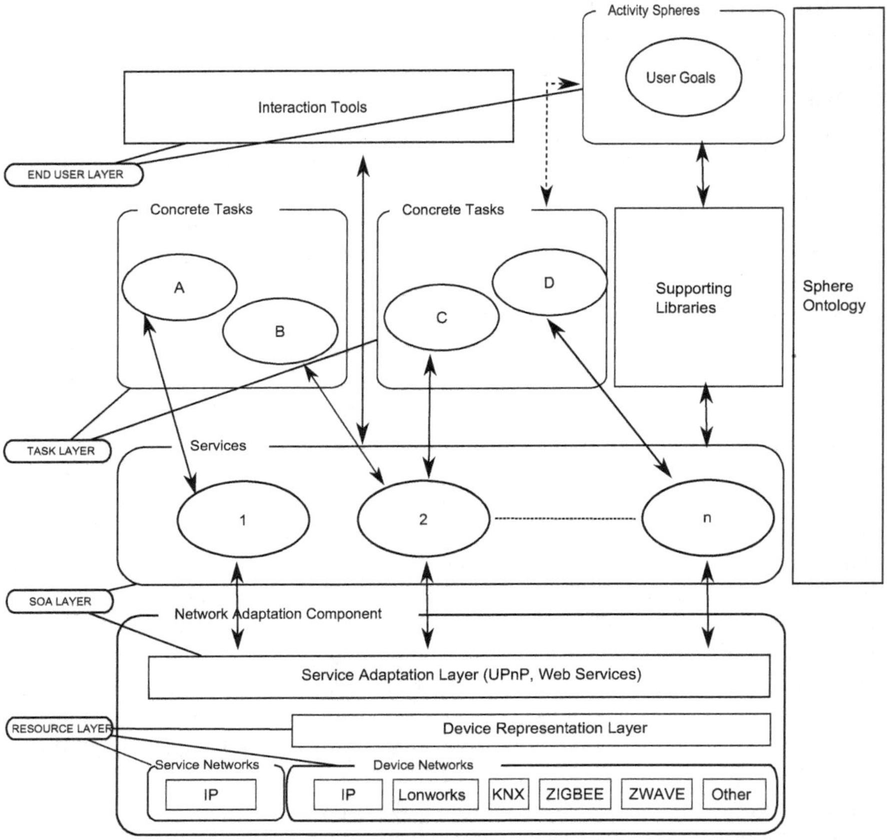

Fig. 2.3 Network Adaptation Concept

All network resources, devices and services, should be provided in an AIE as services. The functionality in these environments is exposed as semantic services which an actor (either a user or an agent) can discover and then compose to form AIE applications. Each service is associated with at least one semantic description which shields the actor from the complexity of the Resource Layer realization and makes it easy for the actor to employ these services in accomplishing interesting and useful tasks. SOA offers one such prospective architecture where every component is either a service provider or a service consumer or both service provider and consumer. SOA may unify AIE processes by structuring large applications as an ad hoc collection of services. Different groups of applications both inside and outside an AIE system can use these services. SOA may use web services standards and web technologies and is rapidly becoming a standard approach for many information systems. However, web services face significant challenges because of particular requirements. Applying the SOA paradigm to an AIE system presents many problems, including response time, support of event-driven, asynchronous parallel applications, complicated human interface support, reliability, etc. The following sections describe the technology that ATRACO has chosen to implement its NA SOA architecture.

In ATRACO, NA is one of the key project dimensions allowing the uniform and transparent access to devices and services present in the networked environment and supporting the realization of activity spheres across a mixture of heterogeneous networks. The NA layer basically consists of two sub-layers; the common Device Representation Layer and the Service Adaptation Layer. The NA Layer is implemented using a mixed architecture, including a centralized home controller able to seamlessly integrate various different devices from different technologies around different home automation and control space and a variety of modules, some of which may be able to work in a distributed architecture. The Device Representation Layer provides an abstraction of the AIE devices to OSGi services that can be used by other AIE components. The service adaptation layer focus is first to provide the Device Representation Layer devices as OSGi services to the other AIE entities. On top of that, OSGi device representations are eventually transformed to UPnP services. Another focus of the Service Adaptation Layer is the UPnP adaptation of IP services participating in an AIE system. With the assumption that the available services are web services it would be straightforward to wrap them as UPnP services or services of any other technology of choice (e.g R-OSGi [18]). It might even be possible to fully automate this procedure in a service agnostic way. UPnP seems to cover most of the aspects that are specified for the Service Adaptation Layer, such as ease of network setup, device discovery, self announcement and advertisement of supported capabilities.

2.5 Network Adaptation Guidelines

An ATRACO system supports adaptation at different levels, such as changing configuration of the ambient ecology, realization of the same activity sphere in different AIEs, realization of tasks in different contexts, as well as dynamic interaction between the system and the user. The design and specification of the NA component is done having the following adaptation guidelines in mind:

- Unified access on devices that belong to different networks (e.g. LON, Z-Wave etc): The proposed NA framework defines common device types that share common variables across networks.
- Interoperability amongst multiple networks: The proposed NA framework allows events from any network to trigger actions to any network.

- Services can use alternative resources for task completion: The proposed NA framework employs a device registry where the available devices can be registered and queried by identifier, property value, constraints, and interface.
- Dynamic discovery for new networks and devices in an Ambient Intelligence (AmI) space ecology: The proposed NA framework features a Device/Driver manager to assure dynamic support for new networks and devices.
- Representation of device resources as services in a SOA middleware in order to be consumed by other AIE components.
- Representation of legacy internet services in order to be consumed by other AIE components.
- Relaying of existing Web Services: An AIE is able to use a web service provided by the NA component without worrying about service availability, accessibility and quality (e.g. NA provides single access to the AIE for a weather report service being able to evaluate availability, accessibility and quality of many existing weather report web services provided by internet sites).

2.6 Network Adaptation Framework

The NA middleware is a technological solution we have used for prototyping a typical AIE which seamlessly blends IP networking with a wealth of multimodal home automation functionality. NA provides uniform access to the controlled devices (including the full range of sensors and actuators) through an adaptation layer mapping all different network domains in the AIE space to the IP level, and some basic services for task execution and event management. Within ATRACO, NA contributes to the realization of activity spheres under the orchestration of the Sphere Manager (SM). NA is able to represent the integrated AIE to the ATRACO ontology level maintaining local device and policy ontologies and collaborating with the Ontology Manager (OM) to respond to queries regarding the state or properties of devices and during ontology alignments to propagate context changes to the sphere ontology.

NA integrates a set of Java tools and components that enable to quickly design, develop and deploy services in an AIE space utilizing the provided OSGi service platform and widely adopted automation technologies. A home controller is used to integrate connectivity with devices of various home control technologies, such as LON PL, LON TP, Z-Wave (RF), X10 etc. The service platform embeds the use of OSGi technology in the home controller, ensuring interoperability with devices in the AIE and an easy way to integrate new services. The different network subsystems are interfaced in a common way through a device representation layer, known as NA-OSGi or ROCob layer, providing unified device representations. The main NA functionality is actually offered through the device representation middleware. User applications may include presentation layer applications, monitoring applications that collect and process data and send them to a backbone server or trusted recipients, home control and pervasive applications.

2.6.1 Why OSGi?

The key reasons for choosing OSGi as the framework to implement the device representation layer of the NA middleware are summarized in the following list:

- Reduced development complexity - Developing with OSGi technology means developing bundles: the OSGi components. Bundles are modules. They hide their internals from other bundles and communicate through well defined services. Hiding internals means more freedom to change later. This not only reduces the number of bugs, it also makes bundles simpler to develop because correctly sized bundles implement a piece of functionality through well defined interfaces.
- Ability to reuse ready made bundles in application development.
- Ease of component installation and management via a standardised API (e.g. using a command shell, a TR-69 or OMA DM management agent, a cloud computing interface etc.). The standardized management API makes it very easy to integrate OSGi technology in existing and future systems.
- Ability of dynamic system updates. Bundles can be installed, started, stopped, updated, and uninstalled without bringing down the whole system.
- Availability of a dynamic service model allowing bundles to find out what capabilities are available on the system and adapt the functionality they can provide making code more flexible and resilient to changes. This requires that the dependencies of components need to be specified and it requires components to live in an environment where their optional dependencies are not always available.
- Simplicity of the OSGi API, powerful security model, portability across different execution environments and wide use.

2.7 Device Representation Layer

NA acts in device representationultiple device types functioning using different technologies. In such environments devices of different types and communication technologies should be interoperable. NA needs a way of controlling and representing seamlessly devices of the same type (e.g a Smart Light), even if they use different communication technology. Moreover, it needs a flexible way of supporting new device types and technologies, through the use of quick, efficient and modular interfaces.

2.7.1 Architecture Description

Figure 2.4 depicts all blocks currently available in the device representation architecture, which constitute a working software release. This architecture can easily be extended to support other technologies, such as X-10 for example, with the development of the corresponding device representation layer drivers. Moreover, the device representation layer API can be easily extended maintaining backwards compatibility, to support new functions in order to cover the functionality of new devices. In the following we include an overview that describes the functionality of each block (generic input/output, internal functionality) and a reference to relevant standards that the block complies to.

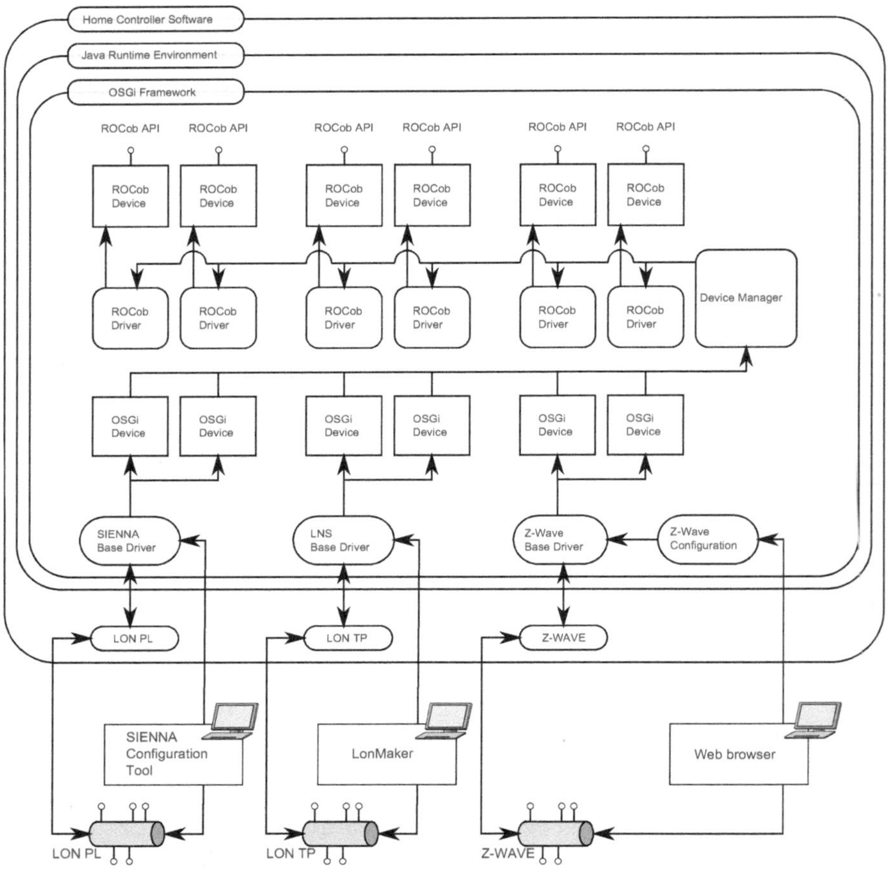

Fig. 2.4 Device Representation Layer Architecture

LON PL Lonworks is a home network protocol supported by the home controller. The physical layer as the PL implies is the powerline network of the house. Lonworks is seven layer protocol (OSI). The LON PL block is a Linux driver for accessing the neuron chip which runs the lower 4 Lonworks layers. All these layers are stacked in a commercial firmware provided by Echelon (the company who invented Lonworks) named MIP (Microprocessor Interface Program). The Linux driver is the software that communicates with MIP and transfers data back and forth from the neuron chip to the microprocessor memory.

LON TP It is also about Lonworks but the physical layer here is the Twisted Pair especially called FTT-10. LON TP block is the same Linux driver like LON PL regarding the source code. At runtime LON TP is a different instance and communicates with a MIP running inside a FTT-10 neuron chip. LON TP and LON PL can coexist in the same home controller.

Z-Wave It is a standard serial Linux driver which allows user programs accessing Z-Wave firmware running on a microcontroller located on a daughter-board which is connected internally on a serial port of the Home Controller. Z-Wave is an RF technology from Zensys that has a five layer protocol stack. The layers are the physical (RF), MAC (CSMA/CA), Transfer, Routing and Application. The first 4 layers are implemented inside the Z-Wave microcontroller. Apart from the protocol layers, the Z-Wave microcontroller provides a serial API so that microprocessor hosted applications have access on the Z-Wave network.

Network Base Drivers The network base drivers are OSGi software elements that basically implement all or part of a control network protocol like Lonworks or Z-Wave. These drivers based on the provided configuration information coming from the network configuration tools, create software representations of the real network devices. These representations are in fact software objects which are called OSGi devices and provide a general interface for other OSGi applications to make actions and take feedback from the real network devices. A device object represents some form of a device. It can represent a hardware device, but that is not a requirement. Each network driver registers the associated device services in the framework in order device manager to find them and attach them to appropriate driver service.

Sienna Especially for the PL network the Home Controller has adopted the SIENNA protocol variation which provides a lightweight and reliable communication protocol for powerline networks. The Sienna Base driver is an OSGi base driver which implements the SIENNA protocol on top of the standard LNS driver. For simplicity the LNS driver doesn't appear as a different block.

LNS The LNS base driver is an OSGi base driver which implements the three upper layers of the Lonworks protocol stack. The LNS base driver is a third party proprietary block provided by Echelon. The LNS base driver has been modified in order to support the hardware LON TP and PL interface on the Home Controller.

Z-Wave The Z-Wave base driver is an OSGi base driver which implements the Z-Wave serial API and allows other OSGi applications to have access over the Z-Wave network but also to information which is locally stored in the Z-Wave microcontroller.

Sienna Configuration Tool The Sienna Configuration tool is a third party proprietary software which comes from Secyourit, through which the installer is able to setup a network of SIENNA LON PL devices. After the installation is completed this tool generates an XML file that contains valuable network information like the names, addresses and the types of the devices. This XML file is further used by the Sienna Base Driver in order to instantiate the sienna OSGi devices.

LonMaker LonMaker is the network configuration tool provided by Echelon for setting up a LNS network. This is a professional tool indented to be used by the installer. After the installation is completed, LonMaker generates an XML file that contains network information like the device names, addresses, network variables etc. This XML file is used by the LNS Base Driver so that the latter to instantiate the LNS OSGi devices.

Z-Wave Configuration The Z-Wave configuration service is responsible to setup a network of Z-Wave devices through a friendly user interface. This interface guides the user via a step-by-step procedure and after the configuration has been completed it transfers all valuable network information to the Z-Wave network base driver.

OSGi Device An OSGi device is a representation of a physical device or other entity that can be attached by a Driver service. OSGi devices that represent physical devices are instantiated by Network Base Drivers.

Device Manager The device manager is responsible for initiating all actions in response to the registration, modification, and unregistration of Device services and Driver services. The device manager detects the registration of Device services and coordinates their attachment with a suitable Driver service. All available Driver services participate in a bidding process. The Driver service can inspect the Device properties to find out how well this Driver service matches the Device service. The highest bidder is selected. The selected Driver service is then asked to attach the Device service. If no Driver service is suitable, the Device service remains idle. When new Driver bundles are installed, these idle Device services must be reattached.

ROCob Driver The ROCob drivers are in principle refinement drivers that translate the network specific device interface to a common interface. This common interface named as ROCob API is used by other OSGi or non OSGi applications in order to communicate with devices in the control networks. Each ROCob driver registers itself in the framework registry expressing its interest for OSGi devices by setting specific criteria. The ROCob driver driven by the Device Manager bids for the device of interest and finally if it is the highest bidder amongst other ROCob drivers it uses the provided device services.

ROCob Device The ROCob device is a representation of a physical device but at a higher abstraction layer compared with the respective OSGi device. ROCob devices are instantiated by ROCob Drivers.

ROCob API The ROCob API is a common interface that hides the specific network interface and gives a general interface for each different device category. For example the API calls for controlling an actuator or receiving events from a sensor are the same either the devices belong to the LON PL network, the LON TP network or the Z-Wave network.

2.7.1.1 The LonWorks Subsystem

The LonWorks Base Driver communicates with the available devices and with the aid of an external Network Configuration XML file it creates the corresponding LonWorks OSGi Base devices. This external file can be produced by a network built with LonMaker with a utility that exports a LonWorks network image to an XML file, suitable for use by the smart home platform.

The LonWorks Base Driver communicates with the available devices and with the aid of an external Network Configuration XML file it creates the corresponding LonWorks OSGi Base devices. This external file can be produced by a network built with LonMaker with a utility that exports a LonWorks network image to an XML file, suitable for use by the smart home platform.

These OSGi Base devices can be used by various clients, such as the ROCob drivers or the Devbrowser bundle. The clients can communicate with the LonWorks Base Devices either with standard LonMark SNVTs and SCPTs provided by the OSGi-stdtypes package or with custom ones.

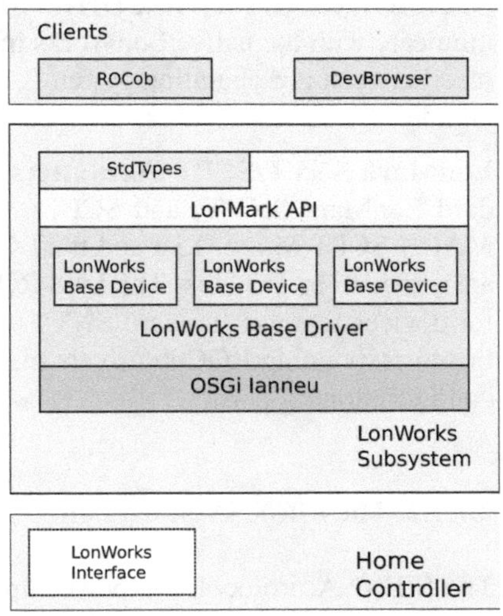

Fig. 2.5 The LonWorks Network Subsystem in the Home Controller

The Lonworks Subsystem is supported in the smart home platform with the LonWorks OSGi stack, which comprises of the following bundles:

LonMark OSGi Device Access API This bundle provides an API with which the various LonWorks ROCob drivers communicate with the LonWorks OSGi Base Devices that are offered by the Base Driver for LonWorks devices bundle. This bundle offers the "language" to read/write network variable values and communicate with the LonWorks devices and LonWorks network interfaces.

LonWorks BaseDriver Bundle This bundle implements the LonMark API and is the heart of the LonWorks support in the home controller. The base driver keeps and maintains the LonWorks network image. Each LonWorks operation is performed through this bundle. Moreover, it supports the LonWorks over IP protocol and the LNS Pass-Through mode which it can be used for network creation. The base driver for LonWorks devices (also called LonWorks bundle) is loaded with a special XML file that holds the information about the LonWorks network. Based on the information that this file provides, the LonWorks bundle creates and registers LonWorks Base OSGi Devices that correspond to the real devices in the LonWorks network. These devices are then used by the ROCob LonWorks Composite Driver in order to dispatch them to the appropriate ROCob drivers and create the relevant ROCob devices.

LonWorks Device Browser This bundle is a debugging tool for base device services and presents a hierarchical explorer tree view of the LonWorks network. If the device has input network variables, the current values of the input network variables are displayed in text boxes (without formatting applied). The input and output network variables can be read, and the input network variables on the device can be written using the Device Browser.

Neuron Network Interface Driver This bundle takes up the responsibility to communicate with the native LonWorks interfaces of the home controller. It is the interface between the operating system's LonWorks interface driver and the LonWorks bundle.

LonMark SNVT/SCPT Formatters This bundle provides the values for the standard LonMark SNVTs and SCPTs, as these are described in detail in the LONMARK SCPT Master List and the LONMARK SNVT Master List documents that are provided by Echelon. The LNS ROCob drivers use this bundle to form messages for devices that use standard SNVTs and SCPTs. However, devices and eventually the corresponding ROCob drivers may use custom NV types and network variable configuration properties.

2.7.1.2 The SIENNA Subsystem

The SIENNA protocol is a power line communication protocol which sits on top of the LonWorks communication protocol. On the home controller, the SIENNA protocol implementation uses the underlying OSGi LonWorks implementation for communicating with the SIENNA Devices and on top of it builds its own logic. The Connected Home SIENNA stack abstracts the physical SIENNA devices into SIENNA OSGi devices. This is performed by the SIENNA Stack Base-driver OSGi Bundle (SSBOB). SSBOB abstracts the physical SIENNA Network in a way that is usable by the end user. In this manner, SSBOB ends up to understand two basic entities:

GE The GE is a group of devices that have the same G and E address values as defined by the SIENNA Protocol. These devices may offer either actuator or sensor type functionality.

GEType The GEType is a group of devices inside a GE that have the same functionality type, actuator or sensor. This means that a GE can contain at most two GETypes. Furthermore, there must be reasonable address assignment between the devices, so that devices with incompatible functionality are not grouped together. If this happens, then a GEType with the safest possible common functionality is created.

The SIENNA stack base driver OSGi Bundle is the interface between the SIENNA devices in the power line network and the upper OSGi layers in the smart home platform. Thus, it creates OSGi services that represent the actual SIENNA or groups of SIENNA devices. On its one end, SSBOB transforms LonWorks messages to SIENNA Messages, both in the incoming and the outgoing direction. For example, LonWorks NV updates that refer to the available SIENNA Devices are translated to the corresponding SIENNA updates and outgoing commands to the sienna devices are transformed to the appropriate LonWork messages. On the other end, SSBOB provides OSGi services to an upper layer OSGi clients, such as the SIENNA Control OSGi bundle. To describe the whole process in the OSGi framework, in the beginning, SSBOB registers the GEs as services. Then, the GETypes can be extracted from the GEs. The GETypes offer the capability to control and receive status updates from the physical SIENNA Devices. The task of extracting the GETypes is performed by the clients of SSBOB. Such clients are for example the SIENNA Control OSGi bundle, which provides a testing tool for the SIENNA Network, and the SIENNA ROCob drivers.

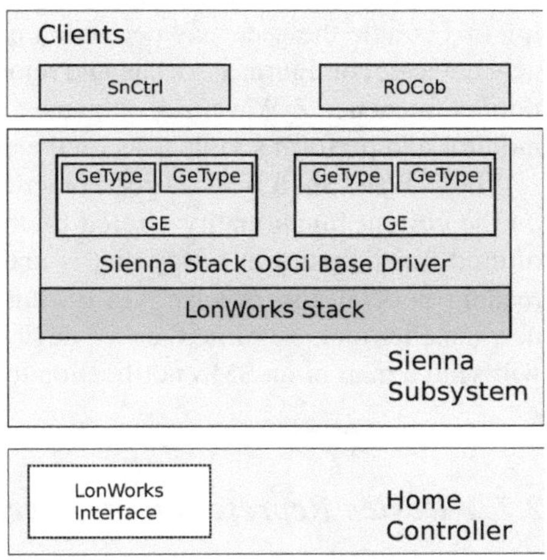

Fig. 2.6 The SIENNA Network Subsystem in the Home Controller

The integration of the SIENNA technology is achieved with the use of the SIENNA Stack Base-driver and SIENNA Control OSGi Bundles.

2.7.1.3 The Z-Wave Subsystem

The smart home platform integrates the Z-Wave technology and provides installation, maintenance, control and monitoring capabilities over Z-Wave devices. The Z-Wave integration is achieved by an embedded Z-Wave interface on the home controller and an extension to the OSGi service platform to provide Z-Wave device services. The architecture of the Z-Wave subsystem is described in the following picture.

The home controller features a Zensys ZM3102 Z-Wave RF module. The Zensys ZM3102 module uses ZW0301 Zensys chips to integrate Z-Wave technology, providing a Serial Application Interface over a custom Zensys serial protocol. The home controller integrates the Z-Wave technology with the Z-Wave Base Driver and the Z-Wave Shell OSGi bundles.

The Z-Wave Base Driver is an OSGi bundle offering application interfaces for creating, managing and controlling Z-Wave networks. It enables the user to create, maintain, control and monitor networks of Z-Wave devices. Using this bundle a developer may provide services for performing new devices registration, configuration, association and removal, as well as most installation and maintenance Z-Wave actions. A developer may also use the Z-Wave bundle in order to provide control, monitor and simple test scenarios execution on Z-Wave devices.

Z-Wave Shell is an OSGi bundle offering both a command-line shell tool over the Z-Wave Base Driver OSGi bundle enabling management and control actions on Z-Wave networks. It enables the user to use the interface of the Z-Wave OSGi bundle for creation, maintenance, control and monitoring of Z-Wave networks. Using this bundle the user may perform a number of tasks by typing commands: add new devices, configure, associate and remove existing ones, as well as most installation/maintenance Z-Wave tasks; the user may also use the Z-Wave Shell to control, monitor and perform simple tests on the devices.

The Z-Wave Shell is also a good practical -yet not complete or detailed- reference on the current functionality offered by the Z-Wave OSGi bundle. It exposes most offered functionality in a way that is understood by a potential developer prior to reading in detail the Z-Wave API documentation: either the developer targets at a new installer tool, user interface for devices control or a simple monitoring tool that will send a mail or an SMS notification to a user.

2.7.2 Device Representation Layer API Specification

The OSGi platform provides the level of required flexibility to add support for new technologies on the runtime, along with other dynamic features. The NA layer we present in the following sections is a set of JAVA classes and interfaces, bind with OSGi services, which are targeting on adding a level of transparency between the higher level services acting on the OSGi layer and the real physical devices. The physical device (OSGi) drivers must implement a number of the NA interfaces, in

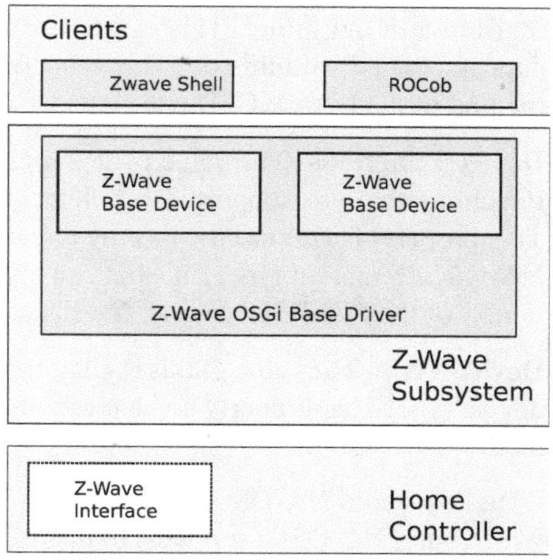

Fig. 2.7 The Z-Wave Network Subsystem in the Home Controller

order that bundles built on top of the NA layer can acquire and use them seamlessly to any underlying physical device technology. The main ingredients of the NA API are the NA Device and Function interfaces. Each OSGi implemented device driver that takes care of the communication specifics between the physical devices and the OSGi framework finally generates and registers to the framework a NA-OSGi Device. A similar term for NA-OSGi is ROCob and we use these similar terms interchangeably in the chapter. The registration must be done in a well defined way (see below). Each registered NA-OSGi Device implementation must implement one or more Function interfaces. Using the methods defined in these interfaces other NA-OSGi aware services will be able to communicate with the NA-OSGi Devices and thus with the physical devices in the underlying network.

2.7.2.1 NA-OSGi Device Interface

A service must implement this interface to indicate that it is a NA-OSGi Device. Services implementing this interface give the system components the opportunity to discover them and retrieve all required information for managing and performing certain actions. Every service that has the intention of being registered as a NA-OSGi Device must conform to the semantics specified by the Device interface of the NA-OSGi API. In detail:

OSGi registration classes This property should be a list containing the NA-OSGi Device interface class name and the class names of all NA-OSGi Function Interfaces implemented by this NA-OSGi Device (see below).

Device Unique ID The value of this registration property denotes the Unique ID among the NA-OSGi Devices registered in the framework. This property is aimed to be used by high level services to distinguish the NA-OSGi Devices. This value should remain the same for every distinct NA-OSGi Device, if it needs to be re-

registered in the future. This is because higher lever services may have generated data associated with this ID. This property can be acquired using the getUniqueId() method of the NA-OSGi Device interface.

Device Functions The value of this registration property is a list of the NA-OSGi Functions that are supported (implemented) by this registered NA-OSGi Device. This property is commonly used by tracker services to filter the search on a specific NA-OSGi Function type. This list can also be acquired by calling method getFunctions() of the NA-OSGi Device interface.

Device Type The value of this registration property denotes the underlying physical device type. This property's value can also be acquired using the NA-OSGi Device method getType().

The registered NA-OSGi Devices need to establish a two way communication between other NA-OSGi services. In the following we describe how other services can handle the NA-OSGi Devices leading to the control of the underlying physical devices and how the NA-OSGi Devices can notify other NA-OSGi services for events like device property updates.

The registered NA-OSGi Devices can be tracked by other services using the defined by the OSGi framework methods. The properties that escort the NA-OSGi Devices during their service registration can be used to help services tracking the desired devices. Once a NA-OSGi Device service is tracked, it can be controlled using the methods of the NA-OSGi Function interfaces that it implements. Then, the service could be casted to the appropriate interface and use its implemented methods.

In the general case, almost every NA-OSGi Device will eventually need a way to notifying other NA-OSGi aware services for device property updates, derived by the underlying physical device (like a lamp turning on), or for any other useful reason. The NA-OSGi Devices use the OSGi r4 EventAdmin specification to deliver events to the other Framework services. These types of events must conform to the following:

Event Topic All the NA-OSGi Device Events must start with the "NA-OSGi" prefix to inform the OSGi Event Handlers that this is a NA-OSGi Event. This can be used for event filtering. The event topic must end with the NA-OSGi Interface name that triggered this event generation. For instance, if an event was generated by a NA-OSGi Device that implements the NA-OSGi Switch Interface, the full event topic would be NA-OSGi/Switch.

A NA-OSGi Device may implement more than one NA-OSGi Interfaces. In that case, the suffix of the OSGi Event Topic will be the Interface name associated with the property, the value of which is published for notification to the Framework. For instance, consider a NA-OSGi Device that stands for a Dimmer physical device and implements the Switch and the LevelAbsolute NA-OSGi Interfaces, to support the on/off and dimming functionality of this dimming physical device. Suppose that this device turns off (i.e. changes its state) and the NA-OSGi Device needs to notify the framework for this change. The property CURRENT STATE that was modified lies within the NA-OSGi Switch Interface. So, the Event topic would be "NA-OSGi/Switch". If the NA-OSGi Device needed to notify for a change in the brightness level of the dimmer device as well, it should create one more event for the property CURRENT LEVEL of the LevelAbsolute Interface with a topic like: "NA-OSGi/LevelAbsolute".

Device Function This property must accompany the NA-OSGi Events. The allowed values for this property derive from the NA-OSGi Function interface. It provides information about the NA-OSGi Interface type that triggered the Event. It has the exact same meaning as the suffix of the OSGi Event Topic of the NA-OSGi Events. Although this information is redundant, it can be used for Property based filtering or alternatively to be used in "switch" statements.

Device Unique ID This is the same property that accompanies the NA-OSGi Device to the framework registration.

In addition to these properties, every NA-OSGi Event must be accompanied with one or more properties that carry the actual information for this event. In the general case, the properties that are used depend on the NA-OSGi Function that triggered the event. These properties are required to exist in the NA-OSGi Event. For instance, the CURRENT STATE property mentioned in the example above is a part of the NA-OSGi Switch Interface. Any other property that could be of interest to the event handlers may be added as well.

2.7.2.2 Function Interface

This interface is not intended to be implemented by any class. Its reason of existence is to provide a number of fields of the available NA-OSGi Function interfaces (that should be implemented by the NA-OSGi Devices), represented as primitive integers along with helper collections that map these integers to strings like friendly names or class names. These collections and fields are meant to help the developer at the NA-OSGi Device registration, event generation and handling. With these functional profiles a NA-OSGi device can represent the functionality of the underlying physical device. The enumeration of the available Function interface representation is the following:

Composite	Interface commonly used for physical devices with multiple endpoints.
Binary Sensor	Interface commonly used for sensor-like devices, like motion sensors etc.
Level Absolute	Interface used for physical devices with level properties like dimmers etc.
Level Absolute Timed	Interface used for physical devices with level properties like dimmers with an extra timing functionality.
Level Relative Control	Interface used for devices with direction semantics, with read only values.

Level Relative	Interface used for physical devices with direction semantics like shades etc.
Level Relative Timed	Interface used for physical devices with direction semantics like shades etc. with an extra timing functionality.
Switch	Interface used for on/off physical devices like lights, simple home appliances etc.
Battery	Interface which is commonly used for representing a physical battery.
Alarm	Interface which is commonly used for alarming devices.
Analog Meter	Interface used for devices that measure analog inputs like pressure and temperature.
Thermostat	Interface which is commonly used for representing a physical Thermostat device.

2.7.2.3 Scene Interface

The scene related interfaces intend to provide a service layer for scene (task) execution. In general, a scene can be considered as a collection of NA-OSGi Devices that will execute a number of actions on these devices when a certain event occurs. An alarm service is included, with which the device representation layer can send alarm events to other NA-OSGi components or higher layers. It makes use of a set of defined alarm levels, types and properties, to cover a wide range of alarm events.

The Scene Interface is used to define the collection of the NA-OSGi Devices that take part in a scene, along with a number of properties per Device that will lead to the appropriate actions on the Device when the Scene is executed. Each Scene implementation has a unique ID among the scenes and holds a collection of NA-OSGi Device IDs. For each of these IDs it holds a Dictionary of properties, the data of which depend on the NA-OSGi Device Function interfaces the specific NA-OSGi Device implements. Moreover, it holds a Function ID per NA-OSGi Device. This Function ID is used from the Scene implementation when the Scene is executed to lead to the appropriate NA-OSGi Device method call on the specific device. At the same time, the properties provided by the Dictionary per NA-OSGi Device are used to determine the method's arguments.

For instance consider a Scene with one NA-OSGi Device that implements the Switch interface and we want this device to turn on upon the Scene execution. The function ID for this device should be "NA-OSGi.Function.SWITCH" and the property within the Dictionary for this device should be: "Switch.CURRENT STATE=Switch.STATE OFF". When this Scene is to be executed, "setState(Switch. STATE OFF)" should be called on this NA-OSGi Device, which will lead to the physical device to turn off.

2.7.2.4 Scene Manager Interface

The implementation of this interface works as a Scene manager service. The role of this service is to hold a collection of scenes provided by other OSGi bundles and execute them upon request. The implementations of this interface should register to the OSGi framework under the class name "NA-OSGi.SceneManager", for other bundles to track and use it.

2.7.2.5 Trigger Interface

For a scene to execute, a certain event must occur. These events are property modifications of the NA-OSGi Devices, like a Binary Sensor at STATE ON, which can be tracked using the OSGi r4 EventAdmin specification as described before. This interface is introduced to be used for that reason. It holds a collection of NA-OSGi Device IDs that take part to the triggering of a scene (or scenes) execution. It also holds a collection of the scene IDs to be executed when a certain event is occurred. It further holds the event that should occur to trigger the scenes' execution. For each of the NA-OSGi Device IDs that take part on this a Dictionary of properties (the data of which depend on the NA-OSGi Device Function interfaces implemented from that specific NA-OSGi Device) is hold. The event is supposed to activate the trigger when these properties of the NA-OSGi Device are met. If just one of the NA-OSGi Devices in the collection of this Trigger meets these property values, the scenes must be executed. Moreover, a NA-OSGi Function ID is held for each NA-OSGi Device in the collection. This is supposed to be used on the NA-OSGi Device event filtering.

For instance, suppose we want a number of scenes to be executed when a Binary Sensor's state changes to STATE ON. This Binary Sensor is described by a NA-OSGi Device implementing the Binary Sensor interface. The Dictionary for this Device in the Trigger should hold the property "BinarySensor.CURRENT STATE=BinarySensor.STATE ON" and the Function ID for this Device should be "Function.BINARY SENSOR".

2.7.2.6 Scene Trigger Interface

The implementations of this interface are used to manage implementations of the Trigger interface. In other words, the implementations of this interface should be OSGi services that hold a collection of Triggers added by other OSGi bundles. The implementations must be registered to the OSGi framework under the class name NA-OSGi.SceneTrigger, for other bundles to track and use to add or remove triggers. They should also take care of the evaluation of the triggers that hold and the scene execution of the scene IDs associated with these triggers.

2.7.2.7 Alert Service

This utility service is designed to provide a simple alerting service to the other NA-OSGi aware services in the framework, based on custom Rules. The alerts make use of the alarm characteristics. This service listens for NA-OSGi Device events and evaluates a series of user defined Rules to finally create the specific NA-OSGi alerts. This service is most commonly used when analog meters are used, like depth meters. For instance in fuel tanks, you may need to have an alert when the fuel level drops below a specific threshold. This alert could be used by a NA-OSGi service to create a notification (e.g. SMS) to the user.

The rules used by the NA-OSGi Alert service are based on the NA-OSGi Device specific properties. The user creates a number of rules based on these specific properties and the evaluation is done by the service by listening for NA-OSGi Device property changes from the NA-OSGi Device events. There are two ways to evaluate a set of properties for a NA-OSGi Device (AND, OR). The NA-OSGi alert service uses the OSGi r3 Configuration Admin Specification to create new Rules. The properties that need to be set in order to create a new Rule include the NA-OSGi Device IDs that take part in this rule, the message that will accompany the alert when generated, the level of the alert, the properties to be evaluated, the evaluation type, as well as a bounce filter and a delay used for avoiding bouncing effects.

2.8 Service Representation Layer

Services are the basis of distributed computing across the Internet. A service consumer locates a service and invokes the operations it provides. As ATRACO has adopted the SOA model each component could be either a service provider or a service consumer or both consumer and provider.

An ATRACO system will use the Network Adaptation component to access network resources. Resources can be either devices or services, but both can be accessed within ATRACO as services. In order to do that, Network Adaptation defines an internal upper layer called service adaptation layer on top of which all resources are viewed and accessed as services.

Devices organized in device types according to their functionality will expose through the service adaptation layer a list of access methods in the form of services. From the service point of view a device can be analyzed as a list of state variables (or parameters) and a list of methods or functions that either read the value or change the value of these state variables.

Other types of services which can be found and accessed in the internet are of significant interest to an ATRACO system and thus they should be provided in the ATRACO environment in a proper shape in order to be consumed from the ATRACO entities. These services could be already implemented as Web Services or may have a completely different architecture and access interface. In either way, the Network Adaptation layer should intervene in order to reshape the service to follow

Table 2.1 Specification Summary of the NA-OSGi API

Interface	Fields and Methods
Alarm	Irrelevant, Level (E), Message (E), NoAlarm, Type (E), Communication Error, Device Inaccessible, Device Specific, Intrusion, Maintenance, Tamper Attempt
AlarmState	getAlarmLevel, getRelevantAlarmTypes
AnalogMeter	Level (E), getAnalogMeterLevel
Battery	Level (E), getBatteryLevel
BinarySensor	Current State (E), State Off (E), State On (E), getState
Composite	getEndpointCount, getEndpointFunctions(index), getEndpoint(index)
Device	Level, Function (E), Functions (S), Type (S), UniqueId (ES), Event Topic, Binary Sensor (S), Door Trap (S), Flood Detector (S), Gas Valve (S), Key Fob (S), Lamp (S), Motion Sensor (S), Shade (S), Smoke Detector (S), Switch (S), Tank Level Meter (S), Thermostat (S), Water Valve (S), Unknown (S), getDefaultName, getFunctions, getType, getUniqueId
Function	Alarm (DE), Analog Meter (DE), Battery (DE), Binary Sensor (DE), Composite (DE), Level Absolute (DE), Level Absolute Timed (DE), Level Relative (DE), Level Relative Control (DE), Level Relative Timed (DE), Switch (DE), Thermostat (DE), Function Classes, Function Names, Count, Device Error, Driver Error, Invalid Args, Invalid State, Network Error, Queued, Stack Error, Success, Transmitted
LevelAbsolute	Current Level (E), Max (E), Min (E), getLevel, setLevel
LevelAbsoluteTimed	setLevel
LevelRelative	startLevelChange, stopLevelChange
LevelRelativeControl	Current Direction (E), Down (E), Stopped (E), Up (E), getLevelChangeDirection
LevelRelativeTimed	startLevelChange
ROCState	getDeviceId, getFunctionId, getProperties
Scene	Actions Count Error, Function ID Error, getActionCount, getDeviceId(index), getFunctionId(index), getProperties(index), getSceneId
SceneManager	addScene, getScene, getSceneIds, removeScene, runScene
SceneTrigger	add, getTrigger, getTriggerIds, remove
Switch	setState
Thermostat	Auto (E), Cool (E), Heat (E), Off (E), Mode (E), Setpoint (E), Temperature (E), getActualTemperature, getOperationMode, getTemperatureSetPoint, setOperationMode, setTemperatureSetPoint
Trigger	getDeviceId(index), getFunctionId(index), getProperties(index), getSceneIds, getTriggerDeviceCount, getTriggerId

[a] E: Event Property, S: Service Property, DE: Device.Function Event Property

the selected ATRACO service architecture acting as a service proxy to existing Web Services and enforcing authentication and security policies.

The NA-OSGi layer provides an abstraction layer for a variety of underlying networking technologies. Looking from the perspective of higher layers, it can be either used directly to implement intelligent environment services or a presentation layer, but it can also act as a common base for extending interoperability with services provided by other well-established technologies, like UPnP and R-OSGi.

The upper layer of NA includes the Service Representation Layer, avoiding that way a direct interface of the service adaptation to different underlying networks.

In an ATRACO environment where NA is used to access devices participating in multiple different networks the integration of the service representation layer with the device representation layer minimizes the complexity of the implementation and allows future additions of networks in a seamless way. This layer converts the available devices from the device representation layer as well as other wrapped legacy services to services of a common technology of choice, e.g. UPnP or R-OSGi. As a result, devices and services can consume or be consumed by other devices and services available in an AIE environment.

The service representation layer supports zero-configuration networking. An automation device from any vendor and supported technology, once registered in the NA component, dynamically announces its name, conveys its capabilities upon request, and learns about the presence and capabilities of other devices or services.

UPnP seems to cover most of the aspects that are specified for the Service Adaptation Layer, such as ease of network setup, device discovery, self announcement and advertisement of supported capabilities. Currently the Network Adaptation framework supports UPnP and integrates it with the OSGi Service Platform. On top of that, a set of NA-OSGi to UPnP wrappers have been developed that eventually transform the NA-OSGi services to UPnP services.

2.8.1 UPnP Adaptation

The UPnP Device Architecture specification provides the protocols for a peer-to-peer network. It specifies how to join a network and how devices can be controlled using XML messages sent over HTTP. The UPnP specifications leverage Internet protocols, including IP, TCP, UDP, HTTP, and XML. The OSGi specifications address how code can be downloaded and managed in a remote system. Both standards are therefore fully complimentary. Using an OSGi Service Platform to work with UPnP enabled devices is therefore a very successful combination allowing the development of OSGi bundles that can interoperate with UPnP devices and UPnP control points.

By using UPnP above the NA-OSGi layer, the functionality of services for standard home control devices (lighting, shades, heating, etc.) can be combined with other spheres of interest, such as audio-visual devices or other web services transformed into UPnP services (network clocks, weather report, news services and others). This combination opens a new wide layer for offered services or complex presentation designs, accommodating and re-innovating the modern lifestyle inside a smart environment.

NA provides a UPnP virtualization of the interfaced physical devices, which eventually belong to heterogeneous non IP networks. Figure 2.8 illustrates a generic view of a networked AIE. UPnP proxies bridge IP networks with non IP networks representing at the same time devices belonging to non IP networks as UPnP entities. Figure 2.9 illustrates the virtualization function of the UPnP proxy. The proxy knows how to communicate with the ZWave microphone. For that reason it uses

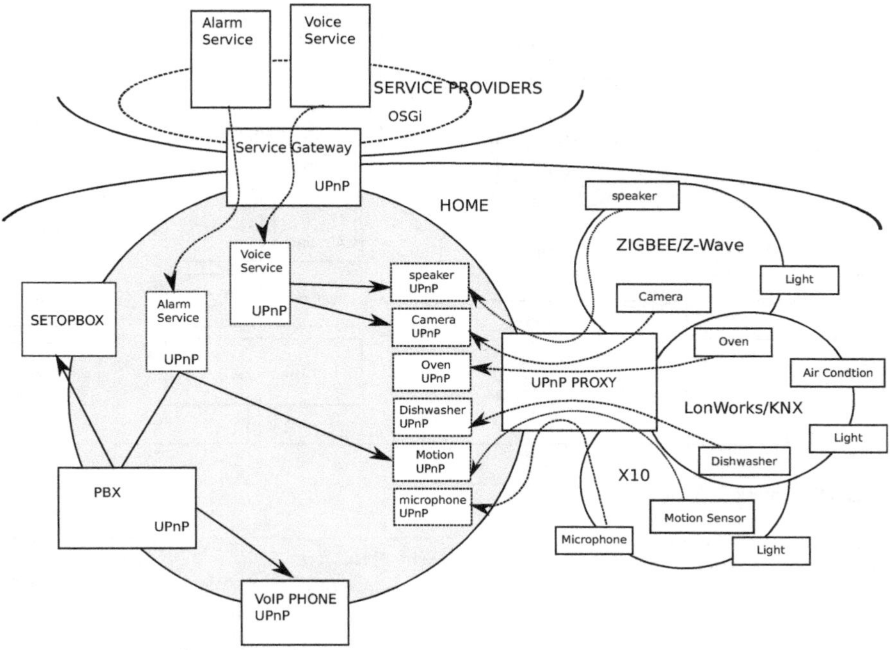

Fig. 2.8 Example of UPnP proxy virtualisation function

a special library that encodes, over the ZWave API, microphone commands to start/stop recording as well as to control the gain and the sampling rate. In order higher perceptual components, such as the voice recognizer, to make use of the microphone, the proxy represents it as a UPnP device exporting appropriate actions for remote invocation (set gain, set sampling, start rec etc). On the other hand taking advantage of the silence detection feature of the microphone, proxy sends appropriate events triggering that way the voice processing at the recognizer side.

Considering the device types, functional profiles and event properties defined in the device representation layer as well as the availability of a standard OSGi UPnP Driver implementation it is straightforward to develop OSGi Wrapper services that export NA-OSGi devices as UPnP devices. The NA layer provides UPnP device services to the AIE network via the UPnP wrappers. Those services expose the relative UPnP device and service description XMLs. Clients that intend to use the offered services can use third party programming APIs, based on the UPnP library on use, e.g. CyberLink (Java), CyberDomo (UPnP), Intel (C#). A side advantage of this procedure is that a single UPnP wrapper can provide UPnP services for a device class, e.g. on/off switch, irrelevant to the technology that the switch uses (LON, Z-Wave, SIENNA), due to the abstraction offered by the NA-OSGi layer. The NA UPnP wrapping architecture is briefly illustrated in Figure 2.10, while Figure 2.11 depicts the NA device layers (in parallel for exemplary LON and ZWave networks), from physical device to UPnP device.

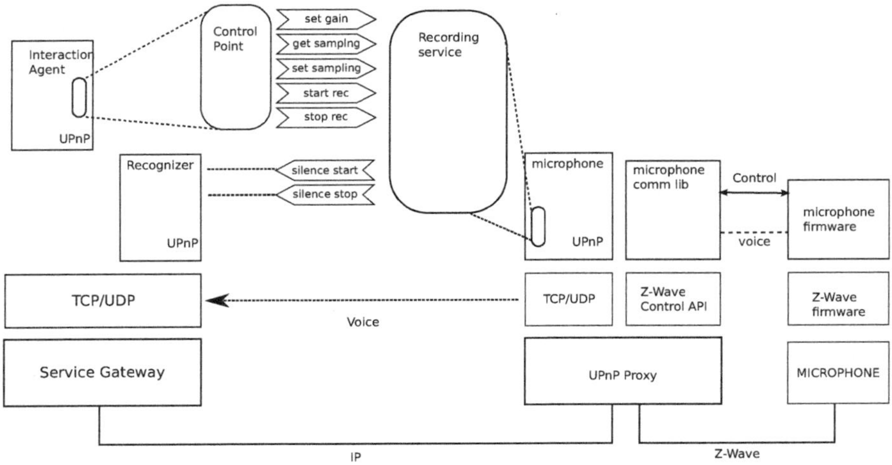

Fig. 2.9 Generic NA view of an AIE

Fig. 2.10 UPnP wrappers overview.

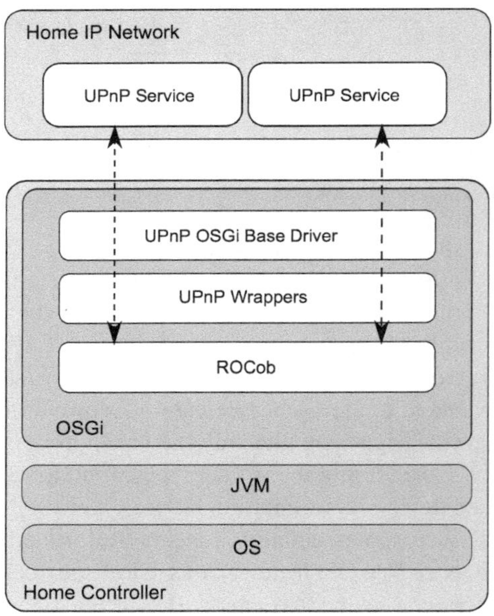

A Wrapper service would usually track the availability of NA-OSGi devices with certain characteristics and then register a UPnP Device service with the appropriate device and service description to represent the underlying device. The registered UPnP Device service is then tracked by the UPnP Base Driver, the bundle that implements the bridge between OSGi and the UPnP networks, and exported to the network as a UPnP device. The registered UPnP Device service should be implemented in such a way that the provided UPnP Service class implementations

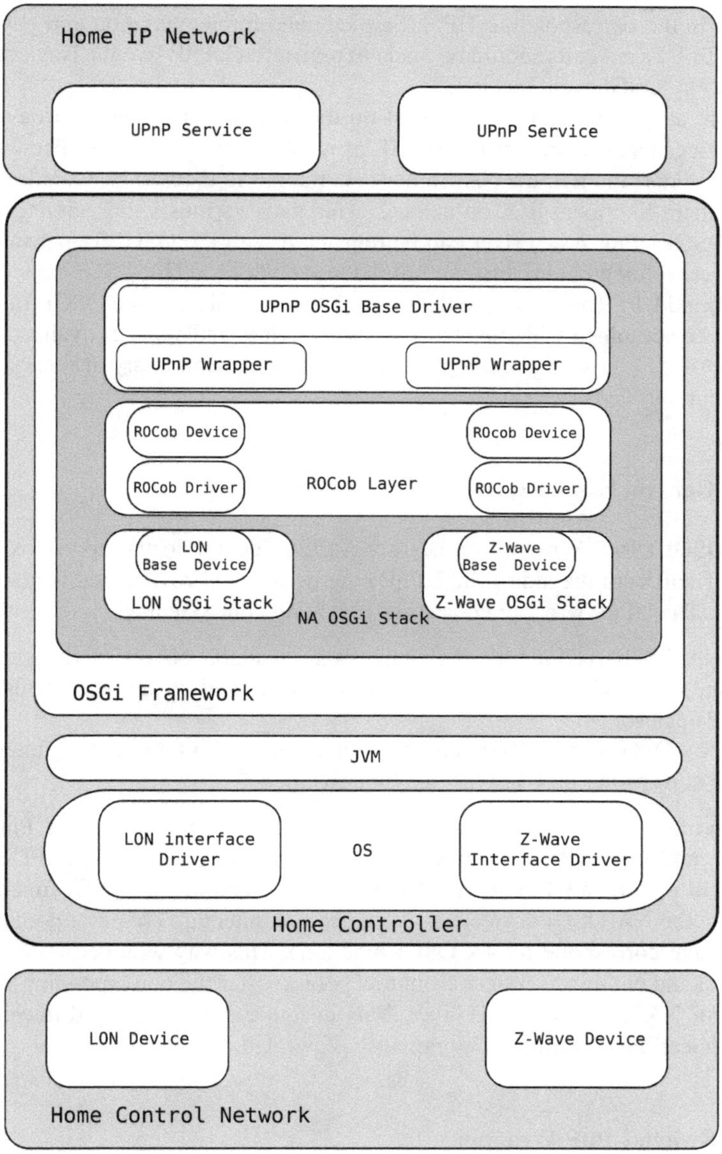

Fig. 2.11 Device adaptation layers, from physical device to UPnP device.

will correspond to the functions supported by the NA-OSGi device. Similarly each UPnP Service should be implemented in such a way that the provided UPnP Action class implementations will correspond to the Java methods defined in the associated NA-OSGi function interface. In the reverse path the NA-OSGi event properties defined in each supported NA-OSGi function should be represented as UPnP State

Variable in the corresponding UPnP Service class implementation and appropriate notify UPnP Event calls should be made to registered UPnP Event Listener services for each NA-OSGi event received.

Device adaptation can also be used on the reverse path to control devices and web services already present in the AIE provided by other sources. The way to do this is to import them in the NA framework, transform them to ROCob devices and operate them like local ROCob devices. This way, various UPnP services tracked by the OSGi UPnP base driver can be registered with the OSGi framework. Then, appropriate refinement drivers can refine these devices to NA-OSGi devices. Since the imported UPnP devices have their NA-OSGi equivalent in the OSGi framework, they can be combined with the rest NA-OSGi devices and exploit advanced features of the NA-OSGi layer. For example, they can be used seamlessly in scenes creation and execution.

2.8.1.1 General Exported UPnP Device Model

A NA-OSGi UPnP Wrapper refines a NA-OSGi device to the appropriate UPnP device. Then, with the aid of the UPnP base driver, this device is advertised in the network. The UPnP wrapper consists of two main components:

- The NA-OSGi to UPnP driver, that tracks the available NA-OSGi devices, selects the suitable devices according to an LDAP filter and creates the corresponding UPnP devices.
- The NA-OSGi to UPnP device, which maps the NA-OSGi device functionality to the correpsonding UPnP device functionality.

For each tracked device a new service is registered implementing the UPnPDevice interface and the registration properties listed in Table 2.2. The UPnPDevice instance will provide a UPnPService for each corresponding NA-OSGi function supported by the NA-OSGi device. The wrapper's architecture creates a pool of UPnP services that correspond to NA-OSGi functions. This way wrapped UPnP devices can select and combine services from this pool to offer the corresponding functionality from NA-OSGi to UPnP layer. This design can lead to rapid development, since replication of common functionality is avoided.

2.8.1.2 Switch UPnP Wrapper

Since the NA-OSGi Switch function matches exactly the UPnP Forum standardized SwitchPower service, the wrapper should eventually export to the network a UPnP device that utlizes this standard UPnP service. Therefore the UPnPService instance should provide two UPnPAction implementations corresponding to *GetStatus* and *SetTarget* actions defined in SwitchPower:1 as well as two UPnPStateVariable implementations corresponding to *Status* and *Target* state variables defined in Switch-

Table 2.2 UPnPDevice Service registration properties

Property Name	Property Type	Property Value
DEVICE CATEGORY	String	UPnP
objectClass	String	org.OSGi.service.upnp.UPnPDevice
UPnP.device.friendlyName	String	A characteristic term for the device plus the value returned by com.inaccessnetworks.rocob. Device.getDefaultName()
UPnP.device.manufacturer	String	inAccess Networks
UPnP.device.modelDescription	String	ROCob - device type - UPnP Wrapper
UPnP.device.UDN	String	A standard or a custom device UDN, according to the device type and the UPnP specification
UPnP.export		Not Required
UPnP.device.type	String	A standard or a custom device type definition, according to the UPnP specification
UPnP.service.id	String[]	Array of Strings with the provided functions, e.g. "urn:upnp-org:serviceId:SwitchPower.001"
UPnP.service.type	String[]	Array of Strings with the supported UPnP service types by the wrapper, e.g.: "urn:schemas-upnp-org:service:SwitchPower:1"

Power:1 (returned by UPnPService methods getActions() and getStateVariables() respectively).

The *GetStatus* UPnPAction.invoke() implementation should call the NA-OSGi Switch.getState() method, while the *SetTarget* UPnPAction.invoke() implementation should call the NA-OSGi Switch setState() method. In both cases the wrapper service must convert NA-OSGi values to UPnP values. Since NA-OSGi Switch function does not have the concept of two state variables, this must be implemented by the wrapper explicitly. The *Status* UPnPStateVariable should have the value returned by the last NA-OSGi Switch.getState invocation or the last NA-OSGi event received. *Target* UPnPStateVariable should just hold the requested value from the last *SetTarget* UPnPAction.invoke() call.

NA-OSGi Switch function defines a single event property Switch.CURRENT STATE and a NA-OSGi event is raised containing this property every time the underlying physical device changes state. Therefore the *Status* UPnPStateVariable is evented and the wrapper bundles should maintain a list of the registered UPnPEventListener implementations (that requested events for this state variable) and call the notifyUPnPEvent callback every time a ROCob event is received. If at least one UPnP control point in the network subscribed for events of this state variable, then the UPnP Base Driver should have an appropriate UPnPEventListener registered and propagate a UPnP Event Notification to the network every time notifyUPnPEvent is called. The *SwitchPower* service description can be found at the UPnP Forum web site [9]. Listing 2.1 presents the UPnP profile for the Switch device in XML description language.

```
<root xmlns="urn:schemas-upnp-org:device-1-0">
```

```
    <specVersion>
      <major>1</major>
      <minor>0</minor>
    </specVersion>
    <URLBase>http://192.168.2.253:4004</URLBase>
    <device>
      <deviceType>urn:com.inaccessnetworks:device:Switch:1</
          deviceType>
      <friendlyName>BinarySwitchZW0000001205</friendlyName>
      <manufacturer>inAccess Networks</manufacturer>
      <manufacturerURL>http://www.inaccessnetworks.com</
          manufacturerURL>
      <modelDescription> Binary Switch UPnP Wrapper</
          modelDescription>
      <modelName>ROCob Binary Switch</modelName>
      <modelNumber>1.0</modelNumber>
      <modelURL>http://www.inaccessnetworks.com</modelURL>
      <serialNumber></serialNumber>
      <UDN>uuid:UPnPROcobZW0000001205</UDN>
      <UPC></UPC>
      <serviceList>
        <service>
          <serviceType>urn:schemas-upnp-org:service:SwitchPower
              :1</serviceType>
          <serviceId>urn:schemas-upnp-org:serviceId:SwitchPower
              :1</serviceId>
          <SCPDURL>/service/0/gen-desc.xml</SCPDURL>
          <controlURL>/service/0/ctrl</controlURL>
          <eventSubURL>/service/0/event</eventSubURL>
        </service>
      </serviceList>
      <presentationURL></presentationURL>
    </device>
  </root>
```

Listing 2.1 The UPnP profile for the Switch device in XML description language.

2.8.1.3 More UPnP Wrappers

Binary Sensor This function is not directly mapped to any standardized UPnP service. Since the binary sensor only sends events with its current state, a UPnP service is needed that can read the binary sensor status or can get notification events when the status is modified. The ideal service to handle this task is a subset of the SwitchPower service. Therefore, a new service has been created that uses one state variable, called *Status*, and a corresponding action to get its value called *getStatus*.

Door Trap UPnP wrapper resembles the functionality of the binary sensor. Their difference lies in the device description. This is necessary to serve better the presentation layer, to allow correct visualization of the devices, as well as to allow expandability in the future.

Motion Sensor UPnP wrapper, just like the door trap wrapper, currently follows the functionality of the binary sensor. This means that it implements only the binarysensor UPnP service. The corresponding profile will be possibly enhanced in the future, when devices with advanced features become available. Such features may be sensor sensitivity and sensor reset timeout.

Illumination Detector UPnP wrapper takes care to represent a NA-OSGi illumination detector in the UPnP layer. This is achieved by utilizing the AnalogMeter UPnP service, which corresponds to the ANALOG METER NA-OSGi subfunction. The AnalogMeter service includes one StateVariable of type float, and an action that returns the value of this variable, called GetAnalogMeter. The measured value corresponds to the percentage of the maximum value. The UPnP device reads the analog meter level value either by polling the NA-OSGi device, using the GetAnalogMeterLevel action, or by serving incoming events that are generated by the NA-OSGi device.

2.8.1.4 Device/Service Control Example

This paragraph presents a device control example in ATRACO (see Figure 2.12). When an ATRACO client needs to use a ZWave Switch connected to the NA component, it will send a UPnP device discovery message to the ATRACO IP network. The OSGi UPnP base driver of the NA component will send the description of the available UPnP devices, including the switch in question, which is handled by the relative OSGi UPnP wrapper. Then the client can query from the UPnP wrapper the available actions of the device, and send for example the ON command to the wrapper. The wrapper, using the ROCob device ends up to send the command to the ZWave Base driver of the NA component. The latter will finally send the ZWave command to the ZWave switch.

2.8.1.5 Device Event Propagation Example

Figure 2.13 depicts device propagation in ATRACO. When an event from a ZWave door trap connected to the NA component occurs, this event is propagated through NA to the various ATRACO clients. The event is first handled by the ZWave Base Driver of the NA component, which propagates it to the corresponding ROCob device. The ROCob device then sends an event to the OSGi framework, which is handled by the relative UPnP wrapper. The wrapper translates the event to a corresponding UPnP event. This event is broadcasted to the ATRACO IP Network and listeners for this event will receive it.

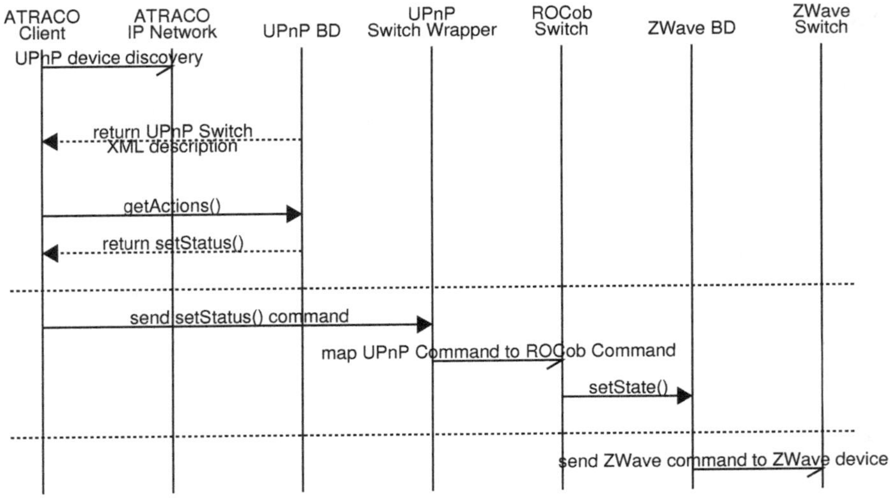

Fig. 2.12 Device control through NA.

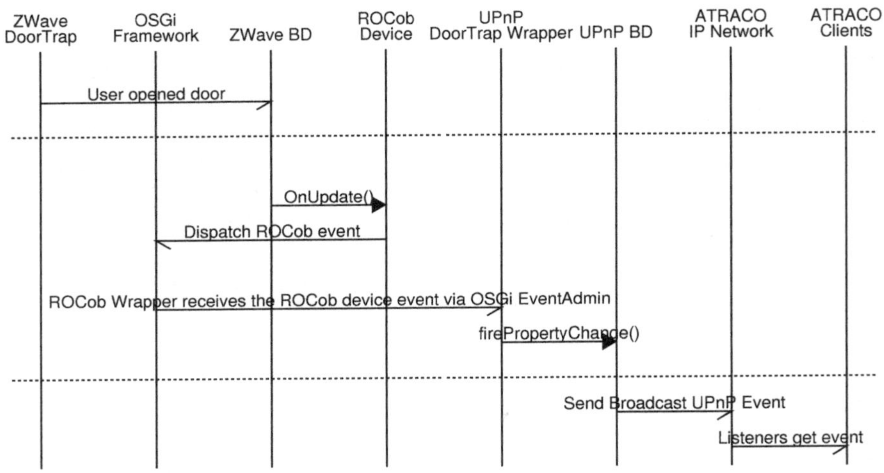

Fig. 2.13 Device event propagation through NA.

2.8.2 Simple Task Execution Manager

The Simple Task Execution Module (STEM) is an OSGi based UPnP service that based on a simple or nested condition evaluation generates a UPnP event about the result of the evaluation and even executes a set of actions on one or more UPnP devices when the evaluation is true. STEM basically operates on UPnP devices/services in a home network. Architectural wise, although STEM could be a simple OSGi service, it follows a network approach and is a UPnP service itself, so that other UPnP services in the same network can use it. It is planned however to provide an OSGi service API in the near future.

The two main concepts of STEM are the scene and the trigger. A scene is a set of actions that can be performed in one or more devices. The actions are described using unmodified UPnP identifiers of the participating devices, so that users of STEM can easily construct them. For functional reasons, each scene has its own unique ID.Thus, it is possible to create, modify and delete a scene.

A trigger is a set of rules, that when the total evaluation turns true an event is generated that notifies that this specific trigger is true. If the trigger is associated with one or more scenes, then these scenes are executed. A rule is set of simple or nested AND and OR conditions. The conditions compare state variable values of a UPnP device with state variable values of the same or other UPnP device or with arbitrary values of the same UPnP data type. This is achieved with a recursive logical evaluation function. Just like the scenes, triggers also have a unique ID that is used for their creation, modification or removal or association/deassociation with a scene.

In order to be able to evaluate conditions, STEM subscribes to the services of the devices that are included in the conditions of the various triggers. When an updated value for a state variable that is used in a condition arrives, STEM re-evaluates the condition and as a result the whole trigger. If the total evaluation turns true, an event containing the trigger id and the status true value is generated and sent to the listeners that have subscribed to STEM's service. If the result of the evaluation is false, an event with the trigger id and the false status value is sent.

STEM by itself is not a UPnP control point, so it is not aware of the available devices. The OSGi UPnP basedriver is responsible for discovering available UPnP devices and updating them or removing lost ones. Using the OSGi API of the OSGi UPnP basedriver, STEM finally becomes aware of the various devices. When a STEM user adds a scene or a trigger, these are validated against the existing UPnP devices. If a device included in a scene or trigger does not exist, or a wrong service or action or state variable is used, then the trigger/scene is discarded. Listing 2.2 shows an example of a scene description XML which handles two binary lights.

```
<xml>
  <scene id="scene1">
    <device id="uuid:UPnPROCobZW 0000000702">
      <service id="urn:schemas-upnp-org:serviceId:SwitchPower
          :1">
        <action id="SetTarget">
          <statevar id="NewTargetValue" type="boolean" value
              ="true"/>
        </action>
      </service>
    </device>
    <device id="uuid:TestLight+004fecde">
      <service id="urn:schemas-upnp-org:serviceId:SwitchPower
          :1">
        <action id="SetTarget">
```

```
          <statevar id="NewTargetValue" type="boolean" value
              ="false"/>
        </action>
      </service>
    </device>
  </scene>
</xml>
```

Listing 2.2 An example of a scene description XML which handles two binary lights.

Listing 2.3 shows an example of a condition description XML; *ge*, *eq* and *nt* are condition operands meaning greater equal, equal and not respectively.

```
<xml>
  <trigger id = "testTrigger">
    <logic operator="OR">
      <condition operator="ge">
        <lcomparator type="upnp" device="uuid:org-upnp:
            testdevice1" service="urn:org-upnp:testservice"
            statevar="level"/>
        <rcomparator type="value" vartype="Integer" value
            ="12"/>
      </condition>
      <logic operator="AND">
        <condition operator="eq">
          <lcomparator type="upnp" device="uuid:org-upnp:
              testdevice2" service="urn:org-upnp:testservice"
               statevar="Status"/>
          <rcomparator type="upnp" device="uuid:org-upnp:
              testdevice3" service="urn:org-upnp:testservice"
               statevar="Status"/>
        </condition>
        <condition operator="nt">
          <lcomparator type="upnp" device="uuid:org-upnp:
              testdevice4" service="urn:org-upnp:testservice"
               statevar="Status"/>
          <rcomparator type="value" vartype="Boolean" value="
              true"/>
        </condition>
      </logic>
    </logic>
  </trigger>
</xml>
```

Listing 2.3 An example of a condition description XML.

Figure 2.14 demonstrates a simple task execution manager example, where a self managed service add triggers that depend on events from UPnP devices A,B and scenes that operate on devices C and D. When events from devices A and B arrive, if the trigger is evaluated true, the associated scene is executed and actions are performed on devices C and D. The task execution manager runs in the OSGi framework, but since it is a UPnP service too, it has also a part in the home IP network.

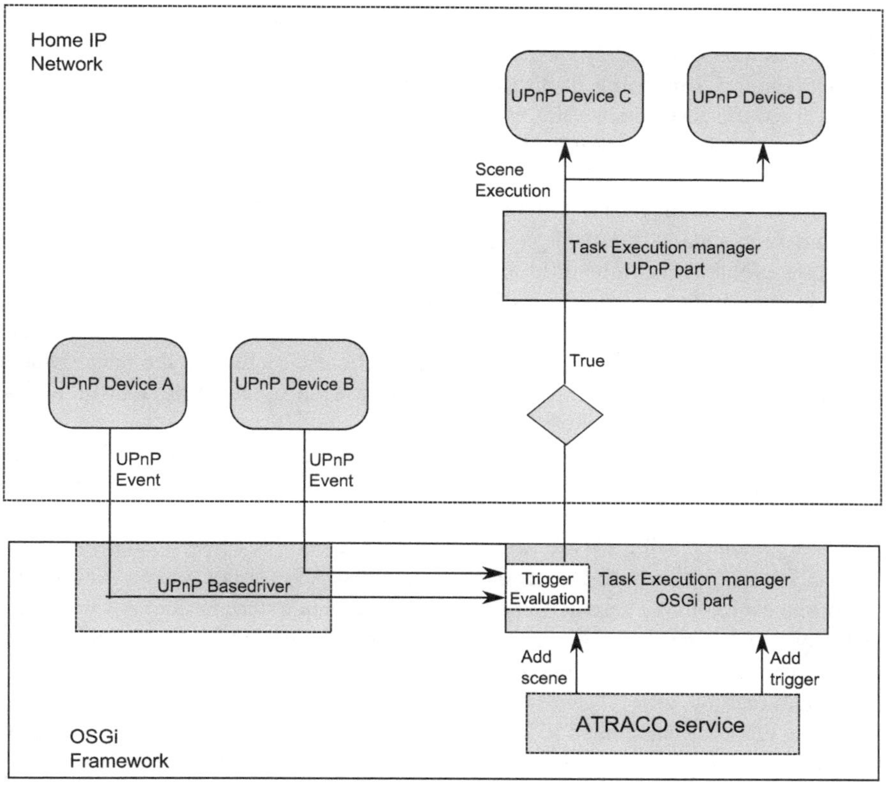

Fig. 2.14 Task Execution Manager Example.

2.9 Network Adaptation in the ATRACO System Prototype

The ATRACO project has implemented a prototype of an ATRACO system in order to test its proposed ambitious architecture. It consists of the main components supporting adaptation specified and prototyped by the project, namely Network Adaptation, Structural Adaptation, User Behaviour Adaptation, Semantic Adaptation, and User Interaction Adaptation, as well as several basic components for controlling the environment (e.g., control of lights, HVAC, music player). Regarding NA, both in terms of device and service adaptation, a set of tools has been created in order to demonstrate the use of NA in real ATRACO environments. This set includes supporting a variety of devices as well as creating all the necessary services that combine the elements found inside an ATRACO ecosystem in order to realize a user goal.

In principle, a user goal, represented as an Activity Sphere (AS), is matched with available device local ontologies, the AS is decomposed into a hierarchy of abstract tasks represented as a workflow, and the Sphere Manager (SM) forms the AS for the specific user goal. The SM connects to device registry to get connectivity data of the devices through the help of the NA component which provides access to home peripherals and a mapping of all different network domains in the AIE to the IP level. Then, the SM initiates the various components (Ontology Manager, Interaction Agent and Fuzzy Task Agent) and performs dynamic service binding in order to execute the workflow. When an action is required to take place in the AIE, like starting the music player or switching on/off HVAC, the SM invokes the NA layer to change the state of the devices and the services. Through the NA, the SM also generates events that can be used by other components of the system, e.g. continuously sending light levels to the FTA adapting the light. The devices involved in an AS communicate over wired/wireless network domains, overlaid with TCP/IP and OSGi/UPnP middleware programmed in Java. The use of Java as the development platform facilitates our system deployment on a wide range of devices including mobile phones and PDAs.

Using the proposed Network Adaptation framework, several device UPnP wrappers for binary sensors, door traps, motion sensors, lights, switches, on/off smart plugs and illumination detectors, as well as services supporting task execution, have been implemented within ATRACO to support the complex trial hosted in the iSpace in Essex demonstrating three activity spheres (AS) for concept validation and prototype evaluation. These activity spheres include an entertainment AS involving tasks for watching TV, reviewing photos, listening to music, reading and playing on a game console, a work AS involving tasks for surfing the internet, writing documents and reading, as well as a sleep AS referring to end of day/sleep activities and involving tasks for watching TV, listening to music/radio and reading. NA represents the physical devices participating in the ASs as ROCob devices (device representation layer) and exports them to the ATRACO ecosystem as UPnP devices (service adaptation layer).

The role of NA is demonstrated in the ATRACO prototype with the transparent communication of devices and services residing in various different physical networks. For instance, X-10 lights connected on a server in the iSpace can be controlled by the Fuzzy Task Agent (FTA) used for artefact and user behavior adaptation, based on the feedback given by a ZWave illumination detector physically connected to the NA component. FTA is agnostic to the protocol used for the control of each individual device, because the devices are presented in a common way thanks to NA.

To provide a clear demonstration of the functionality of NA in an ambient environment such as an ATRACO ecosystem, we examine use cases in different activity spheres. Specifically, we will examine two use cases. The first one demonstrates how a high level preference of a user for energy efficiency is realized transparently through NA inside the context of an activity sphere related to user *entertainment*. The second use case provides the implementation of a a security service during a *sleep* activity sphere by employing actual devices as well as abstract UPnP services that do not correspond to actual devices.

2.9.1 Key Components in Use Cases

Every use case is implemented with the participation of various persons, devices and services. The most obvious of them are the persons that participate and initiate actions or become receivers of the results of adaptation actions performed by the ATRACO system.

Other less obvious components that take place in use case realization are the devices that are used in the context of an AS. This set of components is less obvious because apart from the devices that the users interact directly with, it includes other devices that have an auxiliary role. Moreover, an AS may define multiple use cases, each one of them may use a differentiated set of devices. It is important to note that the characteristics of a device, such as network protocol and physical medium, are transparent to the ambient ecosystem thanks to NA and do not affect the definition and the functionality of the use case in the activity sphere.

A key device in the ambient ecosystem is the home controller, which provides connectivity with devices of different protocols and physical mediums. This device diversity is finally abstracted in the ambient space by using suitable UPnP wrappers that bring all these devices in a common network layer (UPnP).

Apart from the obvious ones, such as the devices and persons involved, some services and functionalities get involved in almost all use cases and are transparent to the user. The most important of them is STEM, which is a network service (UPnP) that enables the execution of a set of actions on one or more devices, when certain conditions are met.

2.9.2 Entertainment Activity Sphere Use Case

The entertainment AS, as implied by its name, demonstrates functionalities and adaptations that take place during an entertainment activity involving particular tasks for watching TV, reviewing photos, listening to music, reading and playing on a game console. For the purpose of this book, we analyze Network Adaptation in the use case of enabling an energy saving capability, in case this is enabled in the user preferences.

The energy saving use case demonstrates how energy can be saved, if this is desired by the user, by combining information and usability of various ATRACO components. The user will be watching TV, and at some time will open the window to let some fresh air in. If energy saving is enabled in the user profile, the ATRACO system should stop the heating from working. Then, when the window is closed, the Interaction Agent will ask the user if the heating should be turned on again.

This energy saving service is realized by using a set of window traps, motor driven curtains and heating equipment inside the ATRACO ecosystem, as well as the *interaction agent* and mostly *STEM*. Additionally there is a software component that provides the *energy saving service* that keeps track of the user's choice of enabling or disabling the energy saving feature and the heating status and also triggers the interaction agent to ask the user if he likes to restart the heating during the *entertainment AS*. The energy saving use case is executed in parallel with the watch TV use case, which additionally involves lights, illumination sensors, various location sensors and a software component providing time and daytime information.

The purpose of the energy saving use case is to demonstrate:

- User behavioral adaptation. The lights are adjusted according to the settings that FTA has progressively learnt for each specific user.
- Sphere adaptation. The system switches from the entertainment *watch TV* sphere to the *energy saving* sphere.
- Application of preferences set via the privacy manager.
- Cooperation of various ATRACO components (FTA, SM, IA, NA)
- *Network Adaptation demonstrating that devices spanning UPnP, x10, LonWorks TP, and Z-Wave networks operate seamlessly in the ATRACO environment and helping to implement user behavioral adaptation and sphere execution.*

The SM becomes aware of the user and his preferences via the privacy manager and sets a condition for the *energy saving* sphere, which is energy saving is enabled, heating is on and window is open. Then it associates with the above condition a scene to turn off heating and send event to set accordingly the appropriate state variable in the energy saving UPnP service. According to the extended use case, the user will turn on the TV and the SM will initiate the watch TV sphere. FTA becomes aware of the active sphere and reads the light sensors data. According to the value it has read, it adjusts the ceiling lights. Optionally, FTA reads the value of a UPnP network clock, and decides whether it is daytime or not. If it is, it closes the curtain in order to prevent the natural light from making watching TV difficult (e.g. too much luminance, possible reflections etc.). Sometime during the watch TV sphere the user may want to let some fresh air in. If the curtain is closed, he uses the interaction agent and controls the curtain (or does this with a relative switch on the wall). Then he opens the window. This sends a *window open* event from the door trap sensor. STEM receives this event, and it checks the condition set. As a result it executes the associated scene as explained before. The SM will receive the heating off event and will place a new condition on STEM, which is heating is off and window is closed and further associates this condition with a scene to ask about enabling the heating. In parallel, the light sensors, if it is day, sense the luminance difference and FTA sets the ceiling lights accordingly. FTA will readjust the dimming value of the ceiling lights if the user closes the curtain again, according to the measured luminance value by the luminance sensors. The user returns to the sofa and continues watching TV. When he decides to close again the window the door trap sensor will send a *window closed* UPnP event. STEM will receive this event and ask IA to display the *Energy Saving UI* on the TV set, which asks the user if the heating should turn on. If the answer is yes, the IA propagates the answer to the SM, which subsequently sends an event to STEM to turn the heating on.

Network Adaptation in this AS is responsible for executing actions on the devices participating in the presented use case, such as turning on/off the heating, opening/closing the motor driven curtain, adjusting the lights etc., as well as translating state changes to UPnP events sent to ATRACO components and services, such as the wireless Z-Wave events sent by the door/window traps and received by the home controller in the ambient ecosystem and via its UPnP wrappers end up as UPnP events, or events conditionally generated by STEM to the UPnP energy saving service by evaluating a set of relative rules defined on STEM by the SM through invoking appropriate STEM UPnP actions.

2.9.3 Sleep Activity Sphere Use Case

The sleep AS, as implied by its name, demonstrates functionalities and adaptations that take place when the user goes to sleep. For the purpose of this book, we analyze the use case of enabling an alarm service.

This alarm service is realized by using a set of door/windows traps and a siren device inside the ATRACO ecosystem, as well as the *interaction agent* and mostly *STEM*. Additionally there is a software component that provides the *alarm service* that keeps track of the user's choice of enabling or disabling the security feature during the *sleep AS*.

Network wise, this is a UPnP service which allows the modification of its state by the user. This change of state can be performed either by using a remote control, such as a Z-Wave keyfob or by a relative user interface, provided that the privacy manager allows the use of these devices or UIs. The first way employs STEM, which connects the actions on the keyfob with the state of the alarm service. This is achieved by translating the ZWave commands issued by the keyfob to UPnP events that are received by STEM, which just afterwards sends the correct UPnP messages to the alarm service. STEM knows which events to use and monitor for this action, by evaluating a set of rules defined by the SM. SM checks the availability of devices, both in terms of physical presence in the ambient space and permission to use by the privacy manager, and sets the relative rules on STEM by invoking the appropriate STEM UPnP actions.

Apart from enabling the alarm service, a new set of rules has to be added, that checks the state of the participating door/window traps (in our example wireless Z-Wave devices) and the alarm service. If the latter is enabled, and at least one sensor generates an *'ON'*, then STEM sends a new event in the ATRACO UPnP network indicating the activation of the alarm service (intrusion), and starts the siren. From a network perspective, wireless Z-Wave events are sent by the door/window traps and received by the home controller in the ambient ecosystem and via its UPnP wrappers end up as UPnP events. Since STEM uses the state of these devices for rule evaluation, it accepts them and evaluates if it should fire the alarm event, again as a UPnP message.

2.10 Lessons Learned

The Network Adaptation component provides the lowest layer of service and device communication upon which the whole ATRACO environment is built. As such it has to tackle requirements and special needs of almost every component. This broad involvement of the NA component in the project was the reason that many lessons have been learned from its development and evaluation.

From a system design point of view, the greatest lesson learned was that widely adopted protocols are not always well implemented. Thus, choosing a technology based among others on adoption and availability of existing implementations does not always ensure that it will provide the fastest and most reliable solution. For example, although both OSGi and UPnP are widely used for a lot of years already, the existing implementation of UPnP for Java and OSGi had a lot of problems. In fact the problems were so many that new UPnP protocol implementations for Java and OSGi have been developed from scratch.

Along the same lines, although software written in Java is widely known to be portable across various operating systems and architectures, it proved that this is not entirely true. Special features of the functionality and implementation of network sockets of each OS running on machines based on different computer architectures (e.g. ARM, x86, x86_64) brought up lots of problems that were very difficult to spot and resolve. Resolution of the problem for a specific pair of architecture and OS very often was breaking the functionality of another. Thus, complex network operations require deep knowledge of the specifics of the underlying OS implementation and it must not be assumed that a simple use of an API provided by the programming language is sufficient.

It is worth mentioning that deviations from the specified functionality were also met in some devices used during the prototype development. This should be expected when working with quite new devices or even not publicly released ones and thus fallback scenarios must exist. Problems of this type were affecting communication with the device, the discovery procedure or could lead to lost events. In some cases that no software workaround was possible, some devices were not used at all. Another interesting issue was that devices with similar functionality, e.g. light sensors, provided input, under the same configuration, with large deviations, requiring the corresponding drivers to be modified in order to weight the incoming data or even to avoid using specific devices in order to achieve robust functionality.

Apart from problems based on deviations from specifications, such as the ones described above, useful conclusions have been drawn regarding integration of multiple components. The first lesson in this area is that quite often integration effort is heavily underestimated. Experience drawn from the ATRACO project has proved that a lot of unexpected problems arise during integration and the more the partners are involved the more the problems appear. The severity of these problems during the project was minimized by constant communication of partners, component unit testing, and following a mixed mode of "big picture" design and step by step design and implementation. Furthermore, excessive use of team development tools and testing of one component from another component's point of view lead to faster development, API clarification and bug resolution. The second great lesson has been learned during the last year of the prototype development, where detailed specification of use cases before implementation helped to identify and resolve quite early component race conditions and limitations.

A very useful conclusion from the evaluation of the system is that testing in the lab cannot provide the same evaluation and testing data compared to the use of the system in real-life conditions. This happens because the time spent in the lab testing a specific part of the system is usually quite limited in comparison with the exposure of the system in real life conditions. Moreover this difference is amplified in real life conditions where usually the system is deployed in larger scale and includes many more devices affecting performance and stability. These observations have changed the way that tests were designed and run and as a result the testing procedure was much more effective.

2.11 Conclusions

Within the ATRACO project we have specified and implemented a Network Adaptation framework in order to provide a set of functions and protocols in a SOA middleware allowing an ATRACO system to interact with the network resources. The goal of the NA layer is to provide seamless use of the devices and services to the ATRACO system and to simplify the access to networks in the AIE. The design and specification process of the NA component involves the consideration of several adaptation guidelines, such as unified access on devices that belong to different networks, interoperability between networks, task completion with alternative resources, dynamic discovery for new networks and devices, representation of legacy internet services and relaying of existing web services. In ATRACO, the NA component meets these requirements through defining common device types across networks, allowing events from any network to trigger actions in any other network, featuring a device registry and device/driver manager as well as providing single access for web services (i.e. a weather report service) able to evaluate the availability, accessibility and quality of multiple existing web services provided by the internet sites.

To cope with the complexity of accessing diverse control networks, a common Device Representation Layer has been introduced. This representation layer is designed in such a way that it handles different networks while hiding the complexity and details of each one of them. For instance, an application which is to be placed on top of the device representation layer should not tell the difference between a Z-Wave lamp and a LonWorks lamp. The device representation layer creates a unique representation of each different device type across networks, which simplifies the evolution of ambient intelligent applications and services. In addition to the devices, the proposed NA framework additionally shapes various other services provided over IP networks (e.g. NTP, VoIP, RTSP etc.) as ATRACO services. The functionality in an ATRACO environment is exposed as semantic services which an actor can discover and then compose to form ATRACO applications. After all, each service is associated with at least one semantic description which shields the actor from the complexity of the resource layer realization and makes it easy for the actor to employ these services in accomplishing interesting and useful tasks.

The NA Layer has been implemented using a mixed architecture, including a centralized home controller that is able to seamlessly integrate various different devices from different technologies around different home automation and control space together with a variety of modules, some of which may be able to work in a distributed architecture. The Device Representation Layer provides an abstraction of the network devices to OSGi services, sharing a common device format and API in OSGi context, that can be used by other ATRACO components. These OSGi services represent the functionality of the underlying physical devices, giving the architectural and application components the opportunity to discover them and retrieve all the required information for managing and performing control actions. The available functional profiles include physical devices with multiple endpoints, binary sensors, level properties like dimmers (with or without timing functionality), direction semantics like shades (with or without timing functionality), direction semantics with read only values, switches, batteries, analog meters, alarms, thermostats etc. Moreover, the Device Representation Layer provides several utilities that intend to provide a service layer for scene execution and alerting. A scene is a collection of OSGi devices that will execute a number of actions on these devices when a certain event occurs. An alarm service allows sending alarm events to other OSGi components or higher layers. It makes use of a set of defined alarm levels, types and properties, to cover a wide range of alarm events. Furthermore, the alert service provides simple alerts to the other OSGi aware services in the framework. It listens to the device events and evaluates a series of user defined Rules to finally create the specific alerts.

The Service Adaptation Layer provides an abstraction of the Device Representation Layer devices as OSGi services so that additional external services can take advantage of the unified abstraction. On top of that, OSGi device representations are eventually transformed to UPnP services. Additionally, it provides UPnP adaptation for IP services participating in an ATRACO system. UPnP seems to cover most of the aspects specified for the Service Adaptation Layer as it provides the ease of network setup, device discovery, self-announcement and advertisement of supported capabilities.

The NA Layer additionally provides an OSGi based UPnP service for task execution, which is called STEM. Based on a simple or nested condition evaluation, STEM generates a UPnP event about the result of the evaluation and even executes a set of actions on one or more UPnP devices when the evaluation is true. In order to be able to evaluate conditions, STEM subscribes to the services of the devices that are included in the conditions of the various triggers. STEM, by itself, is not a UPnP control point, so it is not aware of the available devices. It becomes aware of the existing devices using the OSGi API of the OSGi UPnP base driver. When a STEM user adds a scene or a trigger, these are validated with respect to the existing UPnP devices.

A prototype that realizes the ATRACO architecture has been implemented, which exhibits, although in primitive form, all types of adaptation specified by the ATRACO project, including adaptation at the network level. The system has been deployed in the iSpace at the University of Essex and tested by real users, with promising results. Using the proposed Network Adaptation framework, several device UPnP wrappers for binary sensors, door traps, motion sensors, lights, switches, on/off smart plugs and illumination detectors, as well as services supporting task execution, have been implemented to support the complex trial hosted in the iSpace demonstrating three activity spheres for concept validation and prototype evaluation. NA materializes a UPnP middleware allowing the transparent communication of devices and services residing in various different physical networks. In order to provide a clear demonstration of the functionality of NA in an ambient environment such as an ATRACO ecosystem, we have presented in detail two use cases. The first one demonstrates how a high level preference of a user for energy efficiency is realized transparently through NA inside the context of an activity sphere related to user *entertainment*. The second use case provides the implementation of a a security service during a *sleep* activity sphere.

References

1. Amigo project. URL `http://www.hitech-projects.com/euprojects/amigo/`
2. Atraco project. URL `http://www.atraco.org`
3. Eperspace project. URL `http://www.ist-eperspace.org/`
4. Home2015 programme. URL `http://home2015.i2r.a-star.edu.sg`
5. Music project. URL `http:///www.ist-music.eu`
6. Ngn @ home. URL `http://portal.etsi.org/at/ATNGNSummary.asp`
7. Sense project. URL `http://www.sense-ist.org/`
8. Teaha project. URL `http://www.teaha.org`
9. Upnp switch power service description. URL `http://www.upnp.org/standardizeddcps/documents/SwitchPower1.0cc.pdf`
10. Baker, C., Markovsky, Y., Van Greunen, J., Rabaey, J., Wawrzynek, J., Wolisz, A.: Zuma: A platform for smart-home environments. In: Int. Conf. on Intelligent Environments, pp. 257–266 (2006)
11. Hargrave, B., Kriens, P.: Osgi best practices. In: OSGi Alliance Community Event, pp. 26–27 (2007)
12. Jahnke, J., d'Entremont, M., Stier, J.: Facilitating the programming of the smart home. IEEE Wireless Communications **9**(6), 70–76 (2003)
13. Kumar, R., Poladian, V., Greenberg, I., Messer, A., Milojicic, D.: Selecting devices for aggregation. In: IEEE Workshop on Mobile Computing Services and Applications, pp. 150–159 (2003)
14. Lin, R., Hsu, C., Chun, T., Cheng, S.: Osgi-based smart home architecture for heterogeneous network. In: 3rd Int. Conf. on Sensing Technology, pp. 527–532 (2008)
15. Matsuura, K., Hara, T., Watanabe, A., Nakajima, T.: A new architecture for home computing. In: IEEE Workshop on Software Technologies for Future Embedded Systems, pp. 71–74 (2003)
16. Perumal, T., Ramli, A., Leong, C., Mansor, S., Samsudin, K.: Interoperability among heterogeneous systems in smart home environment. In: IEEE Int. Conf. on Signal Image Technology and Internet Based Systems, pp. 177–186 (2008)

17. Ramparany, F., Boissier, O., Brouchoud, H.: Cooperating autonomous smart devices. In: Smart Objects Conference, pp. 182–185 (2003)
18. Rellermeyer, J., Alonso, G., Roscoe, T.: R-osgi: Distributed applications through software modularization. In: ACM/IFIP/USENIX 8th Int. Middleware Conference (2007)
19. Roalter, L., Kranz, M., Moller, A.: A middleware for intelligent environments and the internet of things. In: 7th International Conference on Ubiquitous Intelligence and Computing, pp. 267–281 (2010)
20. Rose, B.: Home networks: A standards perspective. IEEE Communications Magazine **39**(12), 78–85 (2001)
21. Valtchev, D., Frankov, I.: Service gateway architecture for a smart home. IEEE Communications Magazine **40**(4), 126–132 (2002)
22. de Vicente, A., Velasco, J., Marsá-Maestre, I., Paricio, A.: A proposal for a hardware architecture for ubiquitous computing in smart home environments. In: Int. Conf. on Ubiquitous Computing: Applications, Technology and Social Issues (2006)
23. Wang, Q., Cheng, L.: Awareware: an adaptation middleware for heterogeneous environments. In: International Conference in Communications, pp. 1406–1410 (2004)

Chapter 3
Ontology-based knowledge management in NGAIEs

A. Kameas and L. Seremeti

Abstract The objective of the knowledge architecture of ATRACO is to enhance communication, as well as to ensure effective knowledge sharing among ATRACO components. It is built around ontologies, ontology managers and agents. Every component of an activity sphere uses an ontology to model its local knowledge and state. These ontologies will certainly be heterogeneous, but they must be used transparently in the context of any sphere. Thus, knowledge management in ATRACO is concerned with the alignment of heterogeneous ontologies, in order to produce the sphere ontology, which encodes the sphere knowledge.

In this chapter, we shall describe the ontology management framework developed in the context of ATRACO, which includes:

1. A set of ontologies, that includes Upper Level Ontology, User Profile Ontology, Generic Device Ontology, Privacy Policy Ontology and Interaction Ontology;
2. The Ontology Manager, a software module that manages ontology and provides an interface to the other system components; it also manages the Sphere Ontology; and
3. The Alignment Module, which is responsible for aligning two ontologies and storing the alignment in a machine readable format.

In Information Technology and Artificial Intelligence, ontologies are the structural framework for organizing and representing information, as a set of concepts in a domain and the relations between those concepts. In the context of ATRACO project, three types of ontologies have been developed: domain ontologies, each one describing devices, services, user profiles and agent knowledge bases; ontologies describing tasks and policies; and an upper level ontology describing the basic entities and relationships of the ATRACO world model.

Achilles Kameas
The Hellenic Open University and DAISy research unit, Patras, Hellas, e-mail: kameas@cti.gr

Lambrini Seremeti
The Hellenic Open University and DAISy research unit, Patras, Hellas, e-mail: seremeti@cti.gr

Ontology alignment is the output of the ontology matching process that attempts to find relationships, or correspondences between entities of different ontologies, in order to produce sets of such correspondences between two or more ontologies. During the alignment operation, the original ontologies are kept unaltered, while the alignment results are stored separately from the ontologies themselves. In ATRACO, we applied alignment procedures in order to deal with lexical, syntactic and semantic heterogeneity.

In order to reconcile the intrisically different models of local knowledge expressed via ontologies, a mathematical formalism based on Category Theory serves as a solid foundation, since it allows the coexistence of heterogeneous entities (ontologies) while focusing on relationships

3.1 Introduction

Ambient Intelligent Environments (AIEs) are human activity spaces populated with smart communicating objects, which are able to perceive the environment, act upon it, process and store data, manage their local state, communicate and exchange data. AIEs provide an infrastructure that supports services such as networking, communication, discovery, location and context estimation [19]. The emerging Next Generation of AIEs (NGAIEs) are being designed to inherently exhibit intelligent behavior and adaptive functionality, in order to provide optimized resource usage and support consistent functionality and human-centric operation. Each of NGAIEs "stakeholders", i.e. humans, agents, devices, services, having its own conceptualization of the world, assumes a variety of roles, in order to realize successfully its tasks, within the NGAIE context.

Intelligence, as the primary means to achieve adaptation, will appear at various levels. For example, local resource management may require decision making mechanisms and even embedded intelligent agents. At a system scale, multi-agent systems, using semantically rich descriptions, learning mechanisms and possibly cognitive functions (such as perception, homeostasis etc.) will be embedded in NGAIEs. But intelligence relies on knowledge; thus, different kinds of knowledge will be encoded in NGAIEs, including knowledge about the state of resources, the tasks to be achieved, the preferences of the users and the policies to be realized.

In the general case, NGAIE resources will be heterogeneous, as they will originate from different manufacturers; consequently, they will probably use proprietary, heterogeneous information and knowledge representation schemes. Within NGAIEs, as well as among interacting NGAIEs, multiple sources of heterogeneity can appear:

- Artifacts and other engineered NGAIE components, each of which has a proprietary, usually closed, model of itself and the world
- NGAIEs, public or temporarily private, which have their own models of their resources and services

- Task models, expressed in various domain-dependent notations, as well as dialogue states and interfaces
- Multimedia objects, which usually adhere to the metadata of multimedia standards; the same holds for other types of "intelligent" content
- Networking protocols and, in general, communication schemes, which require specific descriptions of artifacts and services and usually have a restricted closed world model
- People, who have their own individual profiles and ways of perceiving, understanding and accepting technology

Thus, two important goals of NGAIE design are (a) to increase the amount of knowledge that is available to the system and (b) to minimize the inaccuracy of knowledge and the ambiguity regarding the interpretation of the shared information, thus enabling NGAIE components to interact successfully through a common communication channel, despite their heterogeneous representations of the world.

Ontologies can be used to address these issues through the semantics they convey. Ontology is defined as an explicit and formal specification of a shared conceptualization [18]. A "conceptualization" refers to an abstract model of some phenomenon in the world, which identifies the relevant concepts of that phenomenon. "Explicit" means that the type of concepts used and the constraints on their use are explicitly defined. "Formal" refers to the fact that the ontology should be machine readable. "Shared" reflects the notion that ontology captures consensual knowledge, that is, it is not private of some individual, but accepted by a group. Thus, ontology is a structure of knowledge, used as a means of knowledge sharing within a community of heterogeneous entities.

A straightforward solution for achieving interoperability is to develop a commonly accepted upper level ontology for NGAIEs and verify its wide applicability over various contexts. To this end, a few upper level ontologies have been proposed, while standardization efforts by the World Wide Web Consortium have led to the development of the ontology language OWL (Web Ontology Language) [45], which is the evolution of DAML+OIL. However, the development of a single ontology, able to completely and accurately describe the key concepts of such environments, as well as the processes carried out within them, is almost infeasible, because of the engineering difficulty of soundly depicting the countless relationships and properties of the large set of concepts. Thus, we expect that multiple heterogeneous ontologies will be developed autonomously, in order to semantically describe the features and capabilities of different NGAIE components.

Consequently, semantic mediation among these different types of knowledge becomes necessary. There are many approaches and techniques, which are based on ontology operations that can be used to develop mechanisms for interfacing the heterogeneous and possibly inaccurate ontologies of the NGAIE components [9]. These include ontology alignment, mapping and merging.

With ontology mapping, the correspondences between the two ontologies are stored separately from the ontologies and thus are not part of the ontologies themselves. The correspondences can be used, for example, for querying heterogeneous knowledge bases using a common interface, or transforming data between different representations. When performing ontology merging, a new ontology is created, which is the union of the source ontologies. The new ontology, in general, replaces the original source ontologies and captures all their knowledge. The challenge in ontology merging is to ensure that all correspondences and differences between the ontologies are reflected in the merged ontology. The first stage of both ontology mapping and merging is ontology alignment, which identifies the semantically related concepts across multiple ontologies.

In a nutshell, knowledge management in NGAIEs comprises two important strands: ontology engineering and ontology matching. In the following section, we discuss the related work on knowledge representation and management in AIEs, starting from ontology-based ubiquitous computing systems and then focusing on ontology engineering methodologies and ontology alignment approaches. In the third section we present the ontology engineering methodology we developed for the purposes of project ATRACO, followed by a presentation of a set of specific NGAIE ontologies. Then, we present the three-step ontology alignment strategy, which we used in ATRACO. This strategy includes an algorithm that guides the selection of appropriate matchers and an algorithm for deciding the combination of matchers to be used. In the following section we briefly present a mathematical formalism based on Category Theory that can serve as the framework for describing the combination and integration of ontological objects of the ATRACO world. Throughout these sections, we highlight the main ATRACO contribution, an ontology-based knowledge management framework, based on the notion of activity spheres, that permits to overcome the barriers of dynamic nature and heterogeneity intrinsic in NGAIEs. Finally, Section 3.6 concludes the chapter, while Section 3.7 proposes further readings.

3.2 Related Work

The challenging issues associated with the dynamic nature of NGAIEs are complex and mostly related to heterogeneity problems encountered at different levels. In this chapter, we shall discuss only those related to the heterogeneity issues present at the level of knowledge representation and management, which appear in cases where different systems utilize different types of ontologies and ontology operations.

3.2.1 Knowledge representation and management in AIEs

As stated in the Introduction, ontologies can be used to tackle the issues that stem at the level of knowledge representation and management from the various sources of heterogeneity that exist within NGAIEs. Already quite a few research projects have employed ontologies for this task, including Gaia [34], CoBRA [7], CoCA [12], Semantic Spaces [45], SOFIA [4], DRAGO [37] and CAMPUS [36]. These projects have developed middleware infrastructure for NGAIEs and support reasoning between ontology-based components. Most of these systems either develop a common upper level ontology, which serves as a reference source for heterogeneous domain ontologies, or use domain ontologies that have been built using a shared vocabulary, that is, they are a priori linked with a core ontology which describes the key concepts of an NGAIE. Only projects DRAGO and CAMPUS consider completely heterogenous ontologies. In the case of DRAGO, predefined mappings are used to align the knowledge representations provided by different ontologies, while CAMPUS focuses on the automatic ontology alignment between ontologies.

In our research, we follow an approach similar to that in CAMPUS project. But, because we expect that numerous inconsistent ontologies will appear within an NGAIE, and taking into account the kind of heterogeneity present in these, we propose a multi-faceted strategy to achieve reliable ontology alignments. This strategy estimates the similarity of ontologies to be aligned, and based on the outcome, automates to a different degree the alignment process (obviously, the higher the similarity, the more automated the alignment process). In this way, an alignment is guaranteed, although in the worst case, the user may be asked to serve as "trusted third party" and evaluate the proposed alignments. The next two sections are dedicated to a survey of ontology engineering methodologies and ontology alignment approaches.

3.2.2 Ontology engineering

Ontology engineering refers to a set of activities that concern the ontology development and management processes, the ontology life cycle, the methodologies for building ontologies and the tool suites and languages that support them. The six basic aspects to consider when creating an ontology are [18, 30, 39]:

- The content of the ontology
- The application in which it will be used
- The language in which it is implemented
- The methodology which has been followed to develop it
- The software tool used to build and edit the ontology and
- The objective principles for guiding and evaluating ontology design

Several research groups have proposed various methodologies and ontology development environments for building ontologies. In this section, we discuss several methodologies for building ontologies, each one involving different activities, depending on the intended scope, size and level of detail of the ontology under construction. Their success has been demonstrated in a number of applications. After surveying an extensive set of ontology development methodologies [25, 28, 29, 33, 42, 43, 44], we can classify them into two large categories [38]: (1) methodologies focusing on building a single ontology for a specific domain of interest and (2) methodologies focusing on the construction of ontology networks.

On the one hand, the methodologies that focus on building a single ontology presently described in the literature, can be further distinguished in:

- Those aiming at building ontologies from scratch, or by reusing pre-existing ontologies, or by using non-ontological resources, according to the resources that are available to developers
- Collaborative and non-collaborative, according to the degree of participation of the involved ontology engineers, users, knowledge engineers and domain experts, in the ontology engineering process
- Application dependent, semi-application dependent and application independent, according to the degree of dependency of the developed ontology on the final application
- Manually, semi-automatically and automatically constructed ontologies, according to the degree of human involvement in the building process

On the other hand, only one methodology appears in the literature [42] that focuses on building ontology networks. It deals with the definition of different scenarios for building a collection of single interconnected ontologies, related to each other via meta-relations, such as hasPriorVersion, if the ontology to be developed is a new version of an existing one, isExtension, if the ontology to be developed extends another existing ontology, etc. In that sense, it emphasizes on reuse, reengineering and merging of available ontological and non-ontological resources.

By surveying the existing methodologies, one can conclude that there is no methodology that fits all cases. In each case, the selection of the best methodology depends on a set of contextual factors, which include the number of the participants in the ontology development process, the kind of the ontology to be built (domain, Upper Level), the application for which the ontology is planned to be used, the participants' knowledge level of a particular domain, or of the ontology field, their cultural biases, their skills, the tools that are going to be used, the availability of participants and resources (documents, ontologies, thesauri), etc.

In the case of building ontologies for modeling NGAIEs' entities, where various stakeholders, with different skills in knowledge engineering are involved in the ontology building process and the entities under description are complex by nature and especially within the ATRACO project, the above mentioned contextual factors are restricted to:

- Since each ATRACO entity describes an isolated domain of interest, such as a specific device, domain ontologies are built
- Only one ontology developer participates in the engineering process, imposing his own interpretation of the domain. For example, a software developer models a UPnP Device, while a knowledge engineer models a user profile
- Non-ontology, or non-domain experts are involved in the ontology building process, since, for example, each manufacturer provides his conceptual model for a specific device he builds
- Ontological, or non-ontological resources are not easily available, in order to be reused, or reengineered

Based on these considerations, we propose in Section 3.3.1 a methodology for building domain ontologies, which covers the drawbacks of the existing methodologies, while, at the same time, benefiting from their advantages. Our methodology is based on different scenarios, with respect to the skills of the ontology creators and the available resources. In Section 3.3.2, we apply this methodology, in order to develop specific resource and other ontologies.

3.2.3 Ontology alignment approaches

Ontology alignment can be a key issue in knowledge-based open-ended environments, such NGAIEs, as it permits the discovery and representation of links (lexical, syntactic, semantic, etc.) between pairs of ontologies, while keeping the original ontologies unaltered. By using alignments, a network of interconnected ontologies is created, which contains the overall knowledge about the domain, albeit distributed in several ontologies and alignments. In this way, the management of the global knowledge contained in an NGAIE becomes a problem of managing many small pieces of knowledge, which describe independent resources and their collaboration. The management of local resource ontologies is cost-effective, because it can be done in a distributed manner (each entity owns its ontology and operates autonomously within the NGAIE). Then, resource collaboration, at the knowledge level, can be described using ontology alignments, which essentially are links that connect the local resource ontologies.

More formally, the problem of ontology alignment is described as: Given two ontologies O_1 and O_2, an alignment between them is defined as a set of relations (equivalence, subsumption, disjointness) between pairs of entities (classes, properties, instances), belonging to the original ontologies. The application of ontology alignment to the interaction of entities in NGAIEs, leads to a set of specific problems:

- Since in open systems like NGAIEs, it's impossible to know in advance the properties of the entities that will interact, it is impossible to use a priori (precomputed) ontology alignments. Thus the alignment process has to be performed at run time, sometimes with (hard) real time constraints.
- Since the interacting entities may share common purposes, or competences, an alignment between two ontologies can be discovered in most applications, although the ontologies may differ in granularity (detail of description of the same entities).

Several alignment approaches have been proposed to tackle the problem of discovering semantic correspondences between entities of different ontologies. They mainly focus on the following aspects [16]:

- Lexical comparison, which relies only on the labels on the ontological entities
- Structural comparison, which relies only on the structure of the ontologies
- Instance comparison, which compares the instances of each ontological entity

- Comparison based on "background knowledge source"

Many researchers have investigated the problem of ontology alignment, mostly by proposing several ontology alignment tools and matchers (or alignment algorithms) [11, 21, 15, 27, 14] which exploit various types of information in ontologies, that is, entity labels, taxonomy structures, constraints and entities' instances. These tools can be classified into two large categories: those that make use of a single matcher in order to calculate similarities between ontology entities and those which use a family of parallel, or sequential matchers in composition. In the latter category, the similarity between two ontology entities is finally computed by a composite method, such as a weighted aggregation of the similarities obtained by each matcher separately. A challenging issue, while applying these methods, consists in deciding whether a single matcher, or a combination of different matchers, performs better and in what cases, that is, for which kind of ontologies in question. Hence, given a specific pair of ontologies to be aligned, one should define a criterion to determine when a special matcher should be used. Based on this consideration and the specification of ontology alignment problems in NGAIEs, we propose in Section 3.4, an ontology alignment strategy which includes the calculation, during a pre-alignment step, of two similarity coefficients, which estimate whether the resemblance of the ontologies in question is mainly lexical, or structural. Then, depending on their values, an agent charged with the task of the alignment process, can select the application of suitable matchers, in order to establish correspondences between ontology entities.

3.3 Own approach to knowledge representation and management in NGAIES

In this section, we discuss the main challenging issues of the ATRACO framework, which is a framework for the development of NGAIE applications, by considering the creation and management of heterogeneous knowledge, which is an inherent feature of NGAIEs. Firstly, we briefly present the fundamental notion of the ATRACO activity sphere, from a knowledge point of view and the problems that arise during its realization.

The ATRACO project aims at conceptualizing an ambient ecology, that is, a space populated by different entities (devices, services, humans, agents), which are interrelated and in interaction with their environment, in order to realize activity spheres. Each activity sphere is considered as a new ATRACO entity, which can be adapted to different conditions (spatial, temporal, user preference related, environmental). For example, an activity sphere serving for the conceptualization and realization of the "reading" activity of a specific user in his/her private room (spatial condition), can be transferred into a hotel by changing this spatial condition and the devices (facilities like book, desk, chair) and services (light, ambient temperature) provided by the specific environment, or it can be transferred into a "reading" activity of another user, by changing the user preference conditions. Moreover, an ATRACO activity sphere, considered as another ATRACO entity, can be interrelated with other ATRACO entities, that is, other activity spheres, in order to describe and realize more complex activities in pervasive computing environments.

The intuition behind this is that once an elementary ATRACO activity sphere is produced, the knowledge associated with it can be reused, in order to produce new activity spheres. Having this in mind, the context of ATRACO is suitable for realizing elementary ATRACO spheres which can then, either be extended, by adding to the sphere, in an appropriate way, an entity of the ambient ecology, or be embedded into another activity sphere, in order to produce a new more complex one.

In order for a specific ATRACO activity sphere to be successfully realized, ontologies, ontology managers and agents are used. These provide information representation, semantic interoperability and exchange mechanisms, so that the entities/members of an ambient ecology that participate in the specific activity sphere can communicate and collaborate. According to the ATRACO approach, each entity of an ambient ecology (a human, an agent, a device, or a service) maintains locally an ontology, which represents the complete set of knowledge associated with it and which is managed only by the owning entity. As the entities are distinct, their ontologies will certainly be heterogeneous. As they are engineered by different creators and for different purposes, an alignment process is needed for them to come in a mutual agreement. This alignment process aims at semantically relating heterogeneous ontologies, in order for a network of interlinked ontologies to be realized.

Since the building blocks of knowledge of an ATRACO activity sphere are the ontologies of the interrelated entities of the ambient ecology participating in it and their alignments, we can consider each ATRACO activity sphere as a network of interlinked ontologies, which encodes the necessary knowledge for its realization. In that sense, ATRACO spheres, as networks of semantically linked ontologies, are means of sharing and reuse. Sharing refers to the fact that different ATRACO spheres can make use of the same ATRACO resources/entities, that is two, or more ATRACO spheres may use the same entity of the ambient ecology for servicing different purposes. For example, the light service provided by a specific lamp device can be eventually shared by two distinct ATRACO activity spheres that are realized simultaneously, i.e. the "reading" activity performed by an individual or for the "playing cards" activity performed by a group of persons. Reuse means to build a new ATRACO sphere by assembling already built ATRACO spheres. For example, a "cooking pasta" activity sphere may be realized by embedding a "reading" (a recipe) activity sphere into a "cooking" activity sphere.

Our experimental research has shown that in order to deal with knowledge representation and management of a network of interlinked ATRACO ontologies, one has to resolve the following fundamental issues:

1. How to build complete and consistent domain heterogeneous ontologies, which define concepts closely related to distinct AIEs entities?
2. Which are the ontologies that need to be created, in order to represent the knowledge which is efficient and necessary for the applications in NGAIEs to be carried out successfully?

3. How to strategically guide the alignment process of a pair of heterogeneous ontologies in order to discover their semantic linkage?
4. How to align a pair of heterogeneous ontologies?
5. In what extent and in which order can an initial alignment of a pair of ontologies be further populated with alignments, in oder to produce a network of interlinked ontologies?
6. How can, a replacement of a specific ontology within a network of ontologies, be achieved?
7. How to evolve a network of interlinked ontologies, in the case an ontology evolves, undergoes changes, or is replaced by an equivalent or similar one?

The above mentioned challenging issues refer to the ontology engineering with respect to the nature of NGAIEs, as well as to the application of alignment techniques in networks of ontologies, which seems to be a rather unexplored topic so far. Considering this context, a methodology for domain ontology building has been proposed and following it, various heterogeneous ATRACO ontologies have been created; these are presented in Sections 3.3.1 and 3.3.2, respectively. Moreover, thorough experiments have been carried out involving the various ATRACO ontologies, in order to confront the third and fourth issues. Considering the third, we propose two ontology resemblance coefficients, presented in Section 3.4.1.1, which detect whether their similarity is mainly lexical or structural, in order to select the most appropriate combination schema of alignment algorithms to be used, while for the fourth, we propose an alignment strategy, presented in Section 3.4, which involves the use of a "trusted third party", in order for a pair of heterogeneous ontologies to be aligned. The remaining issues are confronted in Section 3.5, within the mathematical framework of Category Theory, which, independently of the language used to represent ontologies and alignments, permits the realization of basic operations in networks of ontologies that comprise the ATRACO activity spheres.

3.3.1 Ontology engineering in ATRACO

Taking into account the considerations of subsection 3.2.2 about ontology engineering issues that emerge in NGAIEs, we designed a task-based ontology engineering process, in order to build the ATRACO ontologies [38].

Initially, we restrict the kind of the ontology to be built, to a domain ontology, since within the ATRACO project, each entity describes a domain, such as a user profile, or a specific device. Moreover, we restrict the number of participants in the ontology development process, to a single novice ontology developer, since, in the context of ATRACO, for example, each manufacturer, who is not an ontology expert, provides his own personal domain ontological model. We further focus on two parameters, which are related to the above mentioned restrictions: the knowledge level of a particular domain the ontology developer has (domain, or non-domain expert) and the strategy followed, in order to build the ontology (from scratch, or by reuse); the strategy depends on the availability of resources (documents, ontologies, group of experts). According to the above mentioned restrictions, the novice ontology developer will have to choose between four scenarios, in order to build his personal domain ontology. In each case, the choice depends on his knowledge level of the particular domain and his access to existing ontologies, related literature, or domain experts. The possible scenarios, as depicted in Fig. 3.1, are:

- 1st scenario: creating a domain ontology from scratch, by a novice ontology developer, who is also domain expert.
- 2nd scenario: creating a domain ontology by semantically reusing existing ontologies, by a novice ontology developer, who is a non-domain expert.
- 3rd scenario: creating a domain ontology from scratch, by a non-domain expert novice ontology developer.
- 4th scenario: creating a domain ontology, by semantically reusing existing ontologies, by a domain expert novice ontology developer.

During the application of each scenario, the ontology developer needs to carry out four mandatory tasks, namely the analysis, definition, selection and evaluation task. These tasks are scattered in all phases of the ontology engineering process, i.e. the specification, conceptualization, implementation and evaluation phase. All these phases appear to be common in the all methodologies available in the literature.

In our approach, we preserve the ordering of the specification, conceptualization, implementation, and evaluation phases, generally adopted by all the methodologies proposed in the literature. Moreover, we further group the activities present in each phase, in four clusters of tasks, each one containing mandatory sub-tasks, according to the selected scenario.

More precisely, concerning these clusters of tasks, the ontology developer needs: (a) To select the available resources, in order to help him conceptualize the domain. These resources include the related literature, existing ontologies and a group of experts of the domain under description; (b) To select the appropriate tool and language, in order to implement his conceptualization; (c) To analyze the selected resources, in order to define what is important for the description of the specific domain, through the competency questions; (d) To evaluate the selected resources, as well as the result of his attempt.

The detailed tasks and the sequential sub-tasks that an ontology developer needs to follow, in order to build his domain ontology, according to the selected scenario, are also depicted in Fig. 3.1. Relations between sub-tasks are denoted by appropriate arrows.

The task of analysis includes sub-tasks, such as analysis of the literature related to the domain, analysis of the classes of the selected ontology and analysis of the properties of the selected ontology. These sub-tasks belong to two different phases, the specification and the conceptualization phase, respectively. In this way we denote the temporal sequence in which these sub-tasks must be carried out. The task of definition includes sequential sub-tasks, such as the definition of the purpose and the domain of the ontology, the definition of a list of competency questions, the definition of the classes, the class hierarchy, the properties and the instances of the ontology.

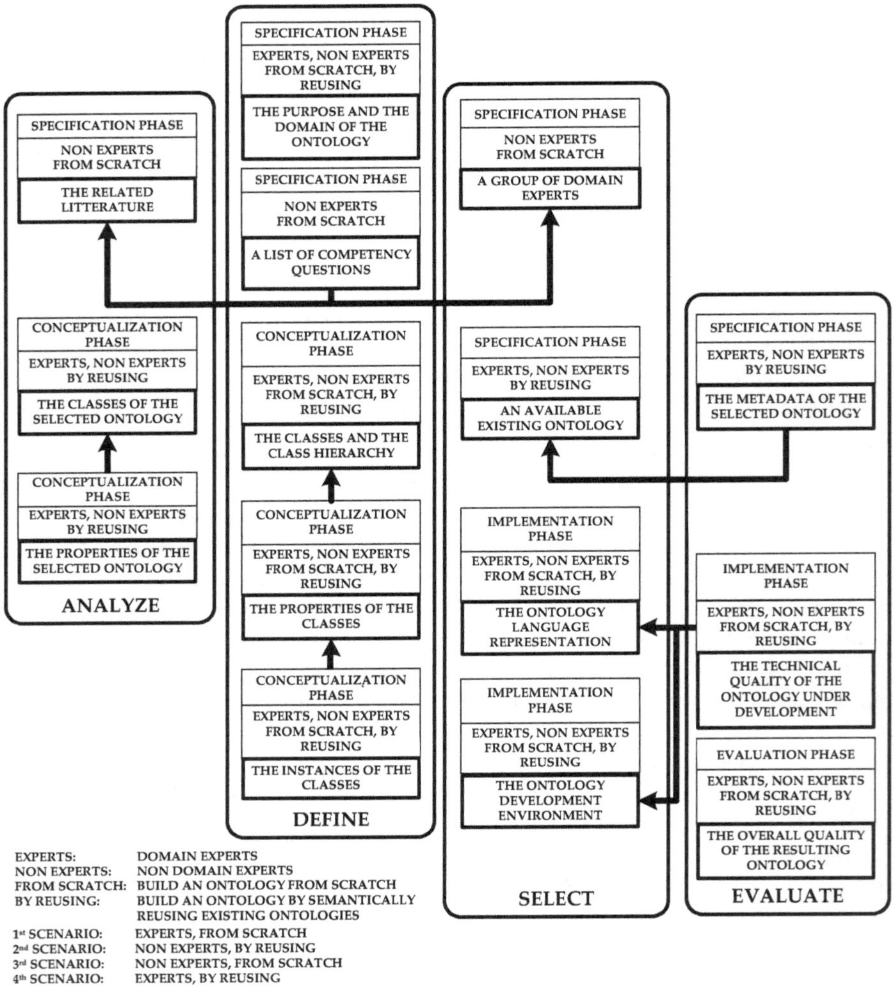

Fig. 3.1 An overview of the possible scenarios in the task-based ontology engineering approach

The selection task includes the selection of a group of experts, the selection of an available ontology, and the selection of the appropriate ontology development environment and the ontology representation language.

The task of evaluation involves sub-tasks that are distributed in the specification phase, in order to evaluate the metadata of the selected ontology, in the implementation phase, in order to evaluate the technical quality of the ontology under construction and in the evaluation phase, in order to evaluate the overall quality of the resulting ontology.

All the above mentioned sub-tasks included in the four mandatory tasks have a temporal ordering, which is denoted by the phase in which they belong. Our aim was to create a methodology for building isolated domain ontologies by a single creator, which is able to:

- Define each task to be carried out precisely, that is, to state clearly its purpose, input and output, the participants involved, the right management of the available resources, the time at which its execution is more convenient and the possible ways of executing it
- Be presented in a prescriptive way, in order to facilitate non-ontology experts
- Be general enough, in the sense that it can help any developer to build his personal domain ontologies, by using any ontology development tool

The tasks proposed in this methodology have been followed in order to build the ATRACO ontologies, which are described in details, in the next section. More precisely, the ATRACO Upper Level Ontology (ULO) was created by following the tasks described in the 1st scenario of the methodology, Device and Service Ontologies (DSOs) were developed by adopting the 2nd and the 4th scenarios, the Privacy Policies Ontology (PPO) was created by following the 3rd scenario, while several User Profile Ontologies (UPOs) were built by adopting the tasks described in the 2nd scenario of the proposed methodology.

Though cost models exist, that aim at predicting the cost involved in developing an ontology, in terms of effort and duration, it is extremely difficult to evaluate an ontology building methodology, since experimentation involves a multitude of uncontrollable conditions. Moreover, it is unlikely that someone would accept to pay twice for building the same complex ontology with different approaches. Thus, we have not conducted such an evaluation for our ontology building methodology, but tailored it to suit the specific contextual factors met in the ATRACO project, as explained in the preceding analysis. In addition, even though we have not evaluated the ontology building methodology that we followed, we have evaluated its output, i.e. the resulting ontologies, through competency questions.

3.3.2 ATRACO ontologies

In ATRACO, each sphere entity (user, agent, device, or service) maintains locally an ontology, which describes its features, capabilities and current state. Each local ontology is managed only by the owning entity. In other words, the local ontology of an entity represents the complete set of knowledge offered about this entity. However, it is expected that these ontologies will be heterogeneous, because they will reflect the conceptualizations of the involved stakeholders (manufacturers, developers, users, experts, etc.). On the other hand, this knowledge must be accessible by all ATRACO entities, in order to support the semantic interoperability among them, so a primary requirement is to define the basic concepts of the ATRACO world model, which are described in the ATRACO Upper Level Ontology (ULO).

Nevertheless, the local ontologies do not have to be semantically compatible with the ATRACO ULO; the task of the project is to align these ontologies by using, when necessary, the ATRACO ULO as a "pivot ontology", or other GFOs (General Foundational Ontologies) as a "third party" background knowledge.

In the context of ATRACO, the following ontologies have been developed:

- The ATRACO Upper Level Ontology (ULO), which serves as a common semantic reference between ATRACO entities. It encodes the basic concepts of ATRACO (i.e. devices, users, agents, services, policies) and their interrelations. Its role is to enhance knowledge and information sharing, between the inherently heterogeneous components of an ambient ecology. Because local ontologies are treated as black boxes (i.e. they can be altered only by the entities they own them), the ATRACO ULO is used by the ATRACO Ontology Manager, in order to optimize the ATRACO Sphere Ontology.
- Device and Service Ontologies (DSOs), which encode, in dissimilar ways, the basic characteristics of each device, as well as the features of the services that they provide. They are maintained by each device.
- Policy Ontologies (POs), which encode entities and rules that describe specific policies, such as user privacy. They are maintained by specific managers in the Ambient Intelligence (AmI) space.
- User Profile Ontologies (UPOs), which encode user traits and preferences. A user can assume different personas, based on context. They are also maintained by specific managers in the AmI space.

Moreover, General Foundational Ontologies (GFOs) are also used for providing background knowledge, whenever this is required.

All these ontologies have been created by following the ontology engineering approach of Section 3.3.1, as described above. In particular, OWL-DL was adopted as the ontology representation language, since it offers maximum expressiveness, while retaining computational completeness and decidability.

3.3.2.1 ATRACO ULO

The basic goal of the ATRACO Upper Level Ontology is to provide a shared referent that will enhance knowledge and information sharing among the ATRACO components (users, agents, devices, services), whenever necessary. Thus, it describes the core terms of ATRACO domain model and their interrelations. Some of the main classes of the ATRACO ULO, as depicted in Fig. 3.2, are:

- ContextEntity, which is the root class of the ontology
- Person, which corresponds to all human entities
- Activity, which refers to a person's current activity
- AbstractPlan, which describes what should be done during a specific activity
- Time, which corresponds to the time at which a specific activity is carried out
- Space, which corresponds to the physical space in which a specific person, physical object, or artifact is placed
- Artifact, which represents a set of devices that offers services
- Agent, which represent a set of all agents in the ATRACO context

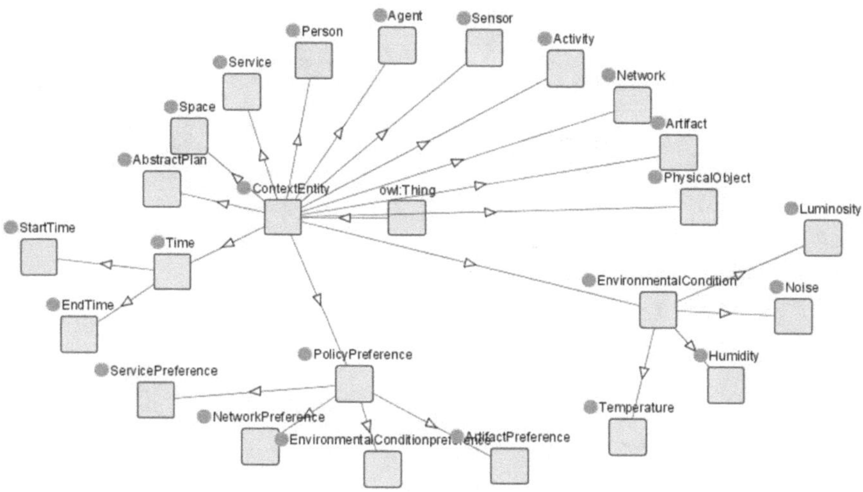

Fig. 3.2 The ATRACO ULO

- Service, which represents a set of services provided by specific devices, or by the network

3.3.2.2 ATRACO DSOs

In the ATRACO scenarios, the following knowledge must be provided: (a) which device offers a specific service, (b) is it permissible, or not, for a user to use a specific service in order to carry out a specific activity, (c) which is the policy of using a specific service, (d) what are the features of a device that are necessary, in order for a user to carry out a specific activity, etc. This kind of knowledge is scattered in the local ontologies of the different autonomous context elements (devices, users) that participate in the execution of an activity sphere; these local ontologies are heterogeneous and may be also inconsistent, or incomplete, as they are developed independently by device manufacturers, software developers, or non-domain experts.

Thus, generally, the ATRACO DSOs have the following characteristics: (a) they are domain ontologies that are short models of the domain under description, (b) they do not have extensive is_a hierarchies, (c) the same concept may be represented at the class level in one ontology and at the instance level in another one, (d) they have complex relations, where classes are connected by a number of different relations, (e) their terminologies, in certain cases, are not identical, even if the ontologies describe the same domain, (f) the modeling principles for them are not well defined and documented. Some examples of the ATRACO DSOs are depicted in Fig. 3.3 and Fig. 3.4.

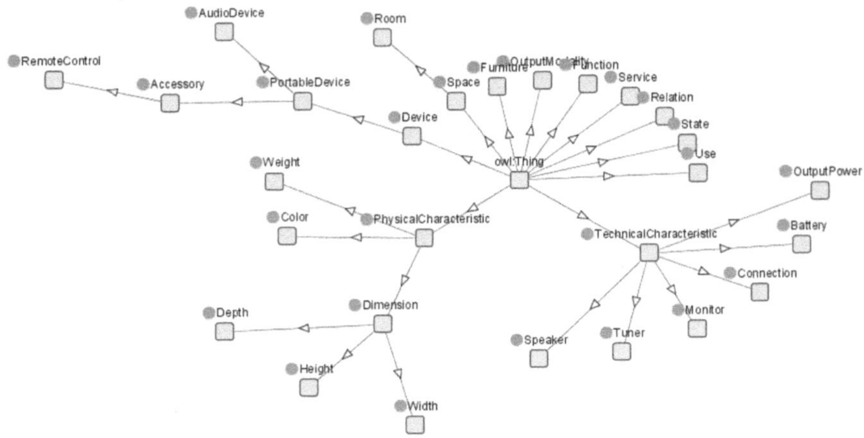

Fig. 3.3 An example of an mp3 player DSO

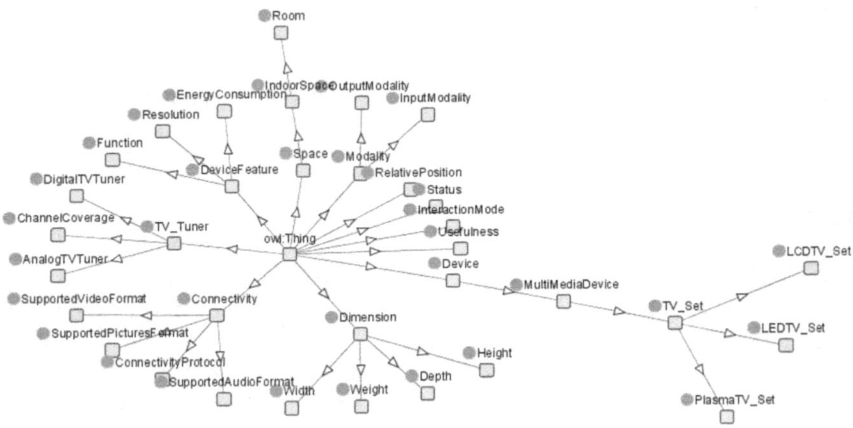

Fig. 3.4 An example of a TV-set DSO

All these ontologies contain information about the features of each device and the services it provides, in a different structure, or in different terminology, as the context of ATRACO states. For example, a Device has DeviceDescription, which contains basic information related to this specific device, such as DeviceName, DeviceType, DeviceVendor and its PhysicalCharacteristics such as Shape, Color, etc.

The Device has also a HardwareDescription, which contains the details of its CPUDescription, ConnectionDescription to the network, its MemoryDescription and UIDescription, as well as SoftwareDescription, which contains the details of the operating system of the device. Another important class in DSOs is the class

Service, which provides the information about the service(s) hosted on the device concerned. An instance of a floor lamp ontology is depicted in Fig. 3.5, where a Lamp with id: Lamp_0234 is described. It has LampGeneralCharacteristics, such as its name, id, etc., LampFunctionalCharacteristics, where the service that it offers is described, LampPositionCharacteristics, such as its location and LampPhysicalCharacteristics, such as height, colour, numberOfBulbs, material, etc. Each Bulb, which is contained in the specific lamp has its own features, such as its type: Fluorescent, shape: Tubular and lifeExpectation.

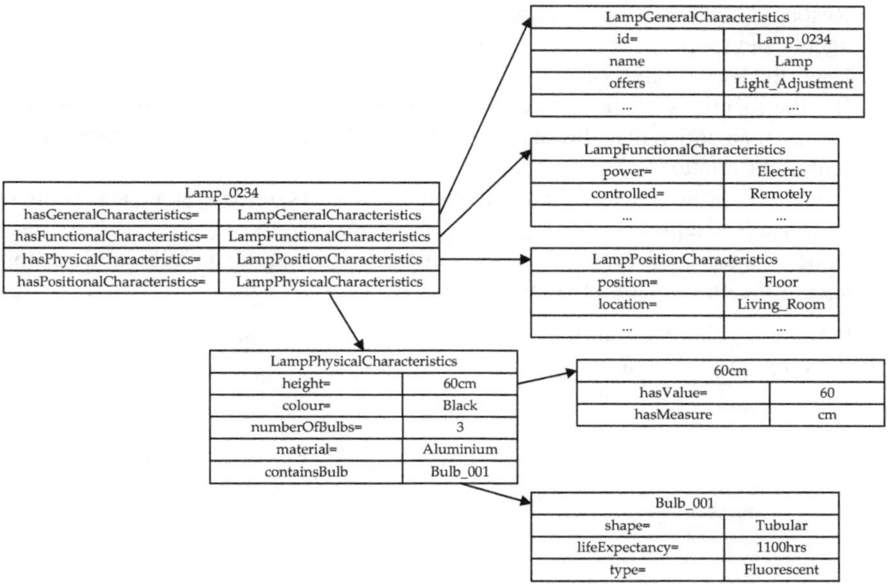

Fig. 3.5 An instance of a floor lamp DSO

3.3.2.3 ATRACO PPO

The Privacy Policy Ontology can be used to define privacy rules, which the various sphere components can use while interacting with each other in the context of a specific activity. Note that the privacy policy is bound to the activity supported by the activity sphere, but not to its particular components. It addresses issues such as: What resource type is needed for a service? What is done with this resource? Who has access to a specific resource? Who processes a user's data? In which way are the data transported? In which way are the data stored? Who are the recipients of the data? Are the recipients humans, or machines? Who is the creator of a specific policy? Which are the mechanisms incorporated in a specific policy? Which policies have authentication, encryption, right, or identification as a policy mechanism? What kind of access does a user have to a specific device?

In order for these questions to be answered effectively, the following basic concepts comprise the main classes of the PPO:

- Mechanism: the set of the mechanisms (actions) that can take place in the context we described and have to do mostly with the data of the system
- PolicyMechanisms: all the mechanisms which are required for the implementation of a policy, such as access control (authentication, authorization and identification)
- Resource: the set of the resources of a ubiquitous computing environment (e.g. agents, devices)
- User: the users in the context
- ID: the identities of the entities of the system. An identity is a list of attribute values of an entity, that allows this entity to be distinguished among other entities within the context
- Policy: all the policies which are defined by the users of the system, according to their needs and preferences
- Time: is a basic concept to describe some actions (e.g. the duration of data storage, or the duration of a policy)

In Fig. 3.6 a graph with all the classes and their subclasses, as they are defined in the ontology, is depicted.

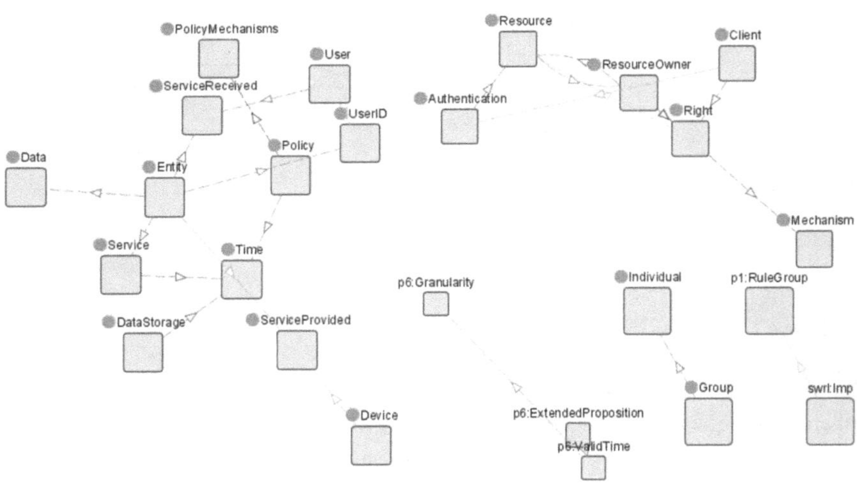

Fig. 3.6 The main classes and their subclasses of the PPO

In order to develop the ontology, the entity classes are connected and hence related to each other, through properties. These are called object properties and provide the available capabilities of each class. Similarly, in order to define in detail the properties and characterize each class, data types (data properties) are created. Some of these properties that have been created, in order to describe the holistic ontology concept, are shown below:

- hasPolicyMechanism: this property connects the classes Policy and PolicyMechanisms. Indicates that any policy created by a user must have at least one policy mechanism for its proper function
- hasDuration: this property relates the classes DataStorage/Policy/Service and Time. It provides the duration of the data storage, or a policy, or a particular service
- BankAccount, HomeAddress, MedicalProfile, PhoneNumber etc.: these properties characterize the class SensitiveData. They indicate which data of a user profile are considered as sensitive
- creatorOfPolicy and policyDescription: these properties characterize the class Policy. They provide information about the user, who has created a concrete policy and the description of this policy, respectively
- identifiesID: this property connects the classes Entity and ID. It provides information about the entity which identifies the identity of a user, or another entity (agent, or device)

The PPO has also been enriched with a set of rules for a better understanding of the basic concepts and also for the representation of privacy principles [31]. Some examples of the rules added, are: (a) the user who has an identifiable ID, which is identified by a resource, always has positive authentication on this resource, (b) the owner of a resource has always positive authentication on this resource, etc.

3.3.2.4 ATRACO UPO

In the ATRACO project, each user has a profile which is described in its User Profile Ontology (UPO). Different users may be involved within the same activity, so heterogeneous ontologies are used in order to semantically describe them. In Fig. 3.7 and Fig. 3.8, some examples of ontologies describing user profiles within the context of ATRACO are depicted.

The creation of these ontologies, as explained in more details in [40], is driven by a set of competency questions, such as: What are the physical characteristics of a specific user? What is the current activity that a specific user is carrying out? What are the preferences of a specific user, concerning a specific activity? What are the preferences of a specific user concerning a specific service that the user uses, in order to carry out a specific activity? What is the general information (name, age, gender, nationality) of a user? What is the contact information (home address, email address, phone number) of a user? Where is a user situated when carrying out a specific activity? When does a user carry out a specific activity? What is the current location of the user? Which is the temperature that a user prefers? What are the user's interests? Does the user belong to a group? What are the light and illumination levels that the user prefers? What are the media options that the user

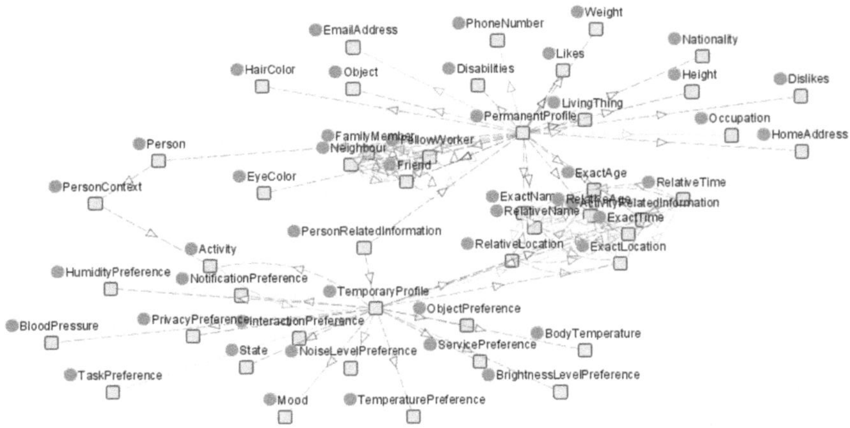

Fig. 3.7 Suki's User Profile Ontology within the ATRACO context

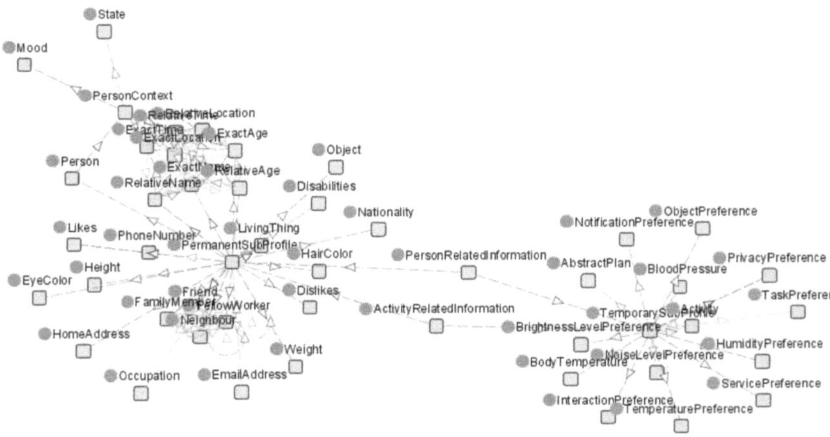

Fig. 3.8 Jona's User Profile Ontology within the ATRACO context

prefers? What modalities does the user prefer for which devices/tasks (e.g. TV <-> remote controller, MP3 player <-> voice)? Which is the profile of a user, according to a specific activity? Which is a user's permanent profile? Which is the user context that is related to a specific activity? What user related information is contained in a specific person's permanent profile? What activity related information is contained in a specific temporary sub-profile? Based on these competency questions, many ontologies with different structure and dissimilar labels can be constructed, but they will all return semantically identical responses to the same queries. We have adopted the following structure of a user profile: each User has a PermanentSub-

Profile and more than one TemporarySubProfile. His PermanentSubProfile contains PersonRelatedInformation, such as GeneralInformation (Gender, Name, Age, Occupation, Nationality); Likes; Dislikes; Disabilities; ContactInformation (PhoneNumber, EmailAddress, HomeAddress); Possessions (LivingThing, Object); SocialInformation (Friend, FamilyMember, FellowWorker, Neighbour); PhysicalInformation (EyeColor, HairColor, Weight, Height). Each TemporarySubProfile is related to a specific Activity and depends on a PersonContext, which contains the Location of the user, the Time at which the Activity is carried out, the State of the user, which means whether the user is Alone, With_Friends, etc., and the person's Mood. A TemporarySubProfile contains ActivityRelatedInformation, such as the person's BiologicalInformation (BloodPressure, BodyTemperature) and the user's Preferences, such as InteractionPreference, NotificationPreference, TaskPreference, ObjectPreference, ServicePreference, PrivacyPreference and EnvironmentalConditionPreference (BrightnessLevelPreference, NoiseLevelPreference, TemperaturePreference, HumidityPreference), which are associated with the Activity that a user carries out. For Privacy related issues, there is a distinction in some sensitive personal data, such as Age, Name, Location and Time. They are distinguished in, for example, ExactAge (e.g. 28 years old) and RelativeAge (e.g. Adult). In that sense, RelativeAge is lessDetailedThan ExactAge.

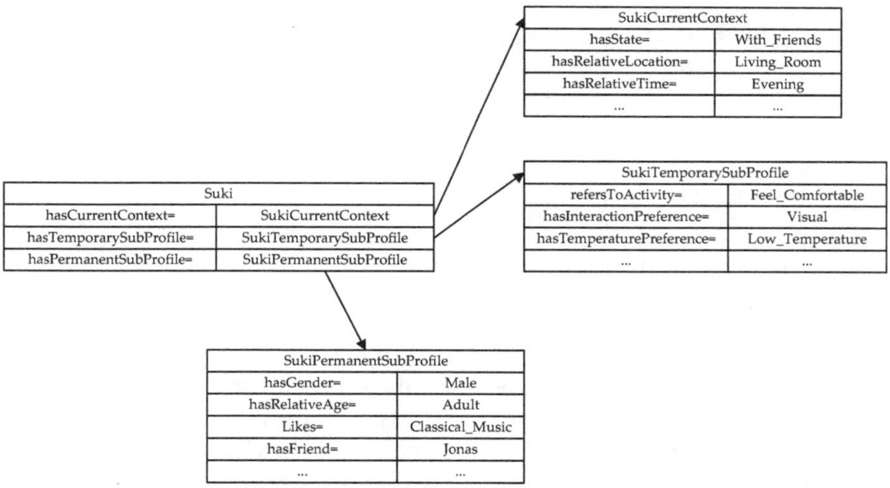

Fig. 3.9 An instance of Suki's Profile Ontology

In order to gain a better understanding of the core structure of this ontology, we exemplify an instance of it, depicted in Fig. 3.9. User: Suki, who has a specific PermanentSubProfile, has the following UserPermanentSubProfile: SukiPermanentSubProfile. He is situated in RelativeLocation: Living_Room, at Time: Evening and he is in State: With_Friends. He is currently carrying out an Activity: Feel_Comfortable. This activity is associated with his TemporarySubProfile: Feel-

ComfortableSubProfile, which contains his TaskPreference: he prefers to Listen-_To_Music, his ServicePreference: TemperatureService, his ObjectPreference: he prefers to use the Air_Conditioning and the TV_Set, his EnvironmentalCondition-Preference: he prefers Low_Temperature, etc.

3.4 ATRACO ontology alignment strategy

The proposed alignment strategy contains three main steps, as depicted in Fig. 3.10:

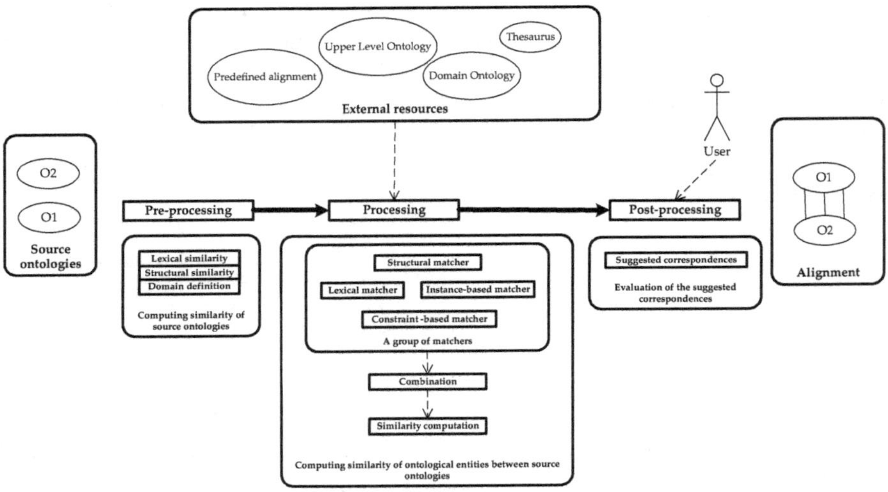

Fig. 3.10 The ATRACO alignment strategy

(1) Pre-processing step, where we try to make an initial estimation, at an ontology level, of whether the ontologies to be aligned appear to be similar, lexically, or structurally. (2) Processing step, where, based on the previous step, we decide about the proper sequence of execution of appropriate matchers (lexical, structural, instance-based, constraint-based algorithms); and (3) Post-processing step, where the final results are evaluated. The above mentioned steps are analyzed in the following subsections.

Once ontology alignments are produced and evaluated, they're further used by the ATRACO Ontology Manager, presented in the respective book chapter, for tasks such as ontology merging, data translation and query answering.

In the ATRACO perspective, an alignment is considered as a meningful object, that is, an an explicit information describing semantic links between different ontologies. Thus, the produced alignments are also shared and reused by different applications within and among NGAIEs.

3.4.1 Pre-processing step

In this first step, given two source ontologies, we consider their particularities (of small, or medium size, of limited hierarchy structure, with entity labels being tightly oriented towards the creators' view of the domain under description) and their similarity characteristics (they have similar labels, or structure), as well as the description of their domains, in order to decide which family of ontology alignment algorithms, or matchers (lexical, structural) to use, whether to combine matchers sequentially, or in parallel, what kind of external resources to use, in order to enhance the alignment process in case of ontologies that differ greatly either lexically, or structurally, or when no instances and constraints are provided, whether the user should be involved, or not in the alignment process, etc.

More precisely, in order for the ATRACO Ontology Manager to be able to execute successfully each alignment process, during this pre-processing step, it must be aware of the source ontologies' particularities, as well as which matchers are suitable for certain input ontologies. We have designed new two similarity coefficients, which we present in the next subsection, in order to detect the lexical, or structural resemblance of the source ontologies. These coefficients are used, in order to select the suitable family of matchers, as well as their way of combination. Their values fall into the range [0, 1]. These similarity coefficients measure similarities at an ontology level and are of two kinds: lexical and structural. The first one examines the relative structure of the two ontologies, based on the comparison of the lengths of all paths leading from the root of each ontology to each of its leaves. The second one, after discovering concepts with identical labels in both ontologies, considers also the relative proximity of these common concepts, inside each one of the ontologies to be aligned.

3.4.1.1 Similarity coefficients for detecting ontologies resemblance

We detail hereafter the computational aspects of the similarity coefficients used in the pre-processing step of the ATRACO alignment strategy.

Structure Similarity Coefficient

Given two ontologies O_1 and O_2, we define the Structure Similarity Coefficient, denoted by $\sigma(O_1, O_2)$, which is a similarity metrics at an ontology level (as opposed to an entity level), with values that range from 0 to 1. The Structure Similarity Coefficient describes the similarity between two ontologies globally (as opposed to local structural similarities between ontology entities), based on their structural resemblance. In order to compute it, one has to follow the constructive procedure described below:

Definition – Structure Similarity Coefficient: Given two ontologies O_1 and O_2, calculate the vectors $\overline{l_1}$, $\overline{l_2}$ having as elements the lengths of all the paths from the root of each ontology, to all its leaves, i.e.,

$\overline{l_1} = [l_{11}, l_{12}, ..., l_{1i}, ...]$, with l_{1i} = length of the path from the root of ontology O_1 to its ith leaf, $i = 1, 2, ..., \#$ leaves of ontology O_1

$\overline{l_2} = \left[l_{21}, l_{22}, ..., l_{2j}, ...\right]$ with l_{2j} = length of the path from root of ontology O_2 to its j^{th}leaf, $j = 1, 2, ..., \#$ leaves of ontology O_2

Let $L = \max\{|\overline{l_1}|, |\overline{l_2}|\}$, with $|\overline{l_i}|$ the dimension of vector $\overline{l_i}$, $i = 1, 2$. Create two new vectors $\overline{a}$, $\overline{t}$, by choosing between the vectors $\overline{l_i}$ $i = 1, 2$ the one that has the biggest dimension and by completing the other vector with leading zeros. Both vectors $\overline{a}$, $\overline{t}$, have dimension L.

If $|\overline{l_i}| > |\overline{l_j}|$, $i, j \in \{1, 2\}$ and $i \neq j$, then $\overline{a} = \overline{l_i}$, $\overline{t} = \left[\overline{0}, \overline{l_j}\right]$, with the dimension of $\overline{0}$ being equal to $L - \min\{|\overline{l_1}|, |\overline{l_2}|\}$.

Compute now a square LxL matrix C, with elements $c_{ij} = |a_i - t_j|$, $i, j = 1, 2, ..., L$. Then, create two new vectors $\overline{r}$ and $\overline{s}$, by appropriately reordering the vectors $\overline{a}$ and $\overline{t}$, as explained hereafter.

Let us consider two sets B and T with cardinalities equal to L and let β_i, τ_i, $i = 1, 2, ..., L$, denote their respective elements. Consider the bipartite graph having as nodes the elements of the sets B and T and containing all possible edges between respective elements of the two sets. The edge linking β_i, to τ_j $i, j = 1, 2, ..., L$, has a weight equal to $c_{ij} = |a_i - t_j|$. One can then always find a square matrix X with dimensions LxL having elements x_{ij}, $i, j = 1, 2, ..., L$, such that the following relations hold:

1. $\forall i = 1, 2, ..., L, \ \sum_{j=1}^{L} x_{ij} = 1$
2. $\forall j = 1, 2, ..., L, \ \sum_{i=1}^{L} x_{ij} = 1$
3. $\forall i, j = 1, 2, ..., L, \ x_{ij} \geq 0$
4. $\sum_{i=1}^{L} \sum_{j=1}^{L} c_{ij} x_{ij}$ is minimized

It can be proven that such elements x_{ij}, $i, j = 1, 2, ..., L$, exist and take either the value 0, or the value 1. If $x_{ij} = 1$, then the ith element of the reordering $\overline{r}$ is $r_i = a_i$, while the jth element of the reordering $\overline{s}$ is $s_j = t_j$. The structural similarity between the two ontologies is finally calculated as the cosine of the angle between the vectors $\overline{r}$ and $\overline{s}$:

$$\sigma(O_1, O_2) = \frac{\overline{r}.\overline{s}}{||\overline{r}||.||\overline{s}||} = \frac{\sum_{i=1}^{L} r_i s_i}{\sqrt{\sum_{i=1}^{L} r_i^2}\sqrt{\sum_{i=1}^{L} s_i^2}}$$

In order to clarify the computation of the Structure Similarity Coefficient, we compute it for the pairs of ontologies of Fig. 3.11 and Fig. 3.12. For the ontologies of Fig. 3.11, we compute the Structure Similarity Coefficient as

$$\sigma(O_1, O_2) = \frac{2 \cdot 2 + 1 \cdot 1 + 1 \cdot 0}{\sqrt{2^2 + 1^2 + 1^2}\sqrt{2^2 + 1^2 + 0^2}} = \sqrt{\frac{5}{6}} = 0.9129,$$

The Structure Similarity Coefficient depicts accurately the similarity of structure between the two ontologies, which becomes apparent when flipping O_1 horizontally. The Structure Similarity Coefficient for the ontologies of Fig. 3.12 is calculated as

$\sigma(O_1,O_2) = \frac{n}{\sqrt{n}\sqrt{n^2}} = \frac{1}{\sqrt{n}}$,

that is, $\sigma(O_1,O_2) \to 0$ as $n \to \infty$.

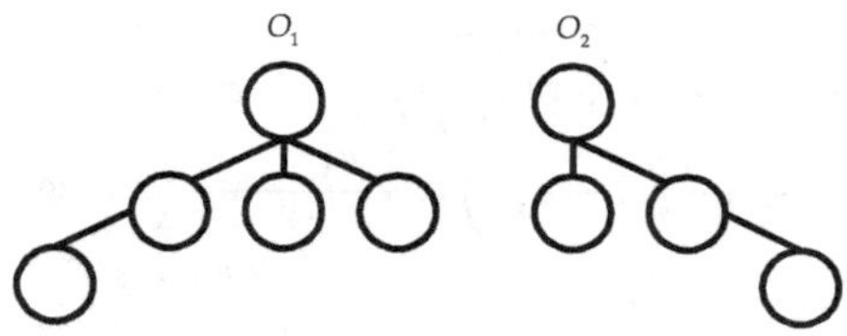

Fig. 3.11 The ontologies of example 1

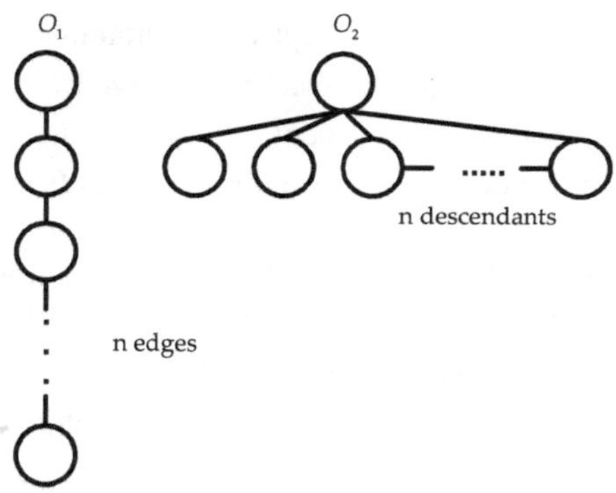

Fig. 3.12 The ontologies of example 2

Lexical Similarity Coefficient

We define the Lexical Similarity Coefficient, at an ontology level, with values ranging from 0 to 1. In order to calculate the Lexical Similarity Coefficient, we consider two factors. The first factor is based on the number of concepts/classes having the same label in both ontologies (inter-ontology factor), while the second one takes into account the relative proximity that these common concepts have among them, inside each one of the ontologies (intra-ontology factor).

Definition – Lexical Similarity Coefficient: Given two ontologies O_i and O_j, $i,j = 1,2,\ i \neq j$, with a number of cc pairs of concepts with the same label, that is, $(\varepsilon_1^{O_i}, \varepsilon_1^{O_j})$, $(\varepsilon_2^{O_i}, \varepsilon_2^{O_j})$,$\cdots$,$(\varepsilon_k^{O_i}, \varepsilon_k^{O_j})$, $i, j = 1, 2, i \neq j$, $k = 1, 2, ..., cc$, respectively, the Lexical Similarity Coefficient is calculated as:

$$\lambda(O_i, O_j) = \frac{\sum_{k=1}^{cc}[1 - \frac{\left|\delta\left(\varepsilon_k^{O_i}\right) - \delta\left(\varepsilon_k^{O_j}\right)\right|}{\max(\delta\left(\varepsilon_k^{O_i}\right), \delta\left(\varepsilon_k^{O_j}\right))}]}{\max(\#conceptsofO_i, \#conceptsofO_j)},$$

$i, j = 1, 2,\ i \neq j$, where the term $\delta\left(\varepsilon_k^{O_i}\right)$ ranks concept $\varepsilon_k^{O_i}$ of ontology O_i, by taking into account how far, in terms of number of edges, the remaining common concepts $\varepsilon_p^{O_i}, p \neq k$ are from concept $\varepsilon_k^{O_i}$ in ontology O_i and is given by

$$\delta\left(\varepsilon_k^{O_i}\right) = \varphi_k^{O_i} + \frac{d\left(\varepsilon_k^{O_i}, n_1\right)}{cc-1}\left(1 - \varphi_k^{O_i}\right)\rho + \frac{d\left(\varepsilon_k^{O_i}, n_2\right)}{cc-1}\left(1 - \varphi_k^{O_i}\right)\rho(1-\rho) + \ldots$$

$$\ldots + \frac{d\left(\varepsilon_k^{O_i}, n_m\right)}{cc-1}\left(1 - \varphi_k^{O_i}\right)\rho(1-\rho)^{m-1} + \frac{d\left(\varepsilon_k^{O_i}, O(n_{m+1})\right)}{cc-1}\left(1 - \varphi_k^{O_i}\right)(1-\rho)^m,$$

where

$\varphi_k^{O_i} = 1 + (\alpha - 1) sgn[(cc-1) - d\left(\varepsilon_k^{O_i}, n_1\right)]$ [1]

with α, $(0 \prec \alpha \prec 1)$ a constant added to the rank of common concept $\varepsilon_k^{O_i}$, due to its lexical similarity to concept $\varepsilon_k^{O_j}, i, j = 1, 2, i \neq j$ and where we define:

n_1 to be the 1-neighborhood of concept $\varepsilon_k^{O_i}$, containing all common concepts $\varepsilon_p^{O_i}, p \neq k$, that are within a distance of exactly one edge from $\varepsilon_k^{O_i}$ in O_i,

n_2 to be the 2-neighborhood of concept $\varepsilon_k^{O_i}$, containing all common concepts $\varepsilon_p^{O_i}, p \neq k$, that are within a distance of exactly two edges from $\varepsilon_k^{O_i}$ in O_i,

$\cdots$,

n_m to be the m-neighborhood of concept $\varepsilon_k^{O_i}$, containing all common concepts $\varepsilon_p^{O_i}, p \neq k$, that are within a distance of exactly m edges from $\varepsilon_k^{O_i}$ in O_i,

$O(n_{m+1})$ to be the remote-neighborhood of concept $\varepsilon_k^{O_i}$, containing all common concepts $\varepsilon_p^{O_i}, p \neq k$, that are within a distance of more than m edges from $\varepsilon_k^{O_i}$ in O_i.

Then,

$d\left(\varepsilon_k^{O_i}, n_q\right), q = 1, 2, ..., m,$

denotes the number of common concepts $\varepsilon_p^{O_i}$, $p \neq k$, that are within a distance of exactly q edges from $\varepsilon_k^{O_i}$ in O_i and

$d\left(\varepsilon_k^{O_i}, O(n_{m+1})\right)$

denotes the number of common concepts $\varepsilon_p^{O_i}, p \neq k$, within a distance of more than m edges from $\varepsilon_k^{O_i}$ in O_i.

$\frac{1}{2} \prec \rho \prec 1$ is a forgetting factor, penalizing more severely the common concepts $\varepsilon_p^{O_i}, p \neq k$ that are more distant from $\varepsilon_k^{O_i}$ in O_i (in more distant neighborhoods).

For clarification reasons, we consider the ontologies of Fig. 3.13 and compute the Lexical Similarity Coefficient of pairs O_1 and O_2 and O_1 and O_3, respectively, by choosing $a = 0.8$ and $\rho = 0.6$. The ontologies O_1, O_2 and O_3 have common

[1] The signum function is defined as: $sgn(x) = \begin{cases} -1 & \text{if } x < 0 \\ 0 & \text{if } x = 0 \\ 1 & \text{if } x > 0 \end{cases}$

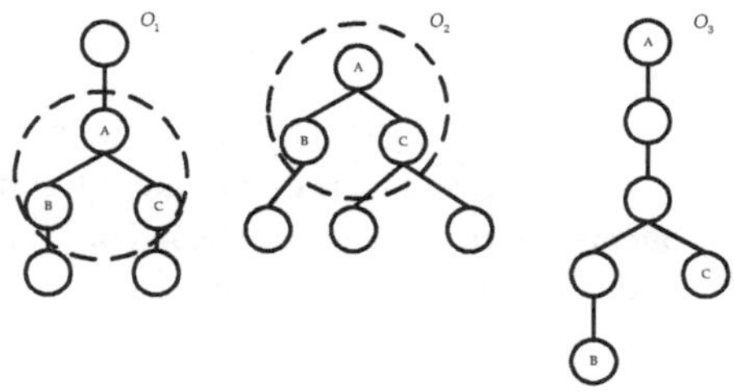

Fig. 3.13 The ontologies of example 3

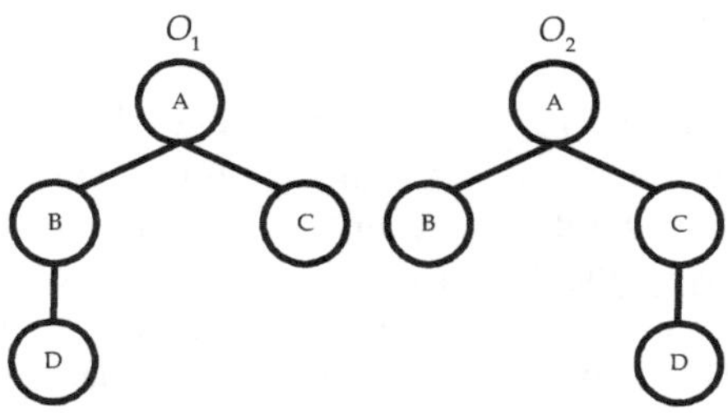

Fig. 3.14 The ontologies of example 4

labels A, B and C . Thus, $cc = 3$ and we choose $m = 2$, limiting ourselves to $1-,2-$neighborhoods n_1, n_2 and $O(n_3)$.

When computing the Lexical Similarity Coefficient between O_1 and O_2, since each common concept distributes in the same way the remaining common concepts in its neighborhoods, in both ontologies, it results that

$\lambda(O_1, O_2) = \frac{1+1+1}{6} = 0.5$

When comparing lexically O_1 to O_3, it is $\delta(A^{O_1})$=1,

$\delta(B^{O_1}) = \delta(C^{O_1}) = 0.8 + \frac{1}{2} \cdot 0.2 \cdot 0.6 + \frac{1}{2} \cdot 0.2 \cdot 0.6 \cdot 0.4 = 0.884$

and

$\delta(A^{O_3}) = \delta(B^{O_3}) = \delta(C^{O_3}) = 0.8 + 1 \cdot 0.2 \cdot 0.4^2 = 0.832$

resulting in

$\lambda(O_1, O_3) = \frac{(1-0.168)+(1-0.0588)+(1-0.0588)}{6} = 0.4524.$

The difference computed is due to the fact that the interrelations among the common concepts A, B and C are preserved in ontologies O_1 and O_2, while in the case of the lexical similarity between O_1 and O_3, our coefficient is calculated to be less than 0.5, depicting the differences in the interrelations among common concepts in these ontologies. The detection of such differences in interrelations among common concepts is essential, since it restricts the problem of polysemy (words that have multiple senses), occurring when comparing ontology entities on the basis of their labels. Indeed, intuitively, groups of common labels in both ontologies, are more probably referring to the same concepts, while distant distinct common labels, may reflect homonyms and thus name different concepts. Another example is depicted in Fig. 3.14, where the ontologies O_1 and O_2 have four common concepts. Here, the common concepts B and C distribute differently the remaining common concepts (A, C and D for B and A, B and D for C), while A and D distribute their respective remaining common concepts in the same way, in both ontologies. The result obtained is $\lambda\left(O_1, O_2\right) = 0.9832$, less than 1, due to the differences in interrelations between the common concepts in the two ontologies. The exact amount of the difference obtained, can be adjusted by a proper choice for the values of the weighting coefficients a and ρ.

3.4.2 Processing step

The output of the pre-processing step, i.e. the values of the structure and lexical similarities that indicate whether the source ontologies resemble mostly structurally, or lexically, respectively, guide the processing step of the ATRACO alignment strategy and more precisely, guide the selection and sequence of use of the appropriate matchers. The alignment process includes several matchers and it is organized sequentially, or in parallel, according to the similarity characteristics of the two source ontologies. These matchers calculate similarities between the ontological entities from the different source ontologies. The higher the similarity of the two ontological entities, the more likely they can be aligned. The matchers can implement techniques based on linguistic matching, structural matching, constraint-based approaches, instance-based techniques and approaches that use auxiliary information, or a combination of the above.

During the phase of linguistic matching, multiple algorithms, or matchers based on linguistic matching, such as Edit Distance and Synonym Matcher through the WordNet synonyms thesaurus [16], are executed. These algorithms make use of textual descriptions of ontological entities, such as names, synonyms and definitions. The similarity measure between entities is based on comparisons of the textual descriptions. The main idea in using such measures is based on the fact that usually similar entities have similar names and descriptions, among different ontologies. More specifically, during this phase we consider establishing correspondences between the entities of the two source ontologies, by using a similarity function and calculating it for the two entities (each one belonging to one of the ontologies). If the returned value is greater, or equal to a threshold set up by a domain expert, then the two entities match. Edit distance, for example, is defined as the number of deletions, insertions, or substitutions required transforming one string into another. The greater the edit distance is, the more different the strings are.

The phase of structural matching uses a number of structural matchers (including Descendant's Similarity Inheritance (DSI) and Sibling's Similarity Contribution (SSC) methods [8]), which use the structure of the ontologies, in order to enhance the alignment results. The algorithms used during this phase rely on the intuition that two elements of two distinct models are similar, when their adjacent elements are similar. In another phase of the alignment strategy, constraint-based approaches are used. In these approaches axioms are used, in order to enhance the previously generated correspondences. For instance, if the range and domain of two relations are the same, this is an indication that there is a correspondence between these relations. If several instances are available in both source ontologies, then instance-based approaches can be used, in order to define similarities between ontological entities. Dictionaries, thesauri representing general knowledge, intermediate domain, or upper level ontologies, or predefined alignments, may be also used to enhance the alignment process.

The choice of the appropriate phases (linguistic, structural, constraint-based and instance-based) and their temporal sequence is driven by the values of the similarity coefficients between ontologies that are calculated during the pre-processing step of the alignment process, according to the following scenarios (see Fig. 3.15):

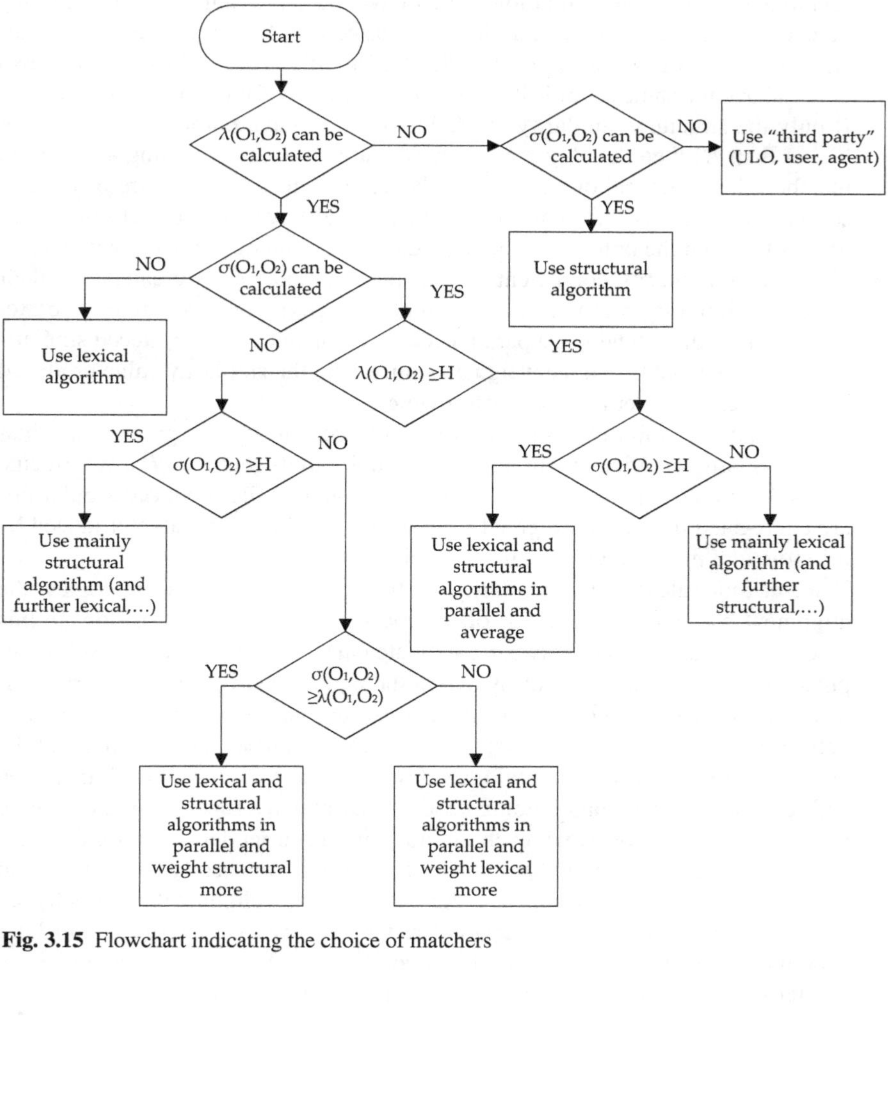

Fig. 3.15 Flowchart indicating the choice of matchers

- If only the label similarity coefficient of the source ontologies is equal, or greater than a threshold H, which is adjusted experimentally, with a value close to 1 (for example $H = 0.9$), then the alignment process should rely only on lexical matchers. For better results, we can further execute a structural, a constraint-based and an instance-based matcher (if any instances are available) sequentially, because, even if some ontological entities have the same label, they may represent totally different concepts. Specifically, starting from a lexical matcher, such as Edit-Distance, the entity pairs that score a high similarity value are further compared using a structural matcher, such as one which uses the is_a and part_of hierarchies. Any correspondences between entity pairs with a low similarity value, according to the lexical matcher, are discarded. Then, the pairs with high similarity values are filtered by using constraint-based information. For example, two relations from the source ontologies that have the same range and domain may be equivalent. Following this, an instance-based matcher may be applied, if instances in both ontologies are available. The intuition behind this is that, if two classes have the same set of instances, it is more likely for them to be equivalent.
- If only the structural similarity coefficient of the source ontologies is equal, or higher than H, then the alignment process starts by firstly applying a structural matcher, then a lexical one and so on, because even if the structure of the two source ontologies is the same, it is difficult to discover the correct correspondences between the ontological entities, due to terminological heterogeneity.
- If the label similarity coefficient of the source ontologies is greater than their structural similarity coefficient, but not high enough (not higher than H), we execute all kinds of matchers in a parallel composition and all the produced similarities are aggregated through a weighted sum, where the similarity value produced by the lexical matcher has the greatest weight.
- If the structural similarity coefficient of the source ontologies is greater than their label similarity coefficient, but not high enough (not higher than H), we execute all kinds of matchers in a parallel composition and all the produced similarities are aggregated through a weighted sum, where the similarity value produced by the structural matcher has the greatest weight.
- If we cannot calculate the similarity coefficients of the source ontologies, the alignment process uses all kinds of matchers in a parallel composition. In this case, the similarity values between candidate pairs of ontological entities is computed, either as the average, or by taking the maximum of all the similarity values that the matchers produce. In the case where no computation of similarity between source ontologies is accomplishable, we additionally try to find out the domain that the two source ontologies describe. Since the meaning (importance and relevance) of a certain ontological entity essentially depends on the specific usage and purpose for which each ontology has been modeled and on different views that creators have on the domain, it is essential to specify the domain that the source ontologies describe, in order to enhance the alignment process by using external resources as a "third party", as we further explain in subsection 3.4.4
- If both the label and structural similarity coefficients of the source ontologies are greater than H, then both lexical and structural matchers are executed in paral-

lel and their average value is used as the similarity of the candidate pairs. Then, entity pairs with high similarity values according to the lexical and structural matchers are filtered by using constraint-based information. Following this, an instance-based matcher may be applied, if instances in both ontologies are available.

In the case of a parallel combination of matchers, their similarities are aggregated through computing their average, or by selecting their maximum value, or through a weighted sum, depending on whether we want, for a given alignment, either all matchers to have the same influence, or require their highest score, or finally want to give more importance to some of them, respectively. The weighted sum is defined as $Sim(e_1,e_2) = \sum_{k=1}^{n} w_k \cdot sim_k(e_1,e_2)$, where n is the number of the combined matchers and sim_k and w_k represent the similarity values between the entities e_1 and e_2 belonging to the two source ontologies, and weights, respectively, for the different matchers. It is also required that $w_k \in [0,1]$ and $\sum_{k=1}^{n} w_k = 1$. If the final similarities obtained after aggregation exceed a threshold, they are selected for the resulting alignment.

3.4.3 Post-processing step

In this step, the produced alignment should be evaluated and further refined, before being stored in RDF\XML format by the Alignment Module of the ATRACO Ontology Manager. In practice, this can be done in three ways: (1) assessing individual correspondences, (2) comparing the alignment to a reference alignment and (3) evaluating the application that uses the alignment. In this step, the user is involved in the alignment process, in order to decide on the final alignment as described in the following sub-section. He accepts, or rejects some of the produced correspondences between ontological entities from the two source ontologies. The user plays the role of the checker, in order for misalignments to be avoided and the role of a helper that enhances the ontology alignment process, by refining suggested correspondences, or adding new matching pairs.

3.4.4 Using a "trusted third party" within the alignment strategy

A "trusted third party" is used within the ATRACO alignment strategy in two ways. First, in order to obtain an indirect alignment, when the direct alignment of a pair of non-overlapping ontologies is impossible and second, in order to evaluate the alignments produced. Considering the first issue, during the processing step of the ATRACO alignment strategy, in order to result in correct correspondences, in the case where the input ontologies describe unrelated domains, the alignment process can rely on ontologies sharing a similar context with the input ontologies. A few approaches [1, 2, 35] have already considered the use of external knowledge, as a way to obtain semantic matching between dissimilar ontologies. Considering the second issue, a "trusted third party" is used to evaluate the produced alignments during the post-processing step of the ATRACO alignment strategy.

A "trusted third party" can be of different types; it depends on the role that it plays within the alignment strategy:

- Use of external ontological resources: as we are in the specific domain of NGAIEs, the missing information during the processing step of the alignment strategy, in case the ontologies to be aligned are very dissimilar, is acquired from existing ontologies sharing a similar context with the aligned ones, or by using WordNet as an external resource, in order to make use of the synonyms that this resource provides
- Manual intervention: at the time being, the automatic alignment of ontologies generates, in most cases, incomplete, or incorrect, or no correspondences at all - it depends on the type of input ontologies (that is, complete, or not, consistent, or not, overlapping, low overlapping, or non-overlapping ontologies). Therefore, we focus on semi-automatic alignment, where the Alignment Module which implements the ATRACO alignment strategy suggests correspondences between concepts of the two ontologies to be aligned and the user, either discards, or follows these suggestions
- Use of agents: another approach that stands between the two approaches mentioned above, is to employ the services of agents that specialize in ontology alignment. These agents can (a) refer to an internal knowledge base of semantic mappings, (b) access public knowledge resources (as suggested in the first item), (c) infer new valid semantic alignments and (d) interact with people to resolve unknown, or ambiguous situations (as suggested in the second approach)

3.5 Using Category Theory as the formalization framework of the network of ATRACO ontologies

In the ubiquitous computing environment of the ATRACO project, the coexistence of intrinsically different models of local knowledge (ontologies), makes the exchange of information difficult. This exchange of information sought, is achieved through the ontology alignment process. Some approaches followed in the literature for the formalization of ontologies and their operations, consist in using the information flow theory [24, 22], Goguen's work on institution theory [26] and Category Theory. Category Theory offers several ways in order to combine and integrate objects and has been used as a mechanism to formalize ontology matching, providing operations to compose and decompose ontologies (alignment, merging, integration, mapping) [23, 20, 46, 6]. Since the basic objects in Category Theory are the relationships between ontological specifications and not the internal structure of a single knowledge representation, it permits to view global operations involving ontologies (like alignment, merging, matching), independently from the languages used to express their local entities and from the techniques used. In the framework of the ATRACO project we have adopted Category Theory as the most appropriate formalization framework. This is because the chosen formalism

- focuses on relationships (categorical morphisms, functors, and natural transformations) and not on entities (categorical objects, and categories)
- allows the coexistence of heterogeneous entities, since it provides the ability to define several categories, according to the kinds of entities to be described (category of ontologies, category of alignments, category of networks of interlinked ontologies), which can be related by the definition of special morphisms (categorical functors)
- offers a set of categorical constructors for creating new categories, by using predefined ones
- provides a means for the combination of categorical objects (colimits can be used to compose them and limits to decompose them), and for the combination of categorical functors (natural transformations)
- provides a multi-level study of its categorical notions, by defining three interrelated levels (the level of categories, the level of functors and the level of natural transformations).

3.5.1 Categorically formalizing ontologies and alignments

In the following, we review the main results of the categorical formalization of ontology operations. The concepts of an ontology are structured in a taxonomy (hierarchy of concepts related by the subsumption relation) and can also be related by more complex non taxonomic relations, like the mereologic ones. Moreover, correspondences can be established between entities (concepts, properties, instances) belonging to two distinct ontologies (this operation is called matching), resulting in a set of correspondences betwen them, called an alignment. Correspondences between entities belonging to different ontologies can be restricted to equivalence only, or could be binary relations of a wider nature, like disjointness, generalization/specification, etc. On the other hand, a category consists of a collection of objects and of a collection of morphisms (or arrows, or maps), each morphism being a relation from a domain object to a codomain object. Each category is equipped with an associative composition operator defined for any composable pair of morphisms (the codomain of one of the morphisms is the domain of the other one) and unique identity morphisms acting as the units of the composition operator. By restricting ourselves to subsumption relations between concepts of the same ontology and to equivalence relations between concepts belonging to different ontologies and in order to have a categorical view of ontologies [6], ontologies are considered as category objects and the morphisms of the category are functions f between a domain and a codomain ontology, mapping concepts, or relations of the domain ontology to respective concepts, or relations of the codomain ontology. The morphisms are such that they preserve any relations between entities in the domain ontology, for example, if c_1 is related to c_2 through the relation r in the domain ontology, then $f(c_1)$ is related to $f(c_2)$ in the codomain ontology by the relation $f(r)$, establishing thus a graph homomorphism between the two ontologies, if the two ontologies are seen as graphs.

The alignment between two ontologies O_1 and O_2, is the task of establishing binary relations between the entities of the two ontologies. In a categorical perspective, an alignment is not expressed as a direct relation between entities belonging to the two ontologies. Instead, each binary relation between entities belonging to O_1 and O_2 respectively, can be decomposed into a pair of mappings from a common intermediate source ontology, O, to the ontologies O_1 and O_2 [23]. The mappings from O to O_1 and from O to O_2, specify how the concepts and relations of O are understood in O_1 and O_2, respectively. This structure, comprising the ontologies O_1 and O_2 to be aligned, the intermediate ontology O and the morphisms f_1, f_2, is called (due to its shape) a V-alignment (see Fig. 3.16); it is a span, in the Category Theory terminology.

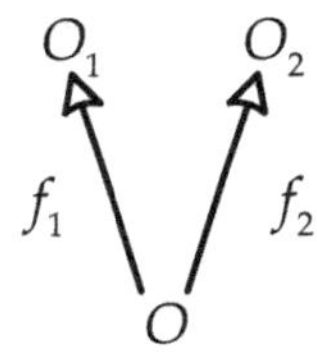

Fig. 3.16 V-alignment

The operation of integrating two aligned ontologies into a single one is called merging and can be accomplished with V-alignments. The ontology resulting from the unification process of merging, embodies the semantic differences of the two ontologies and collapses the semantic intersection between them. Merging of aligned ontologies can be described, in the Category Theory formalization, in terms of a Category Theory construct, called pushout, which is a special case of another construct, called colimit. The pushout involves three objects, the three categories O_1, O_2 and O and the two morphisms of the alignment diagram. The pushout is a new object, a new ontology O' in our case, together with morphisms f_1', f_2', such that

$f_1' \circ f_1 = f_2' \circ f_2$

The commutativity of the pushout diagram in Fig. 3.17, means that components of O_1 and O_2 that are images of the same component in O (that is, the semantic intersection of O_1 and O_2), are collapsed in the resulting ontology O' (mapped to the same entity), which is exactly the definition of the merging operation. That is, the pushout ontology realizes the merging of O_1 and O_2. Moreover, for any other object (ontology) O'' for which the commutativity holds, i.e. for which

$f_1'' \circ f_1 = f_2'' \circ f_2$

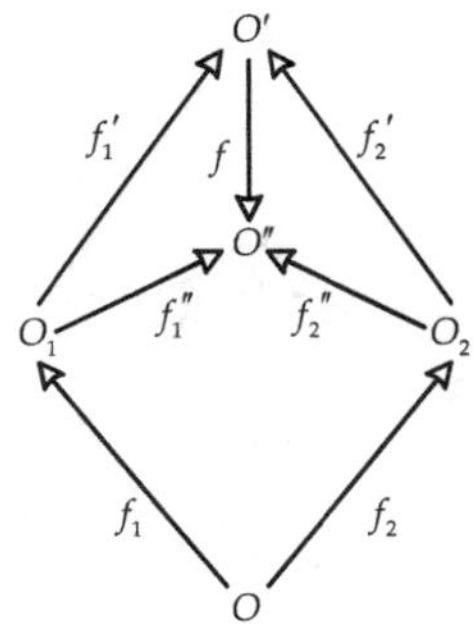

Fig. 3.17 Merging through the pushout construct

there exists a unique morphism f such that
$f \circ f_1' = f_1''$
$f \circ f_2' = f_2''$
that is, the pushout O' is the most compact ontology that can embody the union of O_1, O_2 and possibly comprises collapsed components (i.e., embodies the semantic differences and collapses the semantic intersection).

In Category Theory, dual concepts arise from the process of reversing all the morphisms in a diagram. Thus, the dual concept of pushout is a construct called pullback, which is a particular case of another construct called limit (dual of colimit). The pullback is used in order to formalize the matching operation, by which similarities between ontologies are detected. We start with what is called a Λ-alignment, of the form depicted in Fig. 3.18.

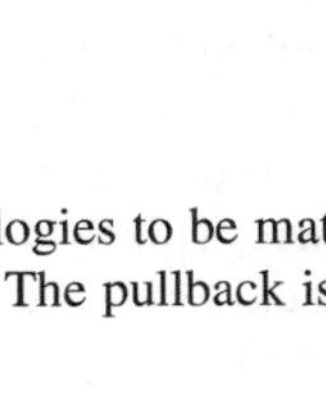

Fig. 3.18 Λ-alignment

Here, O_1 and O_2 are the ontologies to be matched and O is an intermediate ontology that guides the matching. The pullback is a new ontology O', together with morphisms f_1', f_2', such that
$f_1 \circ f_1' = f_2 \circ f_2'$
that is, the pullback O'embodies all information of O_1 and O_2 that is semantically equivalent.

The commutativity of the pullback diagram in Fig. 3.19, means that components of O_1 and O_2 that have the same image in O (i.e., are semantically equivalent), are

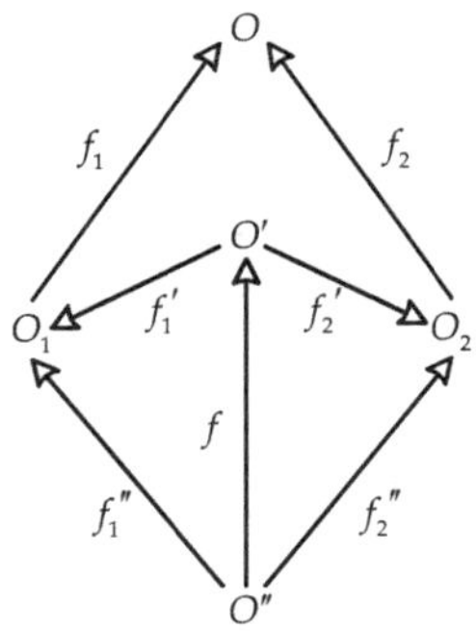

Fig. 3.19 Matching through the pullback construct

images of the same component in O', which is exactly the definition of the matching operation. Thus, the pullback operation realizes the matching of O_1 and O_2. Moreover, for any other object (ontology) O'' for which the commutativity holds, i.e.

$f_1 \circ f_1'' = f_2 \circ f_2''$

there exists a unique morphism f, such that

$f_1' \circ f = f_1''$

$f_2' \circ f = f_2''$

that is, the pullback O' is the biggest ontology that includes all the semantic intersection of O_1 and O_2.

3.5.2 Categorical formalization of activity spheres

In the ATRACO environment, the activity spheres conceptualize an ambient ecology populated by different entities, such as devices, services, humans, or agents, interrelated and in interation with their environment. Once an activity sphere is produced, the knowledge associated with it can be reused, in order to produce new activity spheres. To this end, activity spheres can be extended by the proper addition of ambient ecology entities, or can be embedded into other activity spheres, to produce more complex ones. Thus, activity spheres undergo changes, each time they adapt to different conditions, or each time a reconfiguration of their structure occurs, due to entities entering, or leaving the sphere. This dynamics can be captured by the alignment composition operation. Indeed, if we have an alignment between ontologies O_1 and O_2 and an alignment between ontologies O_2 and O_3, we can compose them and obtain an alignment between ontologies O_1 and O_3, thus depicting relations holding between the entities of ontologies O_1 and O_3.

The operation of alignment composition can be formulated in the categorical framework [46], as the composition of spans in Category Theory, through the use of the pullback construct. In Fig. 3.20, if (O_A, f_1, f_{2A}) is a V-alignment expresing the alignment between ontologies O_1 and O_2 and (O_B, f_{2B}, f_3) is a V-alignment expresing the alignment between ontologies O_2 and O_3, then the ontology O together with the morphisms $f_1 \circ f_A$ and $f_3 \circ f_B$ constitute the composition of the two V-alignments sought, where O, together with the morphisms f_A, f_B is the pullback of the Λ-alignment of O_2 (O_2 with f_{2A}, f_{2B}).

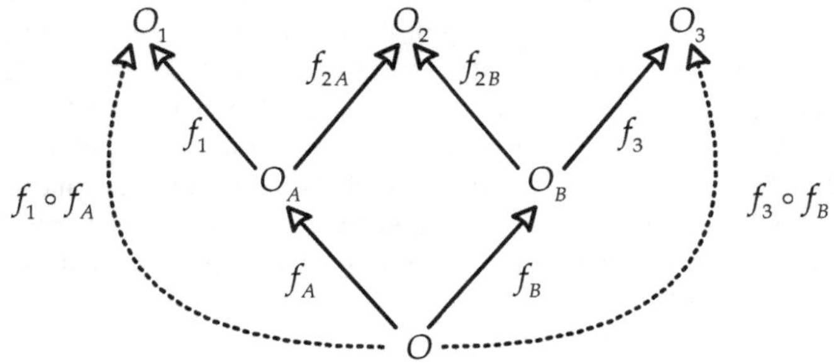

Fig. 3.20 Composition of alignments

In the ATRACO perspective, an activity sphere is a network of already aligned ontologies. In order to define an activity sphere under a categorical perspective, we define a category having ontologies as objects. There exists a morphism between two ontologies, objects of this category, if and only if there exists a V-alignment between them. Thus, composable morphisms in this category, reflect to the composition of V-alignments, which becomes then the main operation needed in the ATRACO perspective.

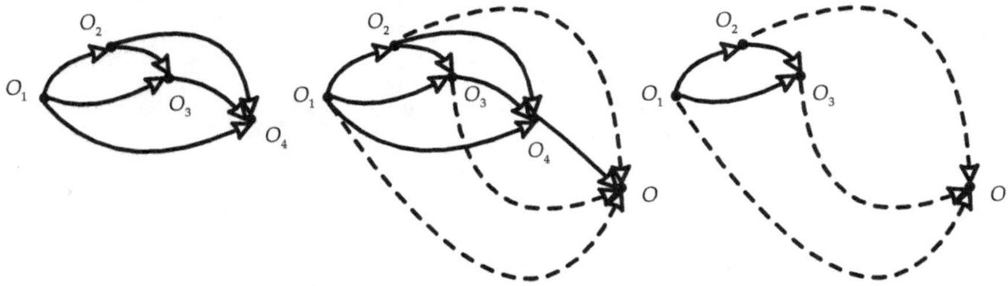

Fig. 3.21 Ontologies entering or leaving an ATRACO activity sphere

More precisely, as depicted in Fig. 3.21, whenever an entity represented by an ontology joins an already established activity sphere of entities represented by a network of already aligned ontologies, it suffices to align it to a single anchor ontology, already participating in the activity sphere. The achor alignment produced, is then composed to already established alignments involving the anchor ontology, producing a batch of composition-generated alignments that remain, even if later on the anchor ontology leaves the activity sphere.

In cases where subsumption, or more elaborate relations, between concepts belonging to different ontologies is to be expressed, since these relations cannot be represented with the vocabulary of any of the two ontologies, it is externalized in an additional new ontology (called bridge ontology), as a bridge axiom. The following diagram depicts the situation, with the original ontologies O_1 and O_2 containing the concepts related via subsumption and the bridge ontology B containing the bridge axioms. The fact that there exist concepts of the ontologies O_1 and O_2 occurring within the bridge ontology, is represented by the two V-alignments between the bridge ontology and the ontologies O_1 and O_2. Thus, the so-called W-alignment is defined (see Fig. 3.22).

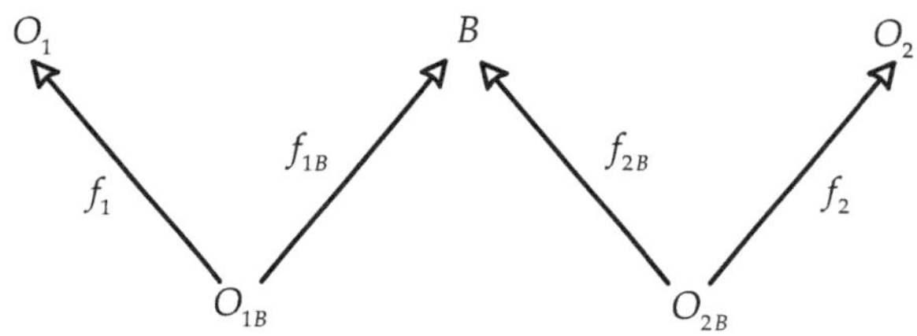

Fig. 3.22 W-alignment

The merging operation in this case, is defined as the colimit of the alignment diagram in Fig. 3.23 and is computed by successive pushouts [46].

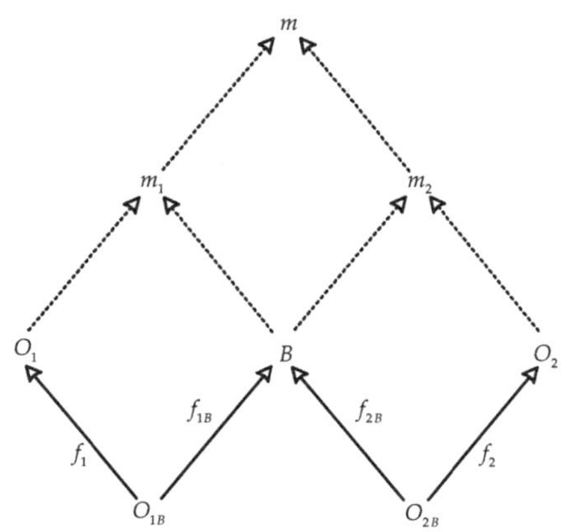

Fig. 3.23 Merging with W-alignments

In a similar way, one can compose W-alignments. If a W-alignment exists between ontologies O_1 and O_2 with bridge ontology B_1 and if also a W-alignment exists between ontologies O_2 and O_3 with bridge ontology B_2, by composing the two W-alignments, it results that a W-alignment exists between ontologies O_1 and O_3, with bridge ontology B, which is obtained if the merging operation is applied to the bridge ontologies B_1 and B_2. The problem of this approach, consists in incorporating in the new bridge ontology B bridge axioms from the ontologies B_1, B_2 and O_2, that might be irrelative to O_1 and O_3.

Another solution to the problem of more elaborate relationships (subsumption, strict inclusion, strict containment, disjointness, overlapping with partial disjointness, temporal relations), between entities belonging to different ontologies, is to enhance the category of ontologies with more elaborate morphisms that denote the relationship that holds between the syntactic entities of the two ontologies (subsumption, strict inclusion etc.) [46, 13]. In this case, when applying the composition operation, if an entity in ontology O_2 has an elaborate relation to entities in the ontologies O_1 and O_3, there is some kind of relation between the two entities in O_1 and O_3. The latter relation depends strongly on the former one. For example, if an entity in O_1 is related to an entity in O_2 with strict inclusion and the same entity in O_2 is related to an entity in O_3 with strict containment, then the entity in O_1 can be related to the entity in O_3 by either of the following relationships: equivalence, strict inclusion, strict containment, disjointness, overlapping with partial disjointness. More specifically, in [13], the formalism of algebras of binary relations has been proposed, in order to solve the problem of expressing the relations between entities of different ontologies in a genaral way, support alignment composition and deduce new alignments from existing ones, via composition tables. In order to capture the dynamic nature of an activity sphere in the case of more general relations between ontological entities, the question remains to construct the appropriate category that captures the structure.

3.6 Conclusion

The knowledge management framework developed in the context of the ATRACO project, was presented in this chapter. We mainly focused on the following axes of research:

- An ontological representation, in order to cope with the dynamic nature and heterogeneity of NGAIEs components, realized in the context of ATRACO with the ATRACO Upper Level Ontology and a set of Device and Service Ontologies, Policy Ontologies, User Profile Ontologies and General Foundational Ontologies
- A task-based ontology engineering process, for the specification, conceptualization, implementation and evaluation of the ATRACO ontologies
- A three step ontology alignment strategy, guided by two similarity coefficients, in order to detect the lexical, or structural resemblance of the source ontologies

- A formalization framework, based on Category Theory, permitting to confront the dynamic nature of ATRACO activity spheres, the constituent networks of interlinked alignments of the ATRACO project

The presented knowledge management framework provides us with theoretical and practical tools so as to overcome the barrier of heterogeneity intrinsic in the ATRACO world, due to the different conceptualizations of its actors.

3.7 Further readings

The reader interested in learning more about the principles and techniques of ontological engineering in any context, is encouraged to consult [41, 17, 3]. To learn more about tools, languages, matchers and approaches on ontology alignment, [16, 10] are the most recent extensive references. Concerning Category Theory, [32] provides a presentation of the basic constructions and terminology of Category Theory and illustrates applications of Category Theory to programming language design, semantics and the solution of recursive domain equations, while in [5] the material covered includes the standard core of categories, functors, natural transformations, equivalence, limits and colimits, functor categories, representables, Yoneda's lemma, adjoints and monads.

References

1. Aleksovski, Z., ten Kate, W., van Harmelen, F.: Exploiting the structure of background knowledge used in ontology matching. In: Ontology Matching (2006)
2. Aleksovski, Z., Klein, M., ten Katen, W., van Harmelen, F.: Matching unstructured vocabularies using a background ontology. In: Proceedings of the 15th International Conference on Knowledge Engineering and Knowledge Management (EKAW'06), Lecture Notes in Artificial Intelligence. Springer-Verlag (2006)
3. Allemang, D., Hendler, J.: Semantic Web for the Working Ontologist: Effective Modeling in RDFS and OWL. Morgan Kaufmann (2008)
4. Anaya, F., Vazquez, J.: Semantic technologies and techniques for interoperable information. In: 1st International Workshop on Semantic Interoperability for Smart Spaces (SISS), in conjunction with IEEE ISCC2010 (2010)
5. Awodey, S.: Category Theory. Oxford University Press (2010)
6. Cafezeiro, I., Haeusler, E.: Semantic interoperability via category theory. In: 26th International Conference on Conceptual Modelling - ER Auckland, New Zeland, pp. 197–202 (2007)
7. Chen, H.: An intelligent broker architecture for pervasive context-aware systems. Ph.D. thesis, University of Maryland, Baltimore (2004)
8. Cruz, I., Antonelli, F., Stroe, C., Keles, U., Maduko, A.: Using agreement maker to align ontologies for oaei 2009 overview, results, and outlook. In: 4th International Semantic Web Conference (2009)
9. Davies, J., Suder, R., Warren, P.: Semantic Web Technologies: Trends and Research in Ontology-based Systems. Wiley (2008)
10. Ehrig, M.: Ontology alignment: bridging the semantic gap. Springer-Verlag New York Inc (2007)

11. Ehrig, M., Sure, Y.: FOAM–Framework for Ontology Alignment and Mapping Results of the Ontology Alignment Evaluation Initiative. In: Integrating Ontologies Workshop Proceedings, p. 72 (2005)
12. Ejigu, D., Scuturici, M., Brunie, L.: Coca: a collaborative context-aware service platform for pervasive computing. In: 4th IEEE International Conference on Information Technology, pp. 297–302 (2007)
13. Euzenat, J.: Algebras of ontology alignment relations. In: International Semantic Web Conference, pp. 387–402 (2008)
14. Euzenat, J., Le Bach, T., Barrasa, J., Bouquet, P., De Bo, J., Dieng-Kuntz, R., Ehrig, M., Hauswirth, M., Jarrar, M., Lara, R.: State of the art on ontology alignment. Deliverable 2.2. 3. Tech. rep., IST Knowledge Web NoE (2004)
15. Euzenat, J., Loup, D., Touzani, M., Valtchev, P.: Ontology alignment with ola. In: 3rd International Semantic Web Conference (2004)
16. Euzenat, J., Shvaiko, P.: Ontology Matching. Springer-Verlag Berlin Heidelberg (2007)
17. Gómez-Pérez, A., Fernández-López, M., Corcho, O.: Ontological Engineering: With examples from the areas of Knowledge Management, e-Commerce and the Semantic Web. Springer (2004)
18. Gruber, T.: Towards principles for the design of ontologies used for knowledge sharing. International Journal of Human-Computer **43**, 907–928 (1995)
19. Heinroth, T., Kameas, A., Pruvost, G., Seremeti, L., Bellik, Y., Minker, W.: Human-computer interaction in next generation ambient intelligent environments. Intelligent Decision Technologies, special issue on Knowledge-based Environments and Services in Human-Computer Interaction **5**(1) (2011)
20. Hitzler, P., Krotzsch, M., Ehrig, M., Sure, Y.: What is ontology merging? - a category theoretic perspective using pushouts. In: 1st International Workshop on Concepts and Ontologies: Theory, Prectice and Applications, pp. 104–107 (2005)
21. Hu, W., Qu, Y., Cheng, G.: Matching large ontologies: A divide-and-conquer approach. Data & Knowledge Engineering **67**(1), 140–160 (2008)
22. Kalfoglou, Y., Schorlemmer, M.: Information-flow-based ontology mapping. In: Ont the Move to meaningful Internet Systems 2002: CoopIS, DOA and ODBASE: Confederated International Conferences, pp. 1132–1151. Springer (2002)
23. Kalfoglou, Y., Schorlemmer, M.: Ontology mapping: the state of the art. The knowledge engineering review **18**(01), 1–31 (2003)
24. Kent, R.: The iff foundation for conceptual knowledge organization. Cataloging and Classification Quarterly **37**, 187–203 (2000)
25. Kotis, K., Vouros, G., Alonso, J.: Hcome: Tool-supported methodology for collaboratively devising living ontologies. In: Workshop on Semantic Web and Databases (2004)
26. Kutz, O., Mossakowski, T., Codescu, M.: Shapes of alignment: Construction, combination, and computation. In: Workshop on Ontologies: reasoning and modularity, WORM-08, ESWC (2008)
27. Li, J., Tang, J., Li, Y., Luo, Q.: RiMOM: A dynamic multistrategy ontology alignment framework. IEEE Transactions on Knowledge and Data Engineering pp. 1218–1232 (2008)
28. Lopez, M., Gómez-Pérez, A., Sierra, J., Sierra, A.: Building a chemical ontology using methontology and the ontology design environment. Intelligent Systems and their Applications, IEEE **14**(1), 37–46 (2002)
29. Nicola, A., Navigli, R., Missikoff, M.: Building an eprocurement ontology with upon methodology. In: The 15th e-Challenge Conference (2005)
30. Noy, N., McGuinness, D.: Ontology development 101: A guide to creating your first ontology. Tech. rep., Stanford Knowledge Systems Laboratory (2001)
31. Panagiotopoulos, I., Seremeti, L., Kameas, A., Zorkadis, V.: Proact: An ontology-based model of privacy policies in ambient intelligence environments. In: Proc. 14th Panhellenic Conf. Informatics (PCI), pp. 124–129 (2010)
32. Pierce, B.: Basic Category Theory for Computer Scientists. The MIT Press (1991)

33. Pinto, H., Staab, S., Tempich, C.: Diligent: Towards a fine-grained methodology for distributed loosely-controlled and evolving engineering of ontologies. In: The 16th European Conference on Artificial Intelligence (2004)
34. Román, M., Hess, C., Cerqueira, R., Ranganathan, A., Campbell, R., Nahrstedt, K.: A middleware infrastructure for active spaces. IEEE Pervasive Computing **1**(4), 74–83 (2002)
35. Sabou, M., d'Aquin, M., Motta, E.: Using the semantic web as background knowledge for ontology mapping. In: Ontology Matching (2006)
36. Seghrouchni, A., Breitman, K., Sabouret, N., Endler, M., Charif, Y., Briot, J.: Ambient intelligence applications: Introducing the campus framework. In: 13th IEEE International Conference on Engineering of Complex Computer Systems (2008)
37. Serafini, L., Tamilin, A.: Drago: Distributed reasoning architecture for the semantic web. The Semantic Web: Research and Applications pp. 361–376 (2005)
38. Seremeti, L., Kameas, A.: A task-based ontology engineering approach for novice ontology depelopers. In: 4th Balkan Conference in Informatics (2009)
39. Seremeti, L., Kameas, A.: Tools for Ontology Engineering and Management. Theory and Applications of Ontology: Computer Applications pp. 131–154 (2010)
40. Seremeti, L., Kameas, A., Panagiotopoulos, I.: An alignable user profile ontology for ambient intelligence environments. In: Proc. 7th International Conference on Intelligent Environments, IE'11 (2011)
41. Sharman, R., Kishore, R., Ramesh, R.: Ontologies : a handbook of principles, concepts and applications in information systems. Springer (2007)
42. Suarez-Figueroa, M.: Neon development process and ontology life cycle - deliverable d5.3. Tech. rep., NeOn Project (European Commission's Sixth Framework Programme under grant number IST-2005-027595) (2007). D5.3.
43. Sure, Y.: Methodology, tools and case studies for ontology based knowledge management. Ph.D. thesis, University of Karlsruhe (2003)
44. Uschold, M., King, M.: Towards a methodology for building ontologies. In: Workshop on Basic Ontological Issues in Knowledge Sharing (1995)
45. Wang, X., Dong, J., Chin, C., Hettiarachchi, S., Zhang, D.: Semantic space: An infrastructure for smart spaces. IEEE Pervasive Computing **3**(3), 32–39 (2004)
46. Zimmermann, A., Krotzsch, M., Euzenat, J., Hitzler, P.: Formalizing ontology alignment and its operations with category theory. In: FOIS'06 (2006)

Chapter 4
Artefact Adaptation in Ambient Intelligent Environments

H. Hagras and C. Wagner

Abstract The paper presents a novel approach to develop strategies that will allow the artefacts to adapt to the uncertainties associated with the changes in the artefacts characteristics, context as well as changes in the user(s) preferences regarding these artefacts in Ambient Intelligent Environments (AIEs). This work is within the framework of an EU funded project entitled ATRACO (Adaptive and Trusted Ambient Ecologies) which aims to contribute to the realization of trusted ambient ecologies in AIEs.

4.1 Introduction

Adaptation is a relationship between a system and its environment where change is provoked to facilitate the survival of the system in the environment. Biological systems exhibit different types of adaptation so as to regulate themselves and change their structure as they interact with the environment.

The dynamic and ad-hoc nature of Ambient Intelligent Environments (AIEs) means that the environment has to adapt to changing operating conditions and user changing preferences and behaviours and to enable more efficient and effective operation while avoiding any system failure. Thus there is a need to provide autonomous intelligent adaptive techniques which should be able to create models which could be evolved and adapted online in a life-long learning mode. These models need to be transparent and easy to be read and interpreted by the normal user to enable the user to better analyze the system and its performance. Such intelligent systems should allow to control the environment on the user's behalf and to his satisfaction to perform given tasks. The intelligent approaches used should have low computational

Hani Hagras
University of Essex in Colchester CO4 3SQ, United Kingdom, e-mail: hani@essex.ac.uk

Christian Wagner
University of Essex in Colchester CO4 3SQ, United Kingdom, e-mail: chwagn@essex.ac.uk

overheads to effectively operate on the embedded hardware platforms present in the everyday environments which have limited memory and processor capabilities. This task based system could be used to control the environment on the user behalf and to his satisfaction to perform given tasks. In addition, the intelligent approaches should allow for real-time data mining of the user data and create on-the-fly updateable models of the user preferences that could be executed over the pervasive network. Moreover, there is a need to provide an adaptive life-long learning mechanism that will allow the system to adapt to the changing environmental and user preferences over short and long term intervals. There is a need also to provide robust mechanisms that will allow handling the varying and unpredictable conditions associated with the dynamic environment and user preferences.

This Chapter will present novel adaptation strategies that will allow the artefacts[1] to adapt to the uncertainties associated with the changes in the artefacts characteristics, context as well as changes in the user(s) preferences regarding these artefacts in AIEs.

Section 4.2 presents an overview of the previous work in adaptation in AIEs. Section 4.3 presents the ATRACO project approach to Artefact Adaptation in AIEs. Section 4.4 provides an overview of the employed Artefact Adaptation approaches while Section 4.5 provide an overview on the Artefact Adaptation design in AIEs. Section 4.6 provides the experiments and results while Section 4.7 provides the conclusions and future work.

4.2 Previous Work in Artefact Adaptation in AIE

Several work have targeted the topic of adaptation in Ambient Intelligent Environments (AIEs) where Davidsson [4] have developed a multi-agent system (MAS) that monitors and controls an office building in order to meet user preferences while conserving energy.

The Oxygen at MIT [13] centres around two rooms containing cameras, microphones, an X-10 controlled lighting system and a multitude of computer vision and speech understanding systems that help the system interpret what people are saying, where they are saying it and what interactions and activities are taking place. The system responds accordingly using speech synthesis when spoken to. The vision system is able to intelligently train itself for a particular environment in less then five minutes using projectors displaying a simulation of someone performing the training. The agents operate in a rather independent intelligence level where each sensor resides in a particular place and uses various local resources.

The goals of the Aware Home [9] agents are to investigate what kind of services can be built on top of an environment that is aware of the activities of its occupants. Besides building models of human behaviour to aid computers in decision making, it aims to support older individuals to 'age in place' by helping them to maintain a

[1] As part of this chapter the word "artefact" is generally used interchangeably with the word "device" and refers to physical devices in the AIE.

good diet, take medication when required, notify family members of their well-being and provide support with everyday tasks. The large number of sensors deployed in the Aware Home range from trip sensors (that detect when a door has been opened or closed) and motion detectors, to higher-fidelity sensors, e.g. embedded microphones and cameras. These sensors, all centrally connected to a computer, are mostly used to determine the current activities, facial expressions, locations and gestures of occupants. The sensors and actuators are fixed and dedicated to performing specific tasks only [1].

The Adaptive Home, a.k.a. the Neural Network Home, explored the "learning user's habits" aspects of an AIE. It was aiming to predict occupancy of rooms, hot water usage and the likelihood that a zone is entered in the next few seconds, using trained feed-forward neural networks. The context information in the project was again mainly comprised of location, but additional state information from rooms like the status of lights or the temperature set by inhabitants was used. Although learning and prediction was done via feed-forward multi-layer perceptrons with the known limitations, it showed that prediction of user locations can help to save resources and support users by learning their behaviour and automating simple tasks.

The UT-Agent at Ajou University in Korea focuses on the use of intelligent agents for ubiquitous smart home environments. An intelligent agent model for AIEs is proposed in which the agents must learn the user preferences in order to assist them, which are represented by user profiles. The proposed UT-AGENT uses case-based reasoning to acquire knowledge about user actions that are worth recording, to determine their preferences and a Bayesian Network as an inference tool to model relationships between them. The UT-AGENT maintains the status of every device present in the home and activates them according to user preference. It generates a sequence of the expected user's query and simultaneously activates the devices with preset preference settings [10].

The MavHome smart home project focuses on the creation of an environment that acts as an intelligent agent, perceiving the state of the home through sensors and acting upon the environment through device controllers [21]. The environment is represented using a Hierarchical Hidden Markov Model and a reinforcement learning algorithm is employed to predict the environmental preferences based on sensors within the environment. Desired actions are proposed for the control of lights within the environment primarily based on motion detection sensors and if the actions are within the bounds of acceptable safety and security policies, they are invoked within the environment. The agent aims to maximise the comfort and productivity of its inhabitants while minimising the energy consumption.

In the University of Essex, the Incremental Synchronous Learning (ISL) approach was developed for embedded agents to realise Ambient Intelligence (AmI) in ubiquitous computing environments [7]. In [6], the AOFIS system was developed to enhance the capabilities of the ISL agents by including mechanisms to extend the original agent based approach to use type-2 fuzzy systems [8]. The type-2 agent generated type-2 FLCs in which type-2 membership functions (MFs), with fixed uncertainties directly modelled and handled the long term environmental uncertainties. Type-2 embedded agents facilitate short term online adaptation of the rules while being robust to long term environmental changes. The agent can incrementally adapt the type-2 FLC rules and MFs in a lifelong learning mode, accommodating for the accumulated long term uncertainties arising from seasonal changes in the environment and the associated changing user behaviour over extended periods of time [8].

Most of the above systems did not manage to allow the artefacts to adapt to the uncertainties associated with the changes in the artefacts characteristics, context as well as changes in the user(s) preferences regarding these artefacts in Ambient Intelligent Environments (AIEs). This will be the focus of our work.

4.3 ATRACO Approach to Artefact Adaptation in AIEs

In the previous subsections, we have shown some of the other work conducted in AIEs and to our knowledge no other work has targeted the Artefact Adaptation (AA) in AIE where AA deals with developing the strategies that will allow the artefacts to adapt to the uncertainties associated with the changes in the artefacts characteristics, context as well as changes in the user(s) preferences regarding these artefacts.

AIEs face huge amount of uncertainties which can be categorized into environmental uncertainties and users' uncertainties. The environmental uncertainties can be due to:

- The change of environmental factors (such as the external light level, temperature, time of day) over a long period of time due to seasonal variations.
- The change of the sensors and actuators outputs due to the noise from various sources. In addition, the sensors and actuators can be affected by the conditions of observation (i.e. their characteristics can be changed by the environmental conditions such as wind, sunshine, humidity, rain, etc.).
- Uncertainties associated with the change in the context and operation conditions.
- Wear and tear which can change sensor and actuator characteristics.

Thus the environmental uncertainties are associated with the change in artefact characteristics and change in the artefact context.
The user uncertainties can be classified as follows:

- Intra-user uncertainties which are exhibited when a user decision for the same problem varies over time and according to the user location and activity. This variability is due to the fact that the human behaviour and preferences are dynamic and they depend on the user's context, mood, and activity as well as the weather conditions and time of year. For the same user, the same words can mean different things on different occasions. For instance the values associated with a term such as "warm" in reference to temperature can vary as follows: depending on the season (for example from winter to summer), context (in the Arctic 15° C might be considered "High", while in the Caribbean it would be considered "low"), depending on the user activity within a certain room and depending on the room within the user home and many other factors.

- Inter-user uncertainties which are exhibited when a group of users occupying the same space differ in their decisions in a particular situation. This is because users have different needs and experiences based on elements such as age, sex, profession, etc. For instance the users might disagree on aspects such as how warm a room should be on any given day.

Thus the user(s) uncertainties are associated with the changes and variations in the user(s) preferences regarding the artefacts and their operation.
Thus it is crucial to employ adequate methods to handle the above uncertainties to enable the system to produce the desired behaviour to perform a given task. In addition, there is a need to produce models of the users' particular behaviours that are transparent and that can be adapted over long time duration and thus enabling the control of the users' environments on their behalf.

Fuzzy Logic Systems (FLSs) are credited with being adequate methodologies for designing robust systems that are able to deliver a satisfactory performance when contending with the uncertainty, noise and imprecision attributed to real world settings [5]. In addition, a FLS provides a method to construct controller algorithms in a user-friendly way closer to human thinking and perception by using linguistic labels and linguistically interpretable rules. Thus FLSs can satisfy one of the important requirements in AmI systems by generating transparent models that can be easily interpreted and analyzed by the end users. Moreover, FLSs provide flexible representations which can be easily adapted due to the ability of fuzzy rules to approximate independent local models for mapping a set of inputs to a set of outputs. As a result, FLSs have been used in AIEs spaces as in [5], [14] and [8].

4.4 An Overview of the Employed Artefact Adaptation in ATRACO

Within the artefact adaption model, the process starts with the high level ontology (termed Sphere Ontology) supplying the artefacts needed to perform a given task. The Sphere Ontology also provides the linguistic labels and the operational ranges of the variables involved with these artefacts. The Sphere adaptation level handles the fault tolerance issues with existing artefacts breaking down or new artefacts being introduced to the system by searching and exposing a suitable replacement artefact if available.

Artefact Adaptation (AA) deals with developing the strategies that will allow the artefacts to adapt to the uncertainties associated with the changes in the artefacts characteristics, context as well as changes in the user(s) preferences regarding these artefacts and their operation. Hence, the artefacts will adapt by adapting the operation values associated with the linguistic labels of the artefacts according to the changes in the artefact characteristics and context. In addition, the artefact will also adapt to the user(s) changes in desire and preferences for the fulfilment of a given task. For example for a temperature sensor which was identified by the sphere ontology to be involved in the task of keeping the temperature comfortable for a given user(s). If the user(s) was using this temperature sensor in kitchen, the temperature values associated with Low, Medium, High will need to be adapted when moved to the living room, hence the artefact will need to adapt these values. On the other hand for a heater actuator, the Artefact Adaptation will involve adapting the heater setting values associated with the linguistic labels Low, Medium, High (supplied by Sphere Ontology) according to the changes in the artefact characteristics and context. For example the heater settings associated with the linguistic labels (Low, Medium, High) will change from Summer to Spring, etc. In addition, the artefact will also adapt to the user(s) changes in desire and preferences for the fulfilment of a given task. Where the heater setting associated with Low will vary from a normal day to day for example involving a party with a large number of people present.

In this way, ATRACO would satisfy one of the main components of the ubiquitous computing and the disappearing Computer visions where the intelligence will not come from one information artefact [20]. Rather intelligence emerges from collections of artefacts interacting and co operating with each other, resulting in new behaviour and new functionality [20]. In Atraco the *artefacts will be able to adapt based on how people use them.* In addition, the objects' coordination with other artefacts will yield new functionalities. Hence, new forms of use will emerge that will enrich everyday life, resulting in a world that is more deeply interconnected.

Each task in ATRACO is associated with a series of devices which are related to this task, for example in terms of "adjusting the light to comfortable levels" this could be a series of lights, light sensors, dimmer switches, etc. AA is achieved by associating each task with a specific Fuzzy Task Agent (FTA) which governs the AA for this task. AA aims to employ the interaction of the user with each individual device for each task in three different stages as follows:

1. User/device interaction capture:
   As the user interacts with the individual devices associated with the given task, the internal parameters set to or received from the device are captured. For example, as part of an "adjust light level to a comfortable level" task, as the user increases the brightness of the lamp, the values of the devices the user is interacting with is recorded. It is important to note that this includes both active interaction (e.g. the lamp) and passive interaction, where we refer to passive interaction as the effect on all sensing devices which are affected by the users action (here - increasing the light level). For the purpose of ATRACO, all sensing devices which are part of a task are considered as passive interaction devices (for example if a timer or clock is associated with the lighting task, then its value will also be recorded during the capturing stage).
2. Interaction data clustering:
   After a sufficient amount of user/device interaction data has been captured. Data clustering is applied for each individual device or groups of devices in order to determine clusters which refer to salient characteristics of the devices' properties. For example in terms of a dimmable lamp, users usually tend to use only a subset of discrete dimming stages such as "off", "dim" (say 50% power) and "bright". It is at this stage that a very important feature of AA is exploited: as the

device characteristics change over time (for example the lamp bulb decreases in brightness), the user responds by altering his use of the specific device (i.e. increases the power to 60% to achieve his preferred level of "dim"). Through clustering, this information is gathered in a completely transparent fashion and rendered useful for the third stage.

3. Adaptive device model creation:
   By employing the information generated in Step 2, an adaptive model of the individual device (which is part of the specific task for the specific user), based on fuzzy sets is created. Additional information which is available such as reliability or accuracy (for example from the manufacturers information (e.g. light bulb life expectancy, output in lumen, Watt, etc.)) can be incorporated at this stage. This third and final stage of AA is the most challenging and research intensive stage as it involves the creation of an adaptive model which generalises over the vast number of user/device interactions while incorporating the ability of handling the device uncertainty (such as temporal variation in its characteristics, wear and tear, etc.).

Recently, type-2 FLSs, with the ability to model second order uncertainties, have shown a good capability of managing high levels of uncertainty. Type-2 FLSs have consistently provided an enhanced performance compared to their type-1 counterparts in real-world applications [3], [8]. A type-2 fuzzy set is characterized by a fuzzy membership function, i.e. the membership value (or membership grade) for each element of this set is a fuzzy set in [0,1], unlike a type-1 fuzzy set where the membership grade is a crisp number in [0,1] [12]. There are two variants of type-2 fuzzy sets - interval valued fuzzy sets (IVFS) and generalized type-2 fuzzy sets (GFS). In an IVFS the membership grades are across an interval in [0,1] and have the third dimension value equal unity. In the case of a GFS the membership grade of the third dimension can be any function in [0,1]. Most applications to date use IVFS due to its simplicity. It has been shown that IVFS based type-2 FLSs can handle the environmental uncertainties and the uncertainties associated with a single user in a single room environment and that type-2 FLSs can outperform their type-1 counterparts [8]. However, no work has tried to approach the challenging area of developing AIEs spaces that can handle the environmental uncertainties as well as the intra- and inter-user uncertainties in an environment.

There are many frameworks for dealing with uncertainty in decision-making including, primarily, those based on probability and probabilistic (Bayesian) reasoning. As an aside, we emphasize that we do not claim that fuzzy-based methods are any better or any more effective than any other uncertainty handling frameworks, rather we claim that some methods are more appropriate in certain contexts. In our experience, fuzzy methods have proven to be more effective than other methods when applied in AmI spaces [5], [14], [8]. This is because the fuzzy methods provide a framework using linguistic labels and linguistically interpretable rules which is very important when dealing with human users.[2]

---

[2] While the Chapter is focused on the creation of adaptive artefact models, the interested reader can find an example of the automatic rule creation based on data for example in [19].

We shall employ the use of type-2 fuzzy logic to model the uncertainties in such AIEs. Consider an example of a central heating system for a living space occupied by two users. In such a situation each user's concept of cold has to be modeled throughout the year. There will be seasonal variations affecting each user's idea of what is cold, an example of intra-user variation. Each individual user will have his notion of what temperatures constitute cold, an example of inter user uncertainty. Modeling either of these uncertainties can be accomplished using a number of existing techniques. The novel challenge with this is to model and cope with the uncertainties created by the dynamic relationship between the interactions of multiple users, each with individual preferences that change over time. For example, Figure 4.1 (a) and (b) shows the use of type-1 fuzzy systems to depict the differences between two users (p1 and p2) concept of cold for the spring/summer and autumn/winter periods. Figure 4.1 (c) and (d) show how interval type-2 fuzzy sets might model each user's concept of cold throughout the year. Figure 4.1 (e) shows how a general type-2 fuzzy set might encompass both the inter- and intra-user uncertainty about what cold is by employing the third dimension, where the different grey levels correspond to different membership levels in the third dimension.

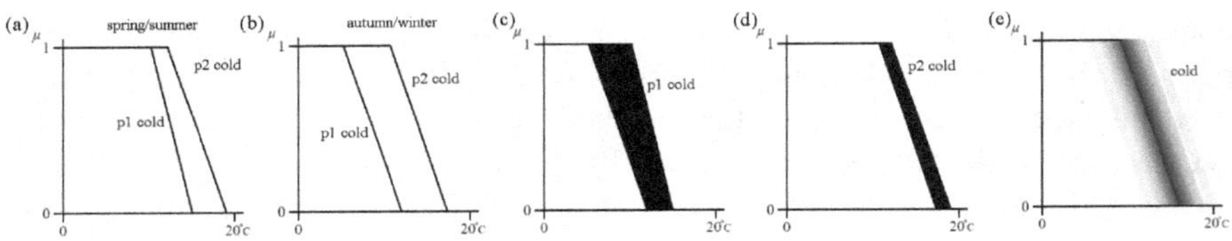

Fig. 4.1 Fuzzy sets modeling the linguistic label cold.

Untill recently, general type-2 fuzzy systems were perceived as very computationally expensive that they do not have any real applications within real time systems like AIEs. However, since the start of ATRACO, UEssex has developed novel theoretical models which allowed the realization of general type-2 fuzzy on real time embedded systems [15], [16],[17], [18].[3]

As mentioned above each task in ATRACO is in turn associated with a series of devices which are related to this task, for example in terms of "adjusting the light to comfortable levels" this could be a series of lights, light sensors, dimmer switches, etc. AA is achieved by associating each task with a specific (AA) agent which governs the AA for this task. AA aims to employ the interaction of the user with each individual device for each task in three different stages (as shown in Figure 4.2).

[3] The current Chapter will focus on the demonstration of the concepts based on type-1 fuzzy sets in order to respect the page limit. The description of the application of general type-2 fuzzy sets will be addressed in a series of publication focusing specifically on the use of type-2 fuzzy logic in AIEs.

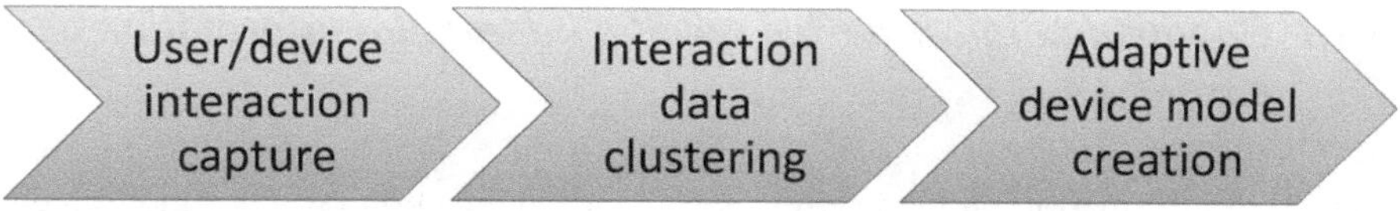

Fig. 4.2 Artefact Adaptation Stages

4.5 Artefact Adaptation Design

As a consequence, within ATRACO and specifically as part of AA, we have chosen fuzzy sets as the basis for the adaptive models of the artefacts/devices within the user environment. Fuzzy sets have a series of capabilities which make them ideal for the application within AA, specifically:

- Fuzzy sets are easily adaptable.
  As fuzzy sets can take a wide variety of shapes and forms (triangular, gaussian, etc.) the parameters of which are easily modified, fuzzy sets are easily adaptable to incorporate new information and provide a very flexible representation for AA.
- Fuzzy sets are easily interpretable.
  Fuzzy logic sets have been designed to model and reflect human reasoning and are generally connected to linguistic labels. For example, a certain device (e.g. a lamp) can be represented using a linguistic variable understandable to human users (e.g. lightlevel) which in turn can be mapped to a series of linguistic labels which are familiar terms for the user (e.g. dim, bright, etc.). Each of these labels can be modelled using a fuzzy set.
- Fuzzy logic sets allow for the modelling and handling of uncertainty.
  Fuzzy logic sets are accepted to have great potential in dealing with real world uncertainty and have been employed in that regard in a wide variety of applications. Further, recent advancements in fuzzy logic theory, (specifically type-2 fuzzy logic theory) have shown great potential in further advances in dealing with uncertainty [8], [12], [15], [16], [17], [18].
- Fuzzy logic sets allow for a seamless integration of AA as part of the FTA. This allows for a seamless integration of the AA into one transparent agent which is referred to as Fuzzy Task Agent (FTA) within the ATRACO project.

The different qualities of fuzzy sets are essential to the successful deployment of AA and as part of ATRACO we are aiming to investigate both the deployment of existing fuzzy set models (i.e. type-1 fuzzy logic sets) as well as to research the application of higher order fuzzy sets such as interval type-2 fuzzy systems [12] and the recently introduced zSlices-based general type-2 fuzzy sets [18].

4.5.1 Capturing of user/device specific interaction information as part of AA

Capturing information on the user/device specific interaction provides the foundation of the AA. While a variety of other methods could be employed to provide information on the device characteristics, for example by asking the user to "program" or to specify the device characteristic, or by retrieving and employing device details released by the device manufacturer (e.g. output in Watts, light beam shape, type of bulb, etc. in terms of a lamp) or simply by subdividing the range of the devices setting into equal subdivisions (i.e. similar to recreating a "dimmer model"), we have opted to capture the user/device interaction as it occurs within the AIE during the normal occupancy of a user.

The reasoning for this approach is mainly two-fold:

1. The ATRACO vision and that of most AIEs in general aims for a transparent operation where the user is unaware of the technology around him in the sense that he/she is not required to take any direct action to understand or control it. The environment should adapt to the user, not the user to the environment. A good counterexample to this paradigm is the video recorder which a whole generation of people in the 1990s purchased in great anticipation but which was so complicated to use and programme. As such, in terms of AA, it is essential that the user is not required to engage directly in the modelling of the devices but that his/her natural interactions (where we define "natural" as the interaction as it would take place in a standard, non augmented home) with the devices is transparently employed to gather the required data.
2. By capturing the users natural interactions with the specific devices, information on the devices' characteristics but also most importantly, information on the users perception and preferences in terms of those devices can be captured. Capturing this additional information allows the AA to not only adapt to the isolated device characteristics but moreover to the device characteristics as they are experienced by the user as part of the current task/sphere. For example users are generally so familiar with dimmer switches that they adjust them without realizing the variation in their sensitivity: for example light dimmer switches often do not produce a visible increase in lighting brightness over the first quarter turn (or more) from the off state. Such information, while obvious to the human user, is generally difficult for an intelligent agent such as the AA to perceive. Nevertheless, by employing the user/device interaction as a source of information, the AA transparently encompasses adaptations undertaken by the user in response to the specific characteristics of a device.

The capturing of user/device interaction information is by design a time intensive activity. The more interactions the AA can record, the more accurate the eventual model of the device characteristics in relation to the current task and user will be. As this is clear, the approach of the AA within ATRACO consists of constantly capturing the user/device interactions thus allowing repeated re-adaptation and updating of the existing device model(s).

With the current section having focussed on the capturing of the user/device specific interaction information, the following section will focus on how to extract a maximum of actual knowledge from this captured information.

4.5.2 *Extraction of salient features for the user/device interaction information as part of AA*

The extraction of salient features from the raw user/device interactions data is a crucial step towards the eventual generation of adaptive models of the artefacts.

In the fields of artificial intelligence and machine learning, one popular technique in the extraction of salient features, also referred to as data mining, is data clustering. Data clustering allows for the segmentation of a series of observations into different subsets. The segmentation process is based on a measure of similarity of the individual observations so that similar observations are assigned to the same subsets.

One of the most popular clustering algorithms is the k-means clustering which was first published in its standard form in 1982 [11]. The basic steps of the algorithms can be summarized as follows:

- Choose the number of clusters, k.
- Randomly generate k clusters and determine the cluster centers, or directly generate k random points as cluster centers.
- Assign each point to the nearest cluster center.
- Recompute the new cluster centers.
- Repeat the two previous steps until some convergence criterion is met (usually that the assignment hasn't changed).

While the k-means algorithms has proved very successful, its very aim and nature of assigning observations to specific sets (an observation is always either a member of cluster/set or no) is a major obstacle in application where the data is subject to large amount of uncertainty such as in AIEs.

It is essential that the salient feature extraction technique chosen as part of the AA addresses the high amount of uncertainty within the captured user/device interaction data (for example because of device wear & tear, subtle variations in user preferences etc...). While k-means clustering is limited in its ability of handling uncertainty because of the reasons mentioned above, we have chosen an enhanced version of k-means clustering which relies on fuzzy logic principles referred to as fuzzy c-means clustering [2].

Fuzzy c-means clustering allows each observation to be member of multiple sets or clusters and to be so at different degrees. This property is essential in dealing with uncertain information as in the context of AA. As part of ATRACO, we are focussing on employing fuzzy c-means clustering while aiming to combine it with zSlices based general type-2 fuzzy logic theory, in particular with a focus on multi-people occupancy of AIEs.

4.5.3 Artefact Adaptation Implementation

4.5.3.1 Instantiation of the AA for a specific task

In ATRACO, the AA is instantiated in the context of a given sphere and a given task. Specifically, AA is instantiated as part of the FTA by the Sphere Manager for a task which has been described by the planner, nevertheless, the AA focuses on individual device/user relations and as such on the individual devices within the task.

During the instantiation process, the Sphere Manager provides a list of devices to the AA which identifies the individual devices which are relevant in the user/environment interaction for this specific task.

Generally a distinction between three types of devices is made:

- Sensor inputs: sensors available within the environment, for example, light sensors, temperature sensors, humidity sensors, etc.
- User input devices: devices which allow the user to interact with the environment, for example switches, dimmers, buttons, voice or touch displays (e.g. via the Interaction Agent), etc.
- Output devices: all devices which actively affect the environment, for example heaters, lights, automated window blinds, etc.

```
<SensorInput>
  <IP>123.456.789.012</IP>
  <UUID>uuid:4711</UUID>
  <FriendlyName>test-phidget-light-sensor:lightservice</
      FriendlyName>
  <ServiceID>urn:schemas-upnp-org:seviceid:lightservice</
      ServiceID>
  <LinguisticVariable>AmbientLightLevel</LinguisticVariable>
  <LinguisticLabels>
    <LinguisticLabel>dark</LinguisticLabel>
    <LinguisticLabel>moderate</LinguisticLabel>
    <LinguisticLabel>bright</LinguisticLabel>
  </LinguisticLabels>
  <Service>setValue</Service>
  <ControlAction>GetLightLevel</ControlAction>
  <ControlArgument>ResultStatus</ControlArgument>
  <Range>
    <From>0</From>
    <To>1000</To>
  </Range>
</SensorInput>
```

Listing 4.1 Example of the XML description of a specific device as passed by the Sphere Manager to the AA (as part of the FTA). Here, a light level sensor is described.

As part of AA, only the Sensor Inputs and the Outputs are considered directly as they represent physical devices which govern the sensing (e.g. what is the current temperature) and the actuation (e.g. heating), while the User inputs are employed to gather the desired settings for the Outputs but are not modelled individually.

For example, in a light controlling task, the list of devices could include a series of light sensors, sun impact sensors, user controllable light switches/dimmers as well as a series of lights and lamps within the environment. All devices communicated from the Sphere manager to the AA are encoded using XML, an example of such a device description is given in Listing 4.1. After the AA has been instantiated as part of the FTA, the given task is resolved which is described in the following Subsection.

4.5.3.2 Task resolution within the AA & execution

Within the AA, as part of the FTA, a task, which can involve a large number of devices is resolved in the following fashion:

- The task is split into a series of subtasks, where each subtask focuses on a single device. (i.e. each sensor input and each output is modelled individually with respect to its input: a sensor is affected by what it is sensing (for example light level in terms of a light sensor), while an output is affected by its setting (expressed through the user input).
- In AA only a single thread of execution is employed which monitors and adapts all devices in a sequential fashion.
- The AA is executed (and continuously running) as part of the FTA.

4.5.3.3 Implementation of the User/artefact interaction capturing within AA.

After the AA has been executed, the user/artefact interactions (in respect to the artefacts/devices which are part of the current task) are monitored and capturing in the form of a data. An example of such a file is given in Figure 4.2.

```
test-phidget-light-sensor-0:0.032;test-phidget-light-sensor-1:0.03;Time:0.0:DMX-Light-6:0.0
test-phidget-light-sensor-0:0.032;test-phidget-light-sensor-1:0.03;Time:0.0:DMX-Light-6:0.04
test-phidget-light-sensor-0:0.043;test-phidget-light-sensor-1:0.042;Time:0.0:DMX-Light-6:0.43
test-phidget-light-sensor-0:0.063;test-phidget-light-sensor-1:0.062;Time:0.0:DMX-Light-6:0.57
test-phidget-light-sensor-0:0.223;test-phidget-light-sensor-1:0.211;Time:0.0:DMX-Light-6:0.97
test-phidget-light-sensor-0:0.475;test-phidget-light-sensor-1:0.224;Time:0.0:DMX-Light-6:01.0
```

Listing 4.2 Example of a small collection of records.

It is important to note that, a record of the states of the devices is captured every time the user interacts with the specific devices, for example, if the user changes the dimmer setting of a lamp, the new setting of the lamp is captured as well as the sensor values related to the current task (i.e. the light sensor values).

In AA, this ensures the connection between the user (interaction) and the specific devices. After a sufficient amount of user/device interactions have been captured (where "sufficient" is dependent on the complexity of the task, the devices, etc.), the accumulated data is processed in order to determine its salient features as described in the following Section.

4.5.3.4 Implementation of the salient feature extraction within AA.

The salient feature extraction has been implemented through the implementation of the fuzzy c-means algorithm [2].

In particular, after the user/device interactions for a specific device have been captured, the fuzzy c-means algorithm is employed in order to find a specified number of cluster centres within the data accumulated for each device. The number of cluster centres is determined by the number of linguistic labels which are to be employed to model the individual device. It is part of the research within ATRACO to determine the best relationship between the number of models and their interpretability in the sense that more individual labels allow for a more precise modelling of a device but also limit the interpretability of the system.

4.5.3.5 Implementation of the adaptive model creation within AA.

The creation of adaptive models within the AA is directly related to the centres of the clusters determined using the fuzzy c-means algorithm as detailed in the previous section (and as such it is directly related to the capturing of user-device interaction data).We are harnessing the advantages in terms of easy interpretability of fuzzy sets by associating each device with a linguistic variable and a series of linguistic labels defining this variable. Associating linguistic labels and thus human-readable concepts with devices and their properties allows for subsequent easy interpretability by the user and thus facilitates trust between the user and the system, a basic requirement for AIEs.

For example, a lamp could be associated with the linguistic variable "Ambient-LightLevel" and the linguistic labels "dark", "ambient" and "bright". Each of these labels in turn is associated with a fuzzy set which is based on the salient information extracted in the previous step which in turn is based on the captured user/device interaction information.

An example of such a linguistic variable (here: AmbientLightLevel which models "Light Sensor 2") is given in Figure 4.3.

Currently, the cluster centres determined in the previous steps are associated with the centres of the fuzzy sets while their endpoints are associated with the cluster centres of neighbouring clusters. In the near future we will aim to employ the novel general type-2 fuzzy system research based on zSlices developed for ATRACO [15], [16], [17], [18].

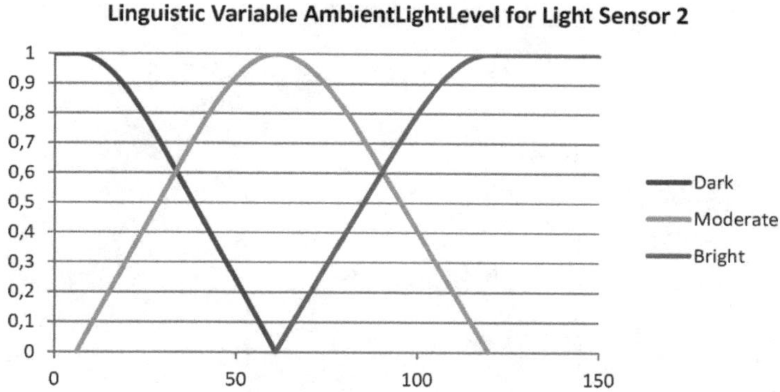

Fig. 4.3 An example of the linguistic variable and its labels associated with a certain device.

4.6 Experiments & Results

We have carried a series of real world experiments in the intelligent apartment (iSpace) at the University of Essex, United Kingdom.

We have described before that the characteristics of a device can vary over time (e.g. wear & tear, outside influences, etc.) and that the user interaction and preferences for the use of a given device is also subject to continuous variation (variation in user preference, "imprecision" of a human user, etc.). Finally, it is clear that different users will interact with individual devices in different ways as part of individual tasks (e.g. adjust lighting). In order to show these different aspects of AA, we have devised a series of experiments which aim to clearly expose, demonstrate and illustrate the AA functionalities in terms of the individual requirements. The aim of the experiments are:

- Show AA in respect to varying user/device interactions.
- Show AA as part of a task in respect to artefact context.
- Show AA as it occurs both in sensing and in actuating artefacts transparently, based solely on user/device interaction.

4.6.1 Experimental Setup

The experiment is set within the iSpace at the University of Essex, UK, specifically within the fully equipped kitchen area shown in Figure 4.4. The kitchen area has been chosen as it provides a realistic testing ground for AA (for example it is subject to a large number of influences, from cooking fumes to strongly varying temperatures, lighting requirements (chopping of food,...), etc.).

Figure 4.4 additionally depicts the individual physical devices which are being employed during the experiments and which are listed below:

Fig. 4.4 The iSpace kitchen area with an overview of devices employed during the experiments.

(a) A touch display allowing full (dimmable) control of the kitchen spot lights.
(b) A light sensor mounted on top of the extraction fan unit. (LightSensor_1)
(c) A light sensor mounted on top of the kitchen cupboard. (LightSensor_2)
(d) A dimmable spotlight illuminating mainly the chopping/food preparation area.

Additional hardware which has been employed consists of a variety of computers running both Windows and Mac operating systems. The AA agent which is run as part of the FTA has been executed on a dual-core laptop running Microsoft Windows Vista®.

The experiments were set in the kitchen area of the iSpace and each experiment (Experiment 1 and Experiment 2) was divided into two phases A and B as follows:

- Phase A
  In Phase A, the artefact (in this experiment the spotlight) is exposed to a series of changes across its whole output range (in this case the output was adjusted from 0-100% in continuous steps of 10%). While this is non-natural in terms of the utilization of a dimmable light by a human user, (as will be discussed further below in this section), it allows us to demonstrate the transparent adaptation of the sensing and actuating devices while also providing a reference point for comparison for Phase B.
- Phase B

In Phase B, the artefact is exposed to a series of changes based on actual changes as they are made by a human user. The difference of these changes to those in Phase A is highlighted and finally, the resulting adaptive models for the devices are shown and compared to those resulting from Phase A.

4.6.2 Instantiation and resolution within the real world environment of the AA experiments

The AA is instantiated as part of the FTA by passing it a set of devices related to the current task. As the number of user inputs/outputs is only one in the case of the experiments presented here. Nevertheless, as it is the adaptation to individual artefacts which we are aiming to demonstrate and illustrate as part of these experiments, showing the adaptation to three physical devices is considered more than sufficient (more devices simply add more complexity and bulk to the document without conveying further information).

The instantiation of the AA is graphically shown in Figure 4.5.

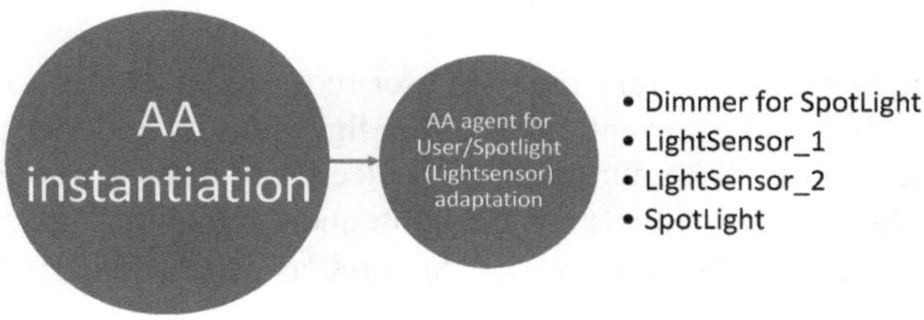

Fig. 4.5 AA instantiation and device association.

After the AA has been instantiated, it is ready for the capturing of the user/environment interactions as described for the individual experiments in the subsections below. The experiments focus on demonstrating and visualizing the following aspects of AA:

- Show AA in respect to varying user/device interactions.
- Show AA as part of a task in respect to artefact context.
- Show AA as it occurs both in sensing and in actuating artefacts transparently, based solely on user/device interaction.

In order to achieve this, AA was executed with the focus on a light adaptation task involving a spotlight, two light sensors and a user input (a touch-screen based dimmer for the spotlight) as shown in Listing 4.3.

4.6.3 Phase 1A

As part of Phase 1A, the user used the touch-screen based dimmer control to increase the output of the spotlight in 10% increments from 0 to 100%. While this is not representative of the standard use of a spotlight within a normally occupied AIE, it allows us to demonstrate the transparent Artefact Adaptation and set a baseline for comparison in the subsequent experiments, both in Phase 2A, in terms of how the different user/device interactions directly influence the resulting (adapted) device/artefact models which considers a drastic change in sensor characteristics.

```
LightSensor_1:3.0;LightSensor_2:4.0;SpotLight:0.0
LightSensor_1:3.0;LightSensor_2:4.0;SpotLight:10.0
LightSensor_1:3.0;LightSensor_2:4.0;SpotLight:20.0
LightSensor_1:3.0;LightSensor_2:4.0;SpotLight:30.0
LightSensor_1:3.0;LightSensor_2:4.0;SpotLight:40.0
LightSensor_1:3.0;LightSensor_2:4.0;SpotLight:50.0
LightSensor_1:13.0;LightSensor_2:14.0;SpotLight:60.0
LightSensor_1:34.0;LightSensor_2:41.0;SpotLight:70.0
LightSensor_1:57.0;LightSensor_2:77.0;SpotLight:80.0
LightSensor_1:75.0;LightSensor_2:111.0;SpotLight:90.0
LightSensor_1:96.0;LightSensor_2:131.0;SpotLight:100.0
```

Listing 4.3 Information captured as part of Phase 1A.

During the experiment, every time the user requested a change to the spotlight output, a snapshot of the current states of the light sensors and the spotlight were taken. This allows for the capturing of the device states directly during user/device interaction which in turn allows for the subsequent modelling in relation to this user/device interaction. The information captured as part of Phase 1A is shown in Listing 4.3.

Table 4.1 Salient Features (approx.) as extracted as part of Phase 1A

	Salient Feature 1	Salient Feature 2	Salient Feature 3
LightSensor_1	4.65	46.22	84.70
LightSensor_2	6.10	60.93	119.41
SpotLight	13.42	48.73	86.14

It is interesting to note when comparing the data in Listing 4.3 to the position of the actual sensors (shown in Figure 4.4) that – as expected, the lighting level recorded by LightSensor_2 exceeds that of LightSensor_1 as LightSensor_2 is more directly exposed to the light beam created by the spotlight. This difference in perceived brightness of different artefacts/devices (even though for example the sensors in this case are identical) is important to note as it illustrates the requirement for the creation of adaptive models which transparently manage the different perceived input/output and allow robust high-level reasoning. The salient feature selection implemented through the fuzzy c-means algorithm was employed to determine three salient features within the data for each physical device. The resulting cluster centres are presented in Table 4.1.

The fuzzy sets modelling the linguistic labels for the individual devices were created based on the salient features extracted in the previous step. As such, the resulting models for the individual devices are shown in Figure 4.6, Figure 4.7 and Figure 4.8 for the LightSensors and the SpotLight respectively.

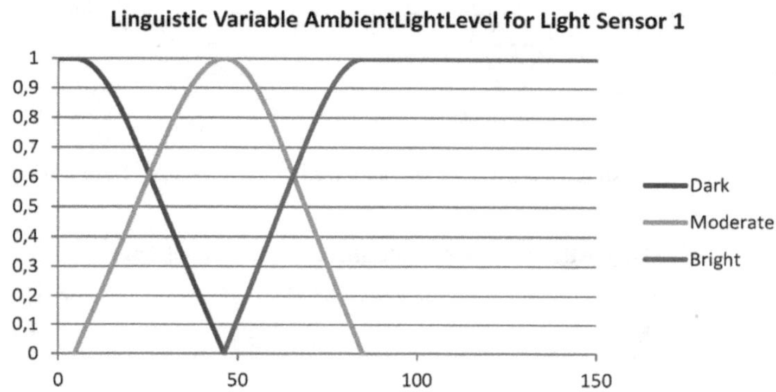

Fig. 4.6 Model for the AmbientLightLevel as perceived by LightSenor_1 employing 3 labels. (Phase 1A)

Considering Figures 4.6 - 4.8, it is clear how the requests in regular intervals have resulted in models which are symmetric over the device ranges. Further, the transparent adaptation to the different contexts (in this case: the different position within the space) of LightSensor_1 and LightSensor_2 can be seen when considering the created models for Dark, Moderate and Bright in both Figures 4.8 and 4.4. Having established the device models for Phase 1A, we now proceed to Phase 1B which illustrates how different user interaction with the devices impacts the resulting models.

In Phase 1B, we repeat the previous experiment executed in Phase 1A but the user requests and preferences in terms of the spot light outputs are changed. The main aim here is to show how different user/device interactions (and thus changes in the user preferences for the use of a given artefact) result in different models for the respective devices.

Additionally, it is worth noting that the set of user requests employed in Phase 1B is more reflective of the actual values requested by real users who generally change the dimming level of lights in their environment until the emitted light level has changed according to their expectations (e.g. if they want a dim environment they will turn the dimmer until the light is dim, and not for example 1/3rd of a turn). When considering Figure 4.7, it is clear that the actual light level in the environment does not change until the spot light output passes 50%-60% (below 50% the power is insufficient to result in appreciable levels of luminance). Thus, the sam-

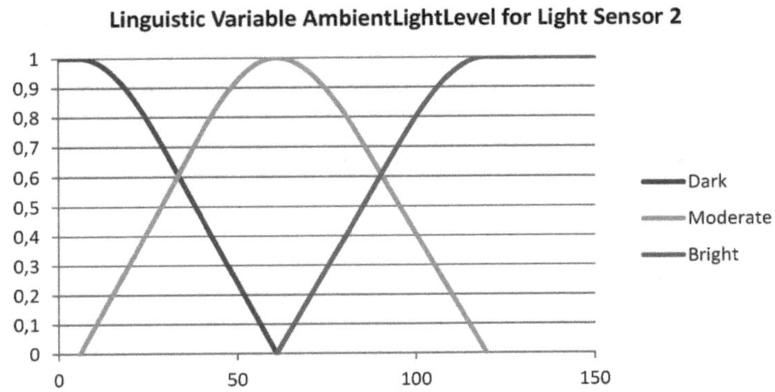

Fig. 4.7 Model for the AmbientLightLevel as perceived by LightSenor_2 employing 3 labels. (Phase 1A)

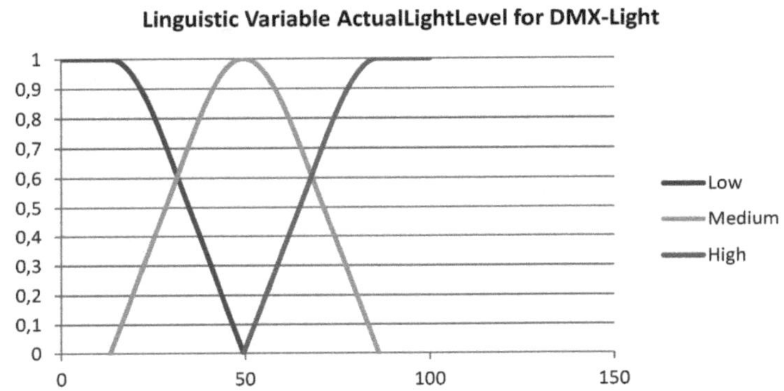

Fig. 4.8 Model for the ActualLightLevel as requested from the SpotLight (Phase 1A)

ples requested by the user as part of Phase 1B vary between 55% and 100% while additionally including 0% (the off state).

```
LightSensor_1:6.0;LightSensor_2:12.0;SpotLight:0.0
LightSensor_1:6.0;LightSensor_2:12.0;SpotLight:55.0
LightSensor_1:16.0;LightSensor_2:32.0;SpotLight:60.0
LightSensor_1:27.0;LightSensor_2:44.0;SpotLight:65.0
LightSensor_1:39.0;LightSensor_2:61.0;SpotLight:70.0
LightSensor_1:50.0;LightSensor_2:82.0;SpotLight:75.0
LightSensor_1:67.0;LightSensor_2:102.0;SpotLight:80.0
LightSensor_1:77.0;LightSensor_2:112.0;SpotLight:85.0
LightSensor_1:87.0;LightSensor_2:124.0;SpotLight:90.0
LightSensor_1:97.0;LightSensor_2:134.0;SpotLight:95.0
LightSensor_1:97.0;LightSensor_2:134.0;SpotLight:100.0
```

Listing 4.4 Information captured as part of Phase 1B.

During the experiment, every time the user requested a change to the spotlight output, a snapshot of the current states of the light sensors and the spotlight were taken. This allows for the capturing of the device states directly during user/device interaction which in turn allows for the subsequent modelling in relation to this user/device interaction. The information captured as part of Phase 1B is shown in Listing 4.4. From Listing 4.4, the requested spotlight levels can be seen to sample the range between 55 and 100% in steps of 5% while additionally including 0% (the off state). As in Listing 4.3, the different positions of the light sensors result in the different perceived light levels as shown.

Table 4.2 Salient Features (approx.) as extracted as part of Phase 1B.

	Salient Feature 1	Salient Feature 2	Salient Feature 3
LightSensor_1	12.59	47.26	88.11
LightSensor_2	23.10	72.05	122.82
SpotLight	0.68	65.00	89.84

The salient feature selection implemented through the fuzzy c-means algorithm was employed to determine three salient features within the data for each physical device. The resulting cluster centres are presented in Table 4.2. The fuzzy sets modelling the linguistic labels for the individual devices were created based on the salient features extracted in the previous step. As such, the resulting models for the individual devices are shown in Figures 4.9, 4.10 and 4.11 for the LightSensors and the SpotLight respectively. Considering Figures 4.9 - 4.11, it can be seen how the increased user demand for values of the spotlight in the range above 55% has resulted in adapted device models both for the light sensors as well as for the spotlight.

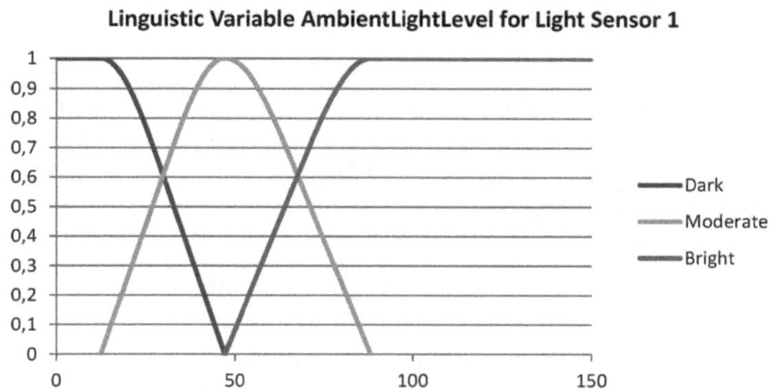

Fig. 4.9 Model for the AmbientLightLevel as perceived by LightSenor_1 employing 3 labels. (Phase 1B)

In order to review the results of the experiments, we review its aims as outlined and discuss the results in relation to each of the objectives below:

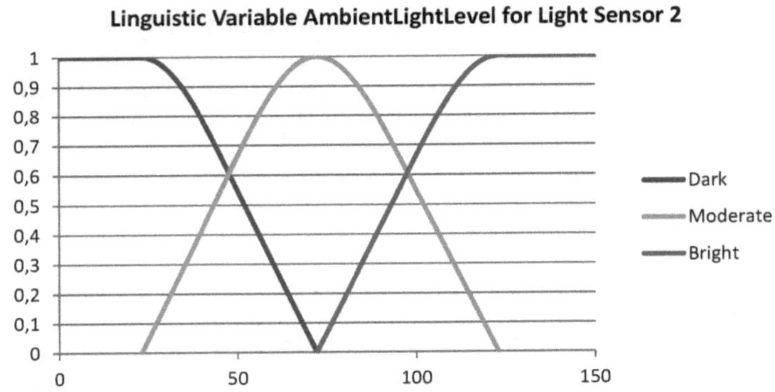

Fig. 4.10 Model for the AmbientLightLevel as perceived by LightSenor_2 employing 3 labels. (Phase 1B)

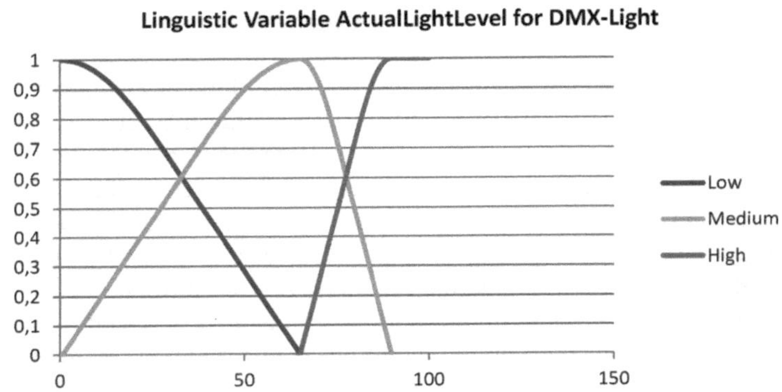

Fig. 4.11 Model for the ActualLightLevel as requested from the SpotLight. (Phase 1B)

- Experiment 1 has shown AA in respect to varying user/device interactions (as a result of the user changing his preferences for the artefact use).
  Considering the models created in Phase 1B (Figures 4.9 - 4.11), and comparing them to the models from Phase 1A (Figures 4.6 - 4.8) a series of changes can be seen, in particular:
  - The models for the light sensors have changed in Phase 1B to reflect the user focus on the higher end of the light level spectrum.
  - The model of the spot light has changed significantly, still encapsulating the off state but showing the increased focus on higher values (skewing to the right).

  The observed changes show how AA adapts the device models according to the user/device interaction (or according to user changing his preferences for the artefact use). As the number of labels remains constant for each device modelled and as the overall range of the devices remains constant, AA adapts the individual

labels to accommodate the values requested (or indirectly generated in terms of the sensors) by the user as good as possible.
- Experiment 1 has shown AA as part of a task in respect to artefact context. Considering the different artefact models for LightSensor_1 and LightSensor_2 (which are identical sensors) both in Phase A and Phase B clearly shows how AA adapts the models of the specific sensors according to their context (in this case the context is mainly affected by their relative position to the spotlight).
- Experiment 1 has shown AA as it occurs both in sensing and in actuating artefacts transparently, based solely on user/device interaction. The adaptive models shown as part of this experiment were created completely transparently, directly employing the captured user/device interaction information as shown in Listings 4.3 and 4.4.

4.7 Conclusions and Future Work

Artefact Adaptation is an integral part of adaptation and evolution within AIEs. As part of ATRACO, AA is addressed in the context of a distributed, ontology based intelligent framework with the eventual aim of supporting a user or multiple users in their home.

In this chapter, we have described the concepts of AA in general and its specific role within ATRACO. We have specified the requirements of AA within ATRACO which involves primary objectives such as the ability to adapt to changes in artefact/device characteristics, to variations in user/device interactions and to changes in the user preferences for the use of a given artefact. The AA also satisfies secondary objectives such as the need to provide interpretable device models and to deal with a variety of strongly heterogeneous devices.

The specified requirements have subsequently been taken forward to create a design for AA within ATRACO. We have established the reasoning for the application of fuzzy sets in order to create artefact/device models which are interpretable, adaptable, easily interpretable by human users.

The design of the individual stages of AA, i.e. the capturing of user/device interactions, the retrieval of salient features and the eventual creation of fuzzy set based device models has been presented and discussed before addressing the implementation of the individual stages. As part of the review of the individual stages of the implementation, we have addressed the importance of the transparent updating of the adaptable device models without affecting higher level reasoning mechanisms such as User Behaviour Adaptation and the ability of AA for this transparent adaptation has subsequently been shown as part of the experimental section.

As part of the experimental section we have provided details on three main experiments which were conducted in order to investigate different aspects of AA. Specifically, we have demonstrated and visualised how the individual requirements on AA are addressed as part of a series of real world (using real devices) experiments conducted in the kitchen area of the iSpace located at the University of Essex,

UK. We have shown in detail as part of a step-by-step approach how changes in the user/device interactions are reflected in the device models as well as how changes in individual devices characteristics are handled as part of AA.

Furthermore, we have provided details on a real world deployment of AA with lay users who interacted with the AA component in a realistic real-world context based around the main living area of the iSpace. As part of this experiment series, we have included the dynamically created device models for three of the participants, demonstrating the ability of AA to generate user-specific device models and to adapt those device models transparently as the characteristics of a device change.

AA provides an essential part of the AIE vision and its viability and potential have been documented as part of this delivery. While the task of AA is highly complex and can be addressed from a variety of directions with a multitude of approaches, involving people from backgrounds ranging from sociology to physics, the current approach to AA which has been chosen as part of ATRACO has been shown to be a viable solution.

While the current version of AA has been shown to be functional, a major challenge which remains is to generate device models which incorporate the uncertainty implied in user/device interactions over time. In other words, while user/device interactions may change, it is currently not possible to model if this change is temporary (for example a mistake by the user) or if it is a definite change in user preference or device characteristics. Further, we are aiming to investigate adaptation to devices over different timescales by different users.

We also aim to investigate the application of zSlices based general type-2 fuzzy sets [18] which are expected to provide advantages in terms of dealing with the uncertainty that the devices within an AIE are faced with while also providing a platform for the modelling of different user/device interaction profiles of different (and multiple!) users.

References

1. Abowd, G., Elizabeth, D., Mynatt, D., Rodden, T.: The human experience. IEEE Pervasive Computing (2002)
2. Bezdek, J.: Pattern recognition with fuzzy objective function algorithms. Kluwer Academic Publishers Norwell, MA, USA (1981)
3. Coupland, S., John, R.: Geometric type-1 and type-2 fuzzy logic systems. IEEE Transactions on Fuzzy Systems **15**(1), 3–15 (2007)
4. Davidsson, P., Boman, M.: Distributed monitoring and control of office buildings by embedded agents. Information Sciences **171**(4), 293–307 (2005)
5. Doctor, F., Hagras, H., Callaghan, V.: An intelligent fuzzy agent approach for realising ambient intelligence in intelligent inhabited environments. IEEE Transactions on Systems, Man and Cybernetics, Part A: Systems and Humans **35**(1), 55–65 (2004)
6. Hagras, H.: Type-2 FLCs: A new generation of fuzzy controllers. IEEE Computational Intelligence Magazine **2**(1), 30–43 (2007)
7. Hagras, H., Callaghan, V., Colley, M., Clarke, G., Duman, H.: A fuzzy logic based embedded agent approach to ambient intelligence in ubiquitous computing environments. IEEE Intelligent Systems Journal (2004)

8. Hagras, H., Doctor, F., Callaghan, V., Lopez, A.: An incremental adaptive life long learning approach for type-2 fuzzy embedded agents in ambient intelligent environments. IEEE Transactions on Fuzzy Systems **15**(1), 41–55 (2007)
9. Kidd, C., Orr, R., Abowd, G., Atkeson, C., Essa, I., MacIntyre, B., Mynatt, E., Starner, T., Newstetter, W.: The aware home: A living laboratory for ubiquitous computing research. In: The 2nd International Workshop on Cooperative Buildings (CoBuild '99) (1999)
10. Kushwaha, N., Kim, M., Kim, D., Cho, W.: An intelligent agent for ubiquitous computing environments: smart home UT-AGENT. In: The Second IEEE Workshop on Software Technologies for Future Embedded and Ubiquitous Systems (WSTFEUS'04). Published by the IEEE Computer Society (2004)
11. Lloyd, S.: Least squares quantization in PCM. IEEE Transactions on Information Theory **28**(2), 129–137 (2002)
12. Mendel, J.: Uncertain rule-based fuzzy logic systems: introduction and new directions. Prentice Hall (2001)
13. Minar, N., Gray, M., Roup, O., Krikorian, R., Maes, P.: Hive: Distributed agents for networking things. In: ASA/MA, p. 118. Published by the IEEE Computer Society (1999)
14. Rutishauser, U., Joller, J., Douglas, R.: Control and learning of ambience by an intelligent building. IEEE Transactions on Systems, Man & Cybernetics – Part A: Systems and Humans **35**(1), 121–132 (2004)
15. Wagner, C., Hagras, H.: A genetic algorithm based architecture for evolving type-2 fuzzy logic controllers for real world autonomous mobile robots. In: The IEEE International Conference of Fuzzy Systems, pp. 193–198 (2007)
16. Wagner, C., Hagras, H.: zSlices – Towards bridging the gap between interval and general type-2 fuzzy logic. In: IEEE International Conference on Fuzzy Systems (FUZZ-IEEE 2008) (IEEE World Congress on Computational Intelligence), pp. 489–497. IEEE (2008)
17. Wagner, C., Hagras, H.: Novel methods for the design of general type-2 fuzzy sets based on device characteristics and linguistic labels surveys. In: The 2009 International Fuzzy Systems Association World Congress and the 2009 European Society for Fuzzy Logic and Technology Conference), pp. 537–543 (2009)
18. Wagner, C., Hagras, H.: Towards general Type-2 Fuzzy Logic Systems based on zSlices. Transactions on Fuzzy Systems (2010)
19. Wang, L., Mendel, J.: Generating fuzzy rules by learning from examples. IEEE Transactions on Systems, Man, and Cybernetics **22**(1), 1414–1427 (1992)
20. Wejchert, J.: The disappearing computer. Information Document, 1st Call for proposals, European Commission, Future and Emerging Technologies (2000). URL `http://www.disappearing-computer.net/mission.html`
21. Youngblood, G., Cook, D., Holder, L.: Managing adaptive versatile environments. Pervasive and Mobile Computing **1**(4), 373–403 (2005)

Chapter 5
User Interaction Adaptation within Ambient Environments

G. Pruvost and T. Heinroth and Y. Bellik and W. Minker

Abstract Ambient environments introduce new user interaction issues. The interaction environment which was static and closed becomes open, heterogeneous and dynamic. The variety of users, devices and physical environments leads to a more complex interaction context. As a consequence, the interface has to adapt itself to preserve its utility and usability. It is no longer reasonable to continue to propose static and rigid interfaces while users, systems and environments are more and more diversified. To the dynamic nature of the interaction context introduced by ambient environments, the user interface must also respond by a dynamic adaptation. Thanks to the interaction richness it can offer, multimodality represents an interesting solution to this adaptation problem. The objective is to exploit all the interaction capabilities available to the system at a given moment, to instantiate and evolve user interfaces. In this chapter, we start by presenting a survey of the state of the art on user interaction adaptation. After, discussing the limitations of the existing approaches, we present our proposals to achieve user interaction adaptation within ambient environments. Then we describe the derived software architecture and the user evaluation it led to. We conclude by some directions for future work.

Gaëtan Pruvost
National Center for Scientific Research (LIMSI-CNRS) BP 133, 91403, Orsay cedex, France, e-mail: `gaetan.pruvost@limsi.fr`

Tobias Heinroth
Institute of Information Technology, Ulm University, Ulm, 89081 Germany, e-mail: `tobias.heinroth@uni-ulm.de`

Yacine Bellik
National Center for Scientific Research (LIMSI-CNRS) BP 133, 91403, Orsay cedex, France, e-mail: `yacine.bellik@limsi.fr`

Wolfgang Minker
Institute of Information Technology, Ulm University, Ulm, 89081 Germany, e-mail: `wolfgang.minker@uni-ulm.de`

5.1 Introduction

During the last years, the use of computers has largely popularized. From kids to seniors, from novices to experts, the ubiquity of computers has impacted a constantly growing variety of users. At the same time, advances in the miniaturization of electronic components have allowed the development of a large variety of portable devices (laptops, mobile phones, personnel digital assistants (PDA), etc.). The devices are various but most of the tasks users want to realize are not device specific. Thus, the *same* task can often be realized through *different* devices, depending on the situation. For instance, it is now common to see people read their e-mails in the bus with a smart phone whereas they would rather use the family computer when they are at home. Nowadays user interfaces (UI) are designed for the specific device they will run on. The same task can be executed on different devices, but manipulating user interfaces that don't have any common point. The users' mobility raises many new interaction situations and for each one, users must go through a new learning phase. Not being able to transfer the skills you have with computers to other interactive devices can be highly frustrating.

Furthermore, while the number of interactive devices is increasing, the capabilities and the richness of interaction should increase too. However, because each device lives in its own closed world, it is not trivial to combine the interaction capabilities of one another into a more synergetic user interface. Users are currently interacting with one independent device at a time. That behavior is the opposite of the notion of ambient ecology described in Section 1.3. When it comes to interaction, the ecology should provide rich interaction with users by combining the capabilities of the different available interaction devices. In this chapter, we will discuss the adaptation of interactive systems in such environments.

In order to clearly explain what adaptation is and what its challenges are, we first need to define a few terms relative to interaction. Taxonomies of interaction generally involve three main concepts: **mode**, **modality** and **media**. The meaning of those terms might slightly change between authors [8] [28] [47] [11]. In our case we adopt user-oriented definitions [6] [69]. A mode refers to the human sensory system used to produce or perceive given information (visual, gestural, auditory, oral, tactile, etc.). A modality is defined by the information structure that is perceived by the user (text, ring, vibration, etc.) and not the structure used by the system. Finally, a medium is a physical device which supports the expression of a modality (screen, loudspeakers, etc.). These three interaction means are dependent. A set of modalities may be associated with a given mode and a set of media may be associated with a given modality.

Before situating our approach compared to similar work, we will define the concept of interaction adaptation – Section 5.1.1 – and the specific challenges it introduces when it comes to Intelligent Environments (IEs) – Section 5.1.2.

5.1.1 Adaptation of user interaction

The interaction adaptation [29] may have several meanings depending on the adopted point of view (user centred [13] [20], target oriented adaptation [66] [14] [37], software architecture adaptation [71], etc.). Among the different points of view, it emerges that interaction adaptation involves the following main concepts:

- Actor: represents the entity responsible for the adaptation task. It can be the user, the designer, the system, etc.
- Components: represent the software entities that will be modified to achieve adaptation. It can be the help system, the kernel core, the task model, the dialog controller, the physical or logical interaction objects [5], etc.
- Time: represents the moment when the adaptation is performed. The adaptation can be *static* (performed during the design phase), *dynamic* (performed at runtime), and sometimes performed between sessions. Certain authors refer to static adaptation by the word *adaptability* while the dynamic adaptation is referred to by *adaptivity* [65] [26].
- Direction: represents the adaptation orientation. The system may adapt its outputs and/or adapt itself to inputs (artefacts adaptation).
- Target: represents the entity we want to adapt to. This is usually denoted by the *interaction context*.

Several definitions exist to describe the notion of *interaction context*. Within the HCI research community the most used definition is the triplet <User, System, Environment> [15] [60] [72]:

- User: the user is described by a profile which informs about their preferences, cultural characteristics, cognitive and sensory-motor capacities, etc. Those can be static (for instance a handicap) and/or dynamic (for instance, the user is not looking at the screen).
- System: the system represents the physical (devices) and logical (software) resources.
- Environment: it represents the physical environment where the interaction is done (luminosity, noise level, etc.)

Some authors include further information such as the current user activity [68] (which can be attached to the user profile), but what must be described in the interaction context depends on what you expect the system to adapt to. Next section explains what is specific in the interaction context of ambient systems.

5.1.2 Ambient specific challenges

Ambient systems are, by definition, integrated in a physical environment. Not only will they be physically located in those environments – like usual personal computers – but they will offer users the opportunity to interact with this environment. The classical desktop interaction takes users into the virtual world, proposing them to interact only with virtual representations - for instance, files. On the contrary, the main application of ambient system is to enhance the real world with digital properties, to bridge the gap between the real world and the virtual world by proposing enhanced interaction with physical objects.

As a consequence, the main differences between classical interaction and ambient interaction are:

- Heterogeneity & Distributivity: The system is not composed of a static set of devices (Computing unit, screen, mouse, keyboard). It is the collaborative reunion of several screens, mice, keyboards, microphones and other devices, some of them offering interactive capabilities.
- Dynamic Media Mobility: Various media can enter and leave the ambient space during interaction (as users will move around taking some devices with them)
- User Mobility: User can move inside the ambient space. We can no longer assume that they are always behind the screen, holding the mouse and the keyboard.

The system needs to combine several media that are not known in advance and which can provide interaction through different modes and modalities. As a consequence, it must be able to provide interfaces that combine those different modalities of interaction. This property is called multimodality – see [35] for a survey of the domain. In general, when developing multimodal systems, the designer knows in advance which modalities it offers, and what is the structure of information that the media can send/receive. In ambient systems however, no assumption can be made about the modalities and media that will be available. In fact, the system must adapt to what is available at runtime and may even have to change during execution if one of the media disappears from the ecology. Adapting to the mobility of devices requires to be able to discover the different media, to model their respective interaction capabilities, and to reason on those capabilities in order to combine them. Besides dynamically appearing/disappearing from the ecology, media are distributed across a network. In classical situations, all the interaction devices are connected to the same computer whereas in ambient systems, it is needed to have protocols that enable the dynamic discovery of those media and the routing of their events towards an entity that can interpret them. Interaction adaptation thus involves to dynamically re-route the flow of data between different media. Last but not least, when users move around in an ambient space, they can expect to seamlessly resume tasks they have started somewhere else. Adapting the interaction to the user mobility requires the possibility to migrate a user interface from one media to another while preserving the current state of the task. Of course, all the features of a task may not be well represented on every device. It is then the role of the system to provide the best compromise between the interaction capabilities of a media and the features that it can provide for a specific task.

In order to explain how such challenges can be addressed, we will first present a state of the art of this research field. We will discuss the different approaches and their respective limits. It will lead us, in Section 5.3, to introduce our vision of interaction adaptation and to present our focus within the ATRACO project. In Sections 5.4 and 5.5, we will demonstrate how we can achieve adaptation on three different levels: Allocation, Instantiation and Evolution. We will first focus on a generic multimodal adaptation system that reasons on the interaction context to allocate and instantiate appropriate user interfaces. Then, we show how the developed concepts can be applied to provide adaptation within a single modality. We will focus on the spoken modality, which offers rich interaction situations in ambient environments, and demonstrate how adaptation can be achieved during on-going dialogue – see Section 5.5.2. Based on this theoretical work, we show in Sections 5.6.1 and 5.6.2 how a software agent responsible of interaction adaptation – called Interaction Agent – has been integrated and evaluated in the ATRACO project. We then conclude by giving some perspectives opened by our work.

5.2 Related Work

In this section, we establish an overview of the state of the art about the adaptation of the interaction to the context in ambient systems. This study of the state of the art gives us a global view of the different dimensions to which an interactive system can adapt. Based on this study, we define some limits of the current systems that we address in ATRACO.

5.2.1 Multimodality, information presentation and model driven approaches

There are many interpretations of the term adaptation. In [38], Kolski and Le Strugeon present a typology of different categories of interaction adaptation. They demonstrate that interaction adaptation ranges from simple adapation of presentation parameters to what they call intelligent agents. Intelligent agents have deep models of users, tasks and planification capabilities. The focus of this study is on how to mapped those adaptation mechanisms to existing interaction models. It results from this study that one research line for smart adaptation of interaction is in using multimodality and in breaking the "couple" composed of one human and one computer; replacing it by a group that includes several computer agents and users.

The use of multimodality for adaptation of interaction has been first explored through the issue of multimodal presentation [11] and the modeling of interaction context [66] [64]. Multimodal presentation focuses on selecting the best communication channels – i.e. the best modalities – to present information, given a certain context. User interface adaptation has been explored notably with the concept of plasticity [70] that tackles the issue of modifying the graphical user interface depending on the inputs, the outputs and the running platform. In ambient systems, two main issues complicate the adaptation of interaction. First, peripherals are heterogeneous and distributed over a network and second, the context of interaction, as defined in [19] is very dynamic and may even change during the interaction. In [17], the concept of Meta-UI is defined in order to represent the set of functions that are necessary and sufficient to control and evaluate the state of an interactive ambient space. A taxonomy is detailed – see Figure 5.1 – that shows 8 axes parted in 3 categories: Quality, Functional coverage and Interaction techniques. The Quality and Functional coverage are not specific to interactive systems, they are also reflected in other adaptation components of the ATRACO system. The category "Interaction techniques" provides a first classification of ambient interaction techniques.

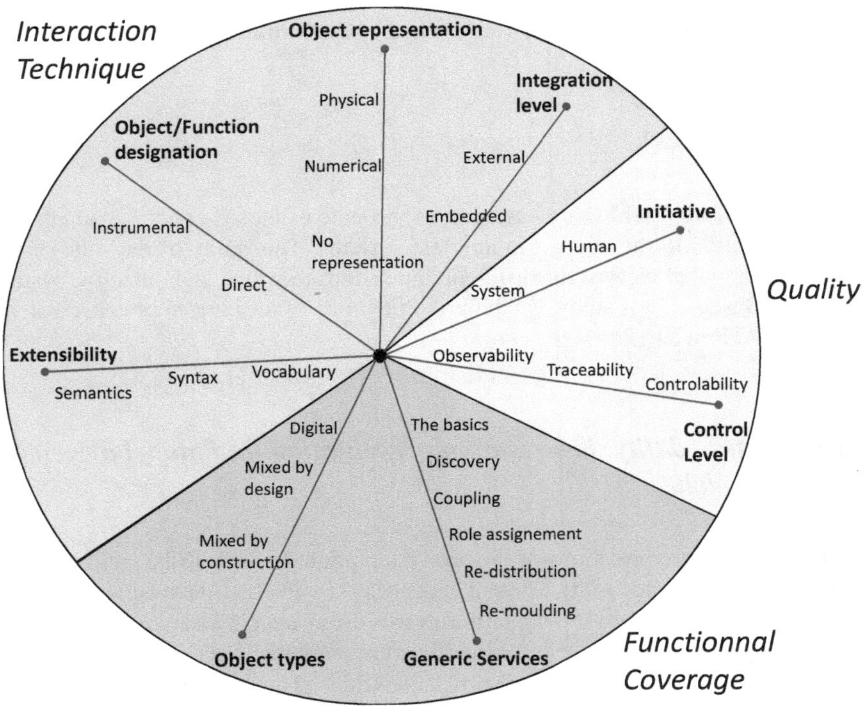

Fig. 5.1 Interaction adaptation taxonomy (courtesy of [17])

Whereas previous works focused on output adaptation, plasticity inspired a more general framework called CAMELEON-RT [4]. This framework explores the different dimensions of interaction adaptation and describes a reference conceptual architecture to implement it both for inputs and outputs. From a technical point of view, this framework highlights the need for an architecture that deals with a multitude of interaction devices. From a conceptual point of view, CAMELEON-RT stresses the importance of using various modalities for improving communication with the user both on efficiency and intuitiveness levels. It provides concepts to further analyse the combination of modalities within a task-oriented approach [45].

As far as information presentation is concerned, a four-level model has been developed named WWHT [59]. This model defines four levels of interaction adaptation: What, Which, How, Then. It presents a rule based system that allows selecting the best communication channels depending on a context model. Using the CARE properties – complementarity, assignation, redundancy and equivalence[18], richness of multimodality is used to improve robustness and expressivity of the system interaction. More details are given on this approach in Section 5.3.1.

The implementation of distributed interactive systems is still an open question though several architectures [57] [67] [3] have been developed. More recently, the OpenInterface European project [62] has developed an architecture that overcomes some technical problems such as network communication and multiple input devices configurations. Next section presents the conceptual and technical issues that remains when it comes to move from a user interface localised on a specific platform to a user interface dispersed in an ambient environment.

5.2.2 Limitations of those approaches

Limits of adaptive interaction systems are strongly connected to the way context is modelled. The interaction context is usually defined as any information relative to a person, a place or an object considered as relevant for the interaction between the user and the system [19]. Research work in this field have analysed key-problems of those dynamic contexts [33] and proposed several solutions based on distributed systems [3] or on uncertainty models [49]. However, no general agreement has been reached on what should be modelled and how to use this knowledge. Those concerns are addressed in the WWHT framework [59]. Yet, another problem with this framework – and the others – is the adaptability of a specific model to a completely new context. Indeed, once the designer has overcome the first issue of choosing which information to represent, he/she needs to connect this information to an information source. This source can be a sensor, another program, the user itself... No standard exists for those information sources and consequently, a model that has been specified for a specific environment might not be usable in another environment. This is the case, for instance, in the WWHT framework. In this framework, the context is described as a set of variables – i.e a name associated to a native value: int, boolean... Reusing a specific model in a new environment would imply binding the new discovered sensors – and other information sources – to the variables of the model. Because no semantic description is associated to those variables, this process cannot be automated. Besides modelling context, the wide variety of channels of communication offered by ambient systems leads to a decidability issue for the interactive system. The interactive system must take a decision as to which modality and which language of interaction it should use to communicate with the user. Contrary to other decision-making systems, there is no such thing as an optimal solution with suboptimal versions of it. As explained by Rousseau in [59], a basic expert system cannot just apply rules to infer the best solutions because the "best" solution will be different from one designer to another. That is why Rousseau introduces an algorithm based on rules to evaluate solutions according to a given behavioural model. This means that behavioural rules are provided by the designer and by applying those rules, the different solutions are granted a mark that serves to rank them. Consequently, our aim is not to provide a tool to select the best modality. It is rather to provide a tool enabling designers – and possibly the user – to define in a generic fashion, what is the best way of adapting interaction. Of course, such a tool would have to apply those specifications at runtime and be able to instantiate the solution that it generated.

It leads us to our last problem: the automatic generation of a user interfaces. In order to automate the process of designing the interaction, it is first needed to formally describe interaction tasks. Some tools have been proposed for that, one of the most well-known is the ConcurTaskTree formalism [51]. Figure 5.2 shows an example of a task described in this formalism. It allows for a multi-level description of the task, its subtasks and the temporal relationships between them. Based on this description, the designer can have a platform independent view of the user interface he/she needs to implement. It helps him/her designing consistently for a same task across several platforms [45].

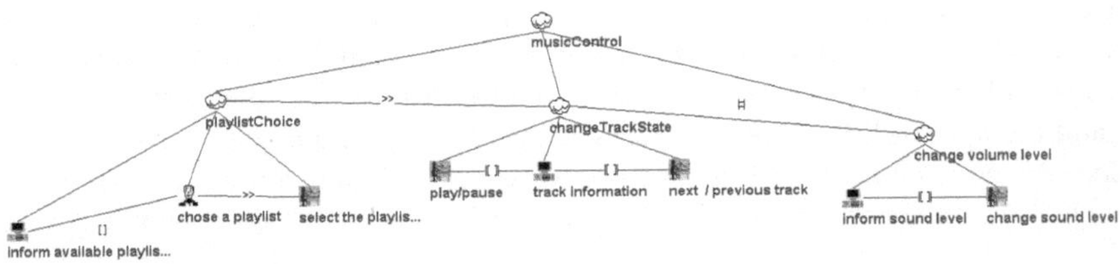

Fig. 5.2 An example of the CTT notation

In the TERESA framework [45], the CTT notation is refined in an Abstract User Interface (AUI) that is composed of interactors which are abstract interaction objects identified by their semantics. This AUI is then refined in a Concrete User Interface (CUI) composed of concrete interaction objects that depends on the target platform. Those concrete interaction objects have attributes such as the size for instance. Finally, the attributes are selected and the CUI is translated into something usable – e.g. a XHTML description, Java code ... – called the Final User Interface (FUI). Similar approaches have designed with other formalisms like UsiXML [43] [63]. Normally, the designer should be necessary for at least refining the CTT into an AUI since this involves understanding the task's semantics. However, tools [44] have been defined in order to partially automate this generative procedure. This approach allows designing a user interface once and then seamlessly instantiate it on devices with different interaction capabilities such as a touch screen or a PDA. Those refinements are done once and for all at the instantiation of the task. Consequently, this approach does not allow adaptation during the task execution. If a modality suddenly becomes unavailable – for instance, if the user receives a phone call, we want to avoid using speech to interact with him/her – then the system needs to switch to another modality during the task, which is not possible with such an approach. Another counterpart for fully top-down approaches is that the user interface that is generated can only be very simple, it is limited to a form-based interaction for graphical user interfaces and to *multiple choices questions* for spoken dialogue.

Even though great progress has been done in the field of model-based HCI, there are still unexplored research directions. The modelling of interaction context, though being at the heart of this field, is still problematic. The dynamicity of the context model has been largely explored but not its reusability on different environments. A model defines a structure that enables the system to choose an appropriate communication channel. But the model instantiation might be foiled by the heterogeneity of the target platforms and by the wide variety of modalities and corresponding languages that are available. Consequently, the top-down generation of a user interface is an interesting approach, but it does not fill all of our needs, in particular the need for rich interaction techniques and for runtime adaptation.

5.3 The ATRACO approach to user interaction adaptation

To resolve the problem of interaction adaptation, different kinds of approaches exist: translation approaches [24], retro-engineering and migrating approaches [16] [9] [50], Mark-up language based interfaces and Model Driven Interfaces (MDI) [63] [23]. As seen in the previous section, most of these approaches try to design several interfaces for a same application, each one being adapted to a different interaction context. This kind of approaches can be used when the different possible interaction contexts are not very numerous and well identified, which is rarely the case in ambient systems. Another possibility consists in integrating adaptation mechanisms that allow the interface to dynamically modify its behaviour in order to stay pertinent with the interaction context [70] [59]. This approach is better suited to our needs as the interaction contexts in the case of ambient ecologies may be numerous and very variable. In Section 5.3.1, we present in more details the WWHT approach. Though this information presentation doesn't solve all our issues, it is a good starting point that guided the design of our interactive system. In particular, we apply it in Section 5.3.2 to specify the different levels of interaction adaptation that are relevant to ambient systems. Finally, in the last section, we detail the global vision of our interaction system and what we have focused on for implementation.

5.3.1 Breaking down the specifications of interaction adaptation

The WWHT approach models the problem of information presentation. It also provides guidelines to establish the needs and to specify the interaction system that supports information presentation. A complete presentation of this model is given in [30] and [59].

The WWHT model relies on four main questions:

- What: what is the information to present?
- Which: which modality(ies) should we use to present this information?
- How: how to present the information using these modalities?
- Then: and then, how to continuously adapt the resulting presentation?

The first three questions (What, Which and How) refer to the initial building of a multimodal presentation while the last one (Then) refers to its future.

What

The starting point of the WWHT model is the semantic information that the system has to present to the user. To reduce the complexity of the problem, WWHT starts by decomposing the initial information into elementary information units. For instance, in the case of a phone call, the information "Call from X" may be decomposed into two elementary information units: the call event and the caller identity.

Which

When the decomposition is done, a presentation has to be allocated to each information unit. The allocation process consists in selecting, for each of them, a multimodal presentation adapted to the current state of interaction context. The resulting presentation is composed by a set of pairs (modality, medium) linked by redundancy/complementarity CARE relations. This process may be complex in particular when several communication modalities are available and/or the interaction context is highly variable. The selection process of presentation means is based on the use of a behavioural model. The representation of this behavioural model may vary depending on the considered system: rules [64], matrices [22], automata [36], Petri Nets [46], etc.

How

When the allocation is done, the resulting multimodal presentation has to be instantiated. Instantiation consists in determining the concrete lexico-syntactical content of the selected modalities and their morphological attributes depending on interaction context. First, a concrete content to be expressed through the presentation modality has to be chosen. Then, presentation attributes (modality-specific attributes like size spatial and temporal parameters) are set. This phase of the WWHT model deals with the complex problem of multimodal generation [1], [55]. Ideally, the content generation should be done automatically. However this is still an open problem for each considered modality such as text generation [56] or gesture generation [12].

Then

This phase addresses the problem of presentation expiration. Indeed, the presentation may be adapted when it is built but there is a risk that it becomes inadequate if the interaction context evolves. This constraint requires the use of mechanisms that allow the presentation to evolve according to the context.

5.3.2 Our vision of adaptation

The WWHT framework addresses the problem of information presentation but not of interaction. Indeed, it focuses on outputs whereas interaction requires both inputs and outputs. Moreover, strong links exist between inputs and outputs [7] which can influence one another. Thus, WWHT cannot be applied as it is to ambient systems. We propose to apply its philosophy and extend it to design an adaptive interaction system for ambient spaces: the Interaction Agent (IA). For this purpose, we assume that the system needs to interact with the user in two main cases:

- *Control tasks:* Those tasks imply the use of user interfaces to provide the user with persistent control of the environment.
- *Dialogue tasks:* Those tasks are used by the system to provide information to users (i.e. information presentation), collect information from them (what are their preferences, what behaviour the system should have) or collaborate/negotiate with them (helping users or finding a compromise between what they want and what the system can do).

We propose a knowledge-based system – the model will be described in Section 5.4 – that addresses three levels of adaptation: *allocation* is the problem of selecting the modalities and devices through which the user interface will be expressed; *instantiation* is the issue of selecting the appropriate parameters for an allocated user interface and *evolution* addresses the evolution of the user interface during interaction. Evolution of a user interface is a very broad problem in itself. It can cover migration – re-allocation – in the case of a context change but also dynamically evolving the dialogue content – refinement. This is typically the case for negotiation or collaboration with the user via natural spoken dialogue.

The IA addresses these three levels of adaptation and answers the four WWHT questions. Both interaction tasks – control and dialogue – are triggered by the Sphere Manager (SM) (see Section 1.3.3) thereby partly answering the first WWHT question: *What?* In ATRACO, the *what* is not restricted to decomposable units of information, we extend it to complex interaction tasks. Like in WWHT, if adaptability is to be considered, the *what* needs to be described in a formalism that affords decomposing it and treating it at different levels of granularity. Our vision is that previously presented formalisms (like CTT) can be used to describe the interaction tasks at different levels – abstract task, sub task... Based on the system state, the SM can trigger such describing trees in the Interaction Agent. The IA should then analyse such a task description to generate a user interface by implementing tasks of different levels in the tree. For instance, the same task can be implemented as a complete graphical user interface on a big screen – i.e. root of the tree – whereas it would be decomposed into a succession of sub-tasks for the small screen of a PDA i.e. the leaves of the tree. The *what* question is thus answered by the SM at the task level – in the activity sphere workflow presented in Section 1.4.

The *Which* and *How* stages are enforced by the IA, in an internal module called the Multimodal Manager (MM). It is responsible for deciding which modes and modalities should be instantiated by the use of which media and how – i.e allocation and instantiation. The MM provides concepts and a rule engine to reason about the possible answers to those questions. It offers a framework for designers to design different behaviours by writing them as a set of rules in a simple language – described in Section 5.4.2.

The MM furthermore addresses these issues in a continuous way during an ongoing task. In this manner it supports evolution, the *then* stage.

5.3.3 Our Focus

The range of research questions that we tackled in previous sections is quite large. They cannot be addressed all at the same time and require long-term efforts from the whole community in terms of plasticity of User Interfaces (UI), interaction context modelling, user modelling, automatic generation of UIs, multimodality... We showed in Section 5.2 that many efforts have already been done in the fields of automatic generation and plasticity. As a consequence, we assume that user interface generation and content generation techniques could be brought into our model later on. We thus chose to restrict ourselves to the problems of interaction context modelling, allocation/instantiation and evolution.

Because we don't know in advance which interaction capabilities the media will provide, user interfaces need to be generated and/or composed at runtime. As we discussed in Section 5.2.2, in order to provide rich user interfaces, we need to free ourselves from the fully top-down approach. Because a fully bottom-up approach would lead to inconsistencies in the flow of interaction tasks [61], we opted for a mixed approach in which a set of already existing interaction objects are combined to generate a user interface. We call those interaction objects Off-the-shelf Interaction Objects (OIO) because they are pre-implemented bundles of code that are reused and composed at runtime. Each of these OIOs can understand a specific interaction language and be manipulated by different interaction techniques. The crucial role of OIOs is further detailed in next sections. We discussed, in previous section, the role of automatic generation and composition of these elementary elements in the generation of a user interface. In our implementation though, we decided not to focus on this problem which has already been investigated – see Section 5.2.2. We simplify the problem by assuming there will always be at least one OIO available for each interaction task and focus our work on selecting and instantiating the OIO in a variable context of available devices[1].

In ambient systems, the heterogeneity of available devices offers rich interaction possibilities. To get the most out of this richness, interaction will be multimodal [8] and rely on various peripherals. The problem of the IA is thus to instantiate each of the interaction tasks through some modalities and choosing which devices to use for that. We do not address the issue of multimodality fusion and fission [40] for the moment and focus on the selection process of the right pairs <OIOs, devices>

[1] The model we propose in following sections was designed to take into account the global vision and will be expanded to include such plasticity in future versions

depending on the context. This approach is inspired by the refining process of Teresa [45]. However, the difference resides in the underlying model. In Teresa, the task model would be refined in abstract interaction objects that can be implemented on any platform. Those abstract interaction objects are elementary interaction objects. Since we wanted a mixed approach for the user interface generation, we designed OIOs to be of arbitrary complexity. An OIO can be a simple button, or it can be a whole widget for music playlist edition. This way, designers have a total freedom on the internal behaviour of the OIO they design, which lets them implement new and/or rich interaction techniques. On the other hand, low-level OIOs could still be composed at runtime to propose a "degraded" – less integrated – user interface.

Tasks are described as multi-level set of subtasks. An OIO can relate to any node of such a hierarchy whereas in Teresa, abstract interaction objects relate to leaves only. This recursive way of considering instantiation allows for instantiation with different granularity depending of the granularity of the task description. Indeed, one of the drawbacks of automatic generation – or computer assisted generation – of user interfaces is that the outcome is often monotonous and offers poor interaction capabilities whereas hand-crafted user interfaces may offer richer and more appealing interactions. Our approach allows for both hand-crafted user interfaces and automatically generated interfaces to coexist.

By taking this approach, we aim at including new hand-crafted interaction techniques right away. We did not implement automatic composition of OIOs since it was out of our focus. However we designed this approach to create a system open and flexible enough to include automatic generation in later releases. Next section details the model that we developed to represent and reason about interaction context.

5.4 Adapting the instance to the context

To address the different levels of adaptation that we exposed, we decided to model the context as an ontology called the Interaction Ontology (IO). The content of this ontology is detailed in Section 5.4.1. It enables describing all the different elements of a user interface – i.e from the user to the media – along with other information about the context. This ontology provides the basic vocabulary the IA will use to gather information about the environment via the Ontology Manager – defined in Section 1.4.3. It will support the description of a specific context in generic terms, which allows designing a rule system that can be seamlessly applied to different environments – see Section 5.4.2.

5.4.1 Modeling distributed and context-dependent HCI

We developed a model that encompasses the description of the user interface along with two facets: the view of the user and the view of the system. It is important to understand and to take into account that whatever is theoretically possible may not be possible in reality because of physical/hardware constraints on the devices and platforms that are available. The system must take into account a grounded description of its capabilities while at the same time trying to serve the high-level needs of the user. Such high-level models have been introduced in Section 5.2. Our attempt at coupling high-level concepts to their more grounded counterparts is presented in the next section. This model is then encoded in an ontology that we describe in Section 5.4.1.2.

5.4.1.1 A conceptual view

Cognitivist approaches conceptualized the problem of human-computer interaction with the notions of mode, modality and medium. Trying to model those concepts and their relationships with OIOs, we realized that they were not sufficient to describe the whole chain of perception/communication. In fact, whereas the relationships between modes and modalities are quite clear (see the notions of primary and secondary modes in [8]) the relations between modalities and media haven't been much investigated. A few decades ago, only one medium could be used for each modality (the screen for graphics or text output, the mouse for pointing input, the keyboard for text input, microphone for voice). Now that other technologies are available (touch screens, gesture recognition, accelerometers…), we do not only discover new modalities, but also new ways of using already existing modalities. A Wiimote[2], for instance, offers different capabilities than a mouse, but it can also be used as a pointing device. The device is different but the language – and its structure – is the same, it still consists in moving a pointer across the screen and clicking. Thus, the user will be able to use and perceive the pointing input modality through a mouse, a touch screen or a Wiimote, with a low cognitive charge. Devices are also subject to variations. Even though the standard mouse has two buttons, some have more. The same observation can be made for keyboards. It is quite obvious to degrade the functionalities of such an extended device to the functionalities of the standard device it derives from. However, the extended capabilities should be used as much as possible. Finally, devices may be wrapped to emulate the behaviour of other devices. The device might not be really adapted but in the case where no suitable device is available, this would prevent the system from just failing and provide a degraded user interface. For instance, if the modality is typing and no keyboard is near the user, we can use a virtual keyboard controlled via the mouse.

---

[2] Wiimote is a sensor for video games that was designed by Nintendo in 2005. It is a remote control that includes vibrators, orientation sensors and screen pointing capabilities.

As we can see, there is no one to one relationship between the modality and the medium. One modality might need several media and one media can support several interaction languages, thus several modalities. By definition [6], modality refers to the structure of the information as it is perceived by the user. However it is also related to the structure of the interaction language used by the system. By modelling directly the relationship between modalities and media, we do not take into account the description of such an interaction language. That is why we developed the concept of interaction language that specifies the different communication acts within a modality. Devices provide inputs/outputs at a very low-level that is dependent on the hardware implementation of the device. This set of native signals is called interaction lexicon. As we have seen with the previous examples, there is a need to describe how the system can move from the interaction lexicon to the interaction syntax, or from the user point of view, how to make a device reflect/impact the interaction through a certain modality. This is the responsibility of modules that we call Mediators. Several kinds of mediators can be envisaged depending on the concerned modality. Indeed, the variety of inputs/outputs is such that there is no global approach to that problem. In graphical user interfaces, it is usually admitted that state machines and their extensions [21] [2] best suits the interpretation of device events. However, this approach is very bad for spoken dialogue, and even worse for natural spoken dialogues – we will see in Section 5.5.2 that spoken dialog is important in ambient environments and how to implement such a mediator. We thus allow different kinds of mediators to cohabit within our system.

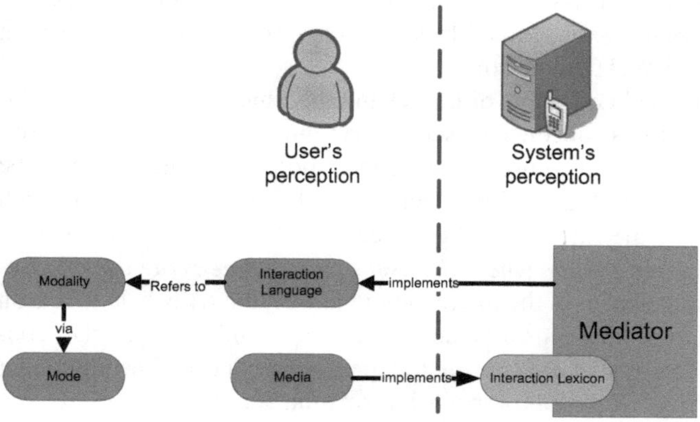

Fig. 5.3 The basic concepts and their relationships to the user or the system

Figure 5.3 summarizes the relationships between the different concepts that we have just presented. We can see that the initial concepts of Mode, Modality and Media represent the user – and designer – facet of the interaction components, whereas the Lexicon and the Mediators refer to a more formal specification that the system is able to interpret in order to instantiate the interaction. Our model thus unifies the system view of interaction with the users' and the designers' ones. Next section presents in details the modelling of those concepts.

5.4.1.2 Modeling rich interaction context with ontologies

The Interaction Ontology[3] is the knowledge base on which the Interaction Agent relies to describe the interaction context. In WWHT, the context was modelled as a simple set of variables, each having a certain range but with no connection between variables and thus no meaning associated to it. Moreover, no standard specification had been given concerning variables that could be found in different applications. For instance, if the noise level has to be part of the context, it could be represented by a variable named noise-level in application A and level-of-noise in application B. Consequently, the behavioural model of application A cannot be reused during the design of application B.

Our approach consists in modelling the context by structuring it in a generic way that includes semantics. This way, two different applications could exchange behavioural model seamlessly since they would rely on the same Interaction Ontology. Even if ontologies were slightly different, alignment methods could be used to try and automatically adapt a behavioural model to a new context of use (see Section **??**). Moreover, standard relationships between concepts are reflected between instances, which allows reasoning. It enables us to make use of the richness of description logics inference in order to describe and apply the election rules. By using semantic web representation (OWL) of interaction context, behavioural models can be more complex and seamlessly adapted from one environment to another which was not the case in the WWHT framework.

Figure 5.4 shows the main concepts of the IO and their hierarchy. The User and the EnvironmentalConditions subconcepts serve as alignment points for other concepts of the Sphere Ontology. The user subconcepts, more particularly, should be aligned with User profile ontologies. This mechanism allows for semantically rich information gathering by the Interaction Agent that is able to ask queries such as *"What is the light level in the room where the user is?"*, *"where is the user?"*. Because those concepts are present in the interaction ontology, the IA is able to reason on them, even though the actual sensor value can only be found in some other concepts of the sphere ontology that are specific to a particular context. The alignment method hides the heterogeneity of the various information sources.

Figure 5.5 shows how the different concepts presented in previous section relate to each other in the ontology. In this figure, some concepts and relations have been hidden to make it more readable. The concepts are spread over three layers. The upper layer represents the designer's and user's view. This view encapsulates the concepts that are used by psychologists and ergonomists to describe the communication: *User*, *Mode* and *Modality*. The lowest layer represents the aim of the system: the *InteractionTask*. In between, the middle layer tries to establish a connection be-

[3] `http://iroom.limsi.fr/atraco/InteractionOntology.owl`

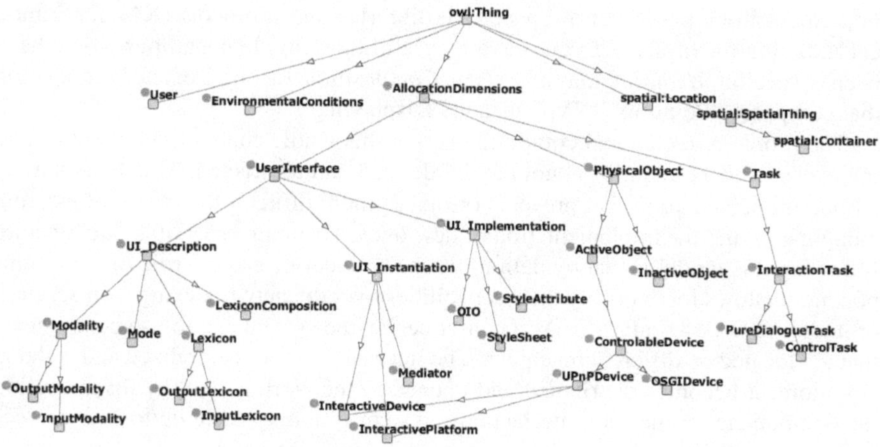

Fig. 5.4 The main categories of the Interaction Ontology

tween the system's needs and the user's perception. The concepts appearing in this layer are hardware devices and software services that can be composed to generate a user interface. Among them, 4 concepts are mostly important: *OIO*, *Mediator*, *Lexicon* and *InteractiveDevice*. The *InteractiveDevice* is a device used for interaction. In this sense, it is a media. As shown in Figure 5.3, the *InteractiveDevice* can send – for input devices – or receive – for output devices – a set of structured events. Such a set of events is called a lexicon. A device is said to "speak" a lexicon. Several lexicons can then be composed in a *Mediator*, a software entity responsible for mixing and interpreting events from a *LexiconComposition*. The *Mediator* can then host *OIOs* that have been designed to implement a specific *InteractionTask*.

At this stage of the design, we can draw a parallel between the usual interaction chain implemented in current graphical user interfaces and our model, which tries to abstract this chain of components. If we take the example of a classical computer, the *InteractiveDevices* are the mouse, the keyboard and the screen. They speak the following *Lexicons*: "Pointing" for the mouse, "Typing" for the keyboard and "Displaying" for the screen. We can informally refine the "Pointing" lexicon here by saying it is composed of 3 events:

- moveTo(x,y)
- right-click()
- left-click()

We could do the same refining for the other lexicons but this is not the point of this comparison – and the list would be too long to be of interest here. Now that we have defined our Lexicons, we would like to implement a language of interaction that we are all used to: WIMP (Windows, Icon, Menu, Pointer). This graphical language is the one used in many mainstream operating systems. It results from the interpretation of mouse and keyboard events to generate drawing events on the screen. Many implementations have been proposed via libraries and protocols (X11 for Linux, GTK, Swing for Java...). In our model, each of those implementations would have been a specific *Mediator* that *affordsLexComposition* for the LexicalComposition that composes "Pointing", "Typing" and "Displaying".

The problem in classical computing is that this whole chain of components and their respective boundaries is not clearly defined and abstracted. This results in an ad-hoc implementation that prevents or makes more difficult the re-use of existing components and the implementation of new ones. The main asset of an architecture following this model is that by delimiting the boundaries and the role of each component, it allows for exposing their capabilities over the network using web-services protocols (SOAP, UPnP, ...). As a consequence, the system can re-use components that are located on different machines. The interaction is not centralized and isolated any more; it becomes distributed and opened to the world. Also, by disentangling the components of the user interface, the system is able to scale up to new devices with unknown *Lexicons*, as long as a specific language has been implemented as a *Mediator* somewhere over the network. Finally, because this model can handle distributivity of the hardware, and reusability of the code via a service-oriented approach, it allows for a certain plasticity of the communication. By composing the *InteractionDevices* with different *Mediators*, we can achieve different interaction styles relying on different modalities or that mix modalities.

Now, as with any distributed system, one question remains: where is the code running? Our vision is that embedded technologies have come to a point where each *InteractiveDevice* could come with an embedded chip that runs the service and the network connection needed to make it work. Said differently, it is technologically possible to create a mouse that would connect via a local network on Wi-Fi and propose on this network a web-service that makes the mouse events available. Still, as we defined them, *OIOs* are pieces of program that realize specific interaction tasks. Those programs have to be run somewhere. For delay reasons and also implementation constraints, it is not trivial to run a graphical user interface on a machine and redirect its output to a screen that is not connected to it. In order to have better results and distribute the computing load, we envision that *OIOs* will be run on different computing units depending on the location of interaction. It will shorten delays and provide more reactivity. In order to provide such flexibility, we need that those computing units provide services to move *OIOs* from one place to another, to deal with several *OIOs* at a time... This set of services is called a meta-HCI [17]. In our model, an *InteractivePlatform* is a computing unit that offers such services. It can be dynamically – or statically – bound to *InteractionDevices*. We call an Interaction Space the pair mixing an *InteractionPlatform* with the set *InteractionDevices* it uses. Figure 5.6 shows the relationships between those concepts. It is interesting to see that the Interaction Space is a concept that matches both the system's view of how components are connected but also the user's view of which devices are "working together". This view helps users anticipate that a specific input device will have impact on a specific output device. Different OIOs can be run in the same Interaction Space, but in order to maintain consistency, an *InteractionDevice* should never be part of two different interaction spaces.

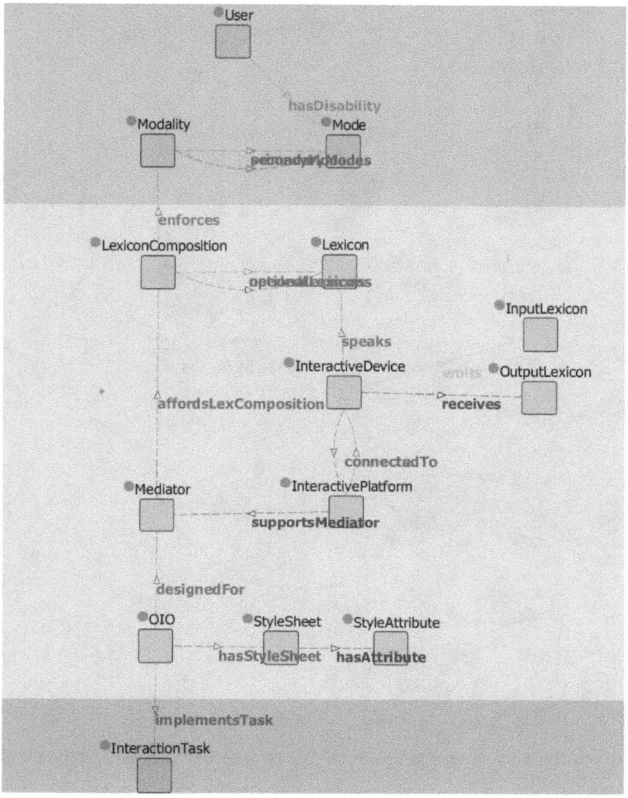

Fig. 5.5 User interface specification in the Interaction Ontology

The knowledge base we presented in this section is of course tightly bound – at the implementation level – to the module of the IA that is responsible for reasoning on the context adaptation, the Multimodal Manager. Next section demonstrates how adaptation can be achieved by reasoning on such a model.

5.4.2 Reasoning for contextual allocation/instantiation

In this section, we aim to describe how the model previously presented can be used to dynamically generate the Interaction Spaces – see Figure 5.6. Having a model of interaction in ambient spaces is not enough. If we want to dynamically adapt, we need to reason over the concepts of this model. Among reasoning techniques, we chose to implement such a reasoner as a rule-based expert system. Rule based expert systems such as Clips [54] provide the powerful expressivity of second order logic – i.e. logic with quantifiers. We first tried to implement our engine as a set

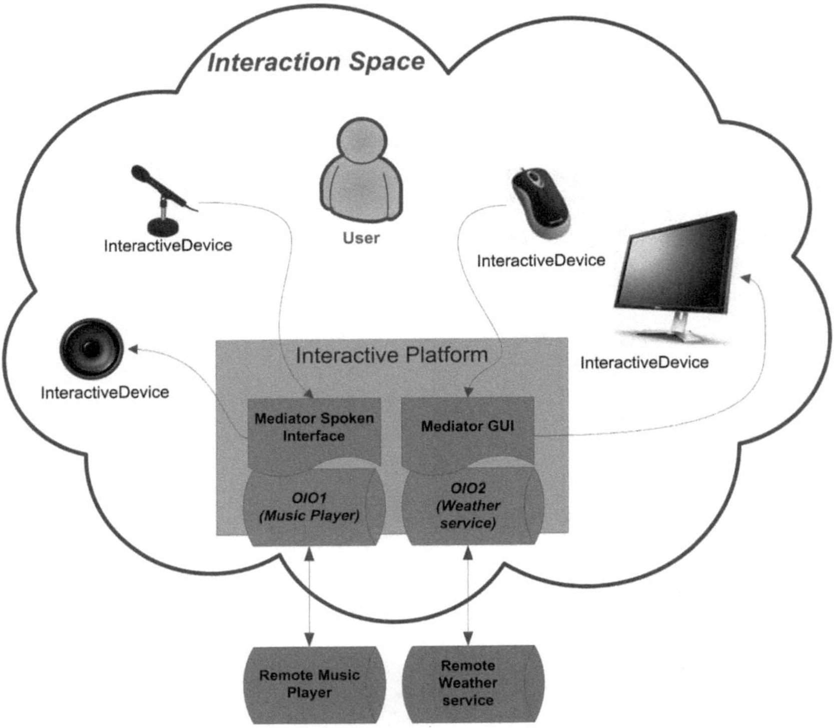

Fig. 5.6 Interaction spaces are formed by composing OIOs, Interactive Devices and Interactive Platforms

of SWRL rules [34]. SWRL has the advantage of being integrated with OWL and other semantic web technologies. Our model being aligned with other knowledge bases from other ATRACO components, SWRL would have been a good way to express rules that could have been aligned with other concepts in the sphere ontology. However, SWRL has a limited expressivity (no support for quantifiers) and lacks flexibility – it is hard, for instance, to add new built-in predicates to the language. Thus it didn't fit our needs for implementing a complex reasoning engine. As a consequence, we implemented a bridge from OWL to Jess [27], a Clips derivate that allows for more complex reasoning.

Our approach consists in three steps. First, contextual information from the ecology is gathered in the Interaction Ontology via the Ontology Manager. This information enables us to specify the interaction context within the Interaction Ontology terms previously described. The use of semantic web formats (OWL) at this phase is motivated by the integration in the more general context of ATRACO. This way, semantic web techniques such as alignment can be used by the Ontology Manager to interpret the heterogeneous knowledge bases from the ecology into terms that the Interaction Agent can understand. In the second phase, we use an expert system to apply rules that determine which are the possible associations of interactive devices.

We call such solutions the candidates – see Section 5.4.2.2. Finally, a set of rules are applied to those candidates to elicit the best option. The behaviour of the interactive ambient system mainly relies on this set of rules. Thus, a consistent set of rules for eliciting the best candidate to interaction is named a behavioural model [59]. Different behavioural models can be described by designers of ambient spaces, the rules being fairly simple to write/modify – see Section 5.4.2.3. This section presents the main rules of our engine that enables eliciting the appropriate Interaction Space.

5.4.2.1 Introduction to Jess/CLips Syntax

Our examples relies on the Jess/Clips common syntax that is described in [54] [27][4]. This language uses a functional style to define 3 main structures:

- Functions
- Facts
- Rules

Facts are a set of predicates that are assumed to be true by the reasoning engine. Those predicates may have a complex structure involving several fields (called slots). Slots can be filled with one data or several (in this case, they are called multi-slots). The strict structure of Facts allows for fast pattern matching and unification.

A rule is composed of two parts separated by symbol "=>": a left-hand side (LHS) that describes a pattern that needs to be matched in order to execute the right-hand side (RHS). RHS is a set of functions that are called which can add new facts, modify/delete existing ones or modify global variables.

When the engine is run, the rules' LHS are matched against the facts using the Rete algorithm [25]. Each rule that has its LHS matched will be executed – or fired – which means that its RHS will be executed with the variables bound in the LHS.

5.4.2.2 Enumerating candidates

In order to construct all the possible Interaction Spaces as described in Figure 5.6, the system must first discover which are the eligible interactive platforms and which mediators it can run on them. Indeed, the use of a specific mediator requires that the interactive devices that are connected to the platform provide the appropriate lexicons – see Figure 5.3. Having gathered information about the interactive spaces, the interactive devices and their lexicons via the Interaction Ontology, the system transforms this knowledge into facts in the reasoning engine.

Based on those facts, the inference engine will try to find which mediator can be instantiated for each interactive platform. For this, it looks at all the devices that are connected to the platform and the lexicons they provide. Each mediator needs a specific composition of lexicons (several compositions of lexicons can match the

[4] More details can also be found on the official jess website
`http://www.jessrules.com`

```
(defrule iplat-mediator-activate
    (InteractivePlatform (name ?ip))
    (supportsMediator ?ip ?mediator)
    (affordsLexComposition ?mediator ?lc)
    (forall (neededLexicons ?lc ?lexicon)
        (connects ?ip ?device)
        (InteractiveDevice (name ?device))
        (speaks ?device ?lexicon))
=>
    (assert (usesMediator (platform ?ip)
                (mediator ?mediator)
                (lexComposition ?lc))))
```

Listing 5.1 Rule that activate valid mediators on Interactive Platforms

```
(deftemplate Candidate
    "A candidate for task instantiation"
    (slot name)
    (slot targetTaskName) (slot oio) (slot platform)
    (slot mediator) (slot style)
    (slot valid) (slot validity-reasons)
    (slot salience (default 0))
    (slot pros-reasons) (slot cons-reasons))
```

Listing 5.2 The template defining Candidate facts

needs of one mediator). For instance, the GUIMediator needs either the displayOutput+mouseInput+KeyboardInput or the displayOutput+touchscreenInput to be able to run. Listing 5.1 demonstrates how mediators are activated on each InteractivePlatforms.

After inferring which mediators can be used on which platform, the system will create every valid candidate. A candidate is the combination of an OIO[5], a mediator, an Interactive Platform and its Interactive Devices. Each candidate is also associated with a style that allows for modifying the morphological attributes of the OIO (for instance, the text size, the voice to be used in speech synthesis...). [6]

Slots *valid*, *validity-reasons*, *salience*, *pros-reasons* and *cons-reasons* are used to grade the candidate. Their use is detailed in the next subsection.

Once mediators are activated, the creation of a candidate is quite simple, it relies on enumerating every possible associations. The LHS of Listing 5.3 declares the constraints that a Candidate must fill in order to be valid. The TargetTask predicate is a placeholder for all the instances of interaction tasks that are required by the activity sphere. Interaction tasks are generic tasks like *controlling the lights* whereas a

[5] i.e. the bundle of code to be executed within a specific mediator

[6] In our implementation, available OIOs and styles are statically specified in the Interaction Ontology, but they could also be specified and retrieved from a remote library if necessary.

```
(defrule candidates-creation
    (TargetTask (name ?ttask) (interactionTask ?itask))
    (OIO (name ?oio))
    (InteractivePlatform (name ?ip))
    (Mediator (name ?med))
    (StyleSheet (name ?style))
    (hasStyleSheet ?oio ?style)
    (implementsTask ?oio ?itask)
    (designedFor ?oio ?med)
    (usesMediator (platform ?ip) (mediator ?med))
=>
    (bind ?cName (generateId "c"))
    (assert-candidate ?cName ?ttask ?oio ?ip ?med ?style))
```

Listing 5.3 Rule creating every possible candidates

TargetTask represents the context-dependent instances of such tasks that are needed, like *controlling the lights in the living-room.*

Listing 5.3 represents how interaction tasks can be implemented via a unique OIO. However, as discussed in Section 5.3.3, it would be interesting to describe the whole interaction task as a CTT and use this task decomposition to derive a structure involving a composition of several OIOs. Work is still in progress to adapt our model to involve multiple-OIOs composition.

Once all possible candidates for a specific target task have been enumerated, we need to select the most appropriate ones.

5.4.2.3 The behavioural model

In order to implement the allocation level of adaptation, the most suitable candidates need to be identified. There is no unique way to solve this problem, it is rather a matter of design decisions to encourage a certain kind of behaviour from the system. As a consequence, we designed the model to be flexible and give designers the possibility to specify the intended behaviour of the system. Thus, we propose a flexible framework to design various behavioural models. This framework relies on the rule syntax presented earlier and a few set of predefined functions that allow granting marks to the different candidates.

Based on the facts encoded in the Multimodal Manager, the designers can apply rules to favour a certain candidate or forbid the use of some of them. The slot *validity* – see Listing 5.2 – is used to forbid the use of some candidates. For instance, in Listing 5.4, a rule is written to disable candidates that make use of a mode that is a user's disability. Such a rule would prevent the system from using speech synthesis when the user is deaf for instance.

There are drawbacks in disabling a candidate altogether. If the candidate was the only one available, then even if it is bad, it might be better than nothing. Of course for deaf people, using speech synthesis will never make sense, whatever the

```
(defrule disableInvalidModes
    (User (name ?theUser))
    (hasDisability ?theUser ?disability)
    ?c <- (Candidate (name ?nc)
    (oio ?oio) (platform ?ip))
    (designedFor ?oio ?mediator)
    (forall (usesMediator (platform ?ip)
                (mediator ?mediator)
                (lexComposition ?lc))
        (enforces ?lc ?modality)
        (primaryMode ?modality ?disability))
    =>
    (forbid ?c "Implies the use of user's disability"))
```

Listing 5.4 Rule disabling candidates using a user's disability

```
(defrule preferTactileModality
    ?c <- (Candidate (name ?nc) (mediator ?mediator))
    (affordsLexComposition ?mediator ?lexComp)
    (enforces ?lexComp tactile)
    =>
    (approve ?c "User prefers the tactile mode"))
```

Listing 5.5 Rule improving the salience of candidates that use the tactile modality

situation is like. Thus it would be safe to use the forbid function. However, in other situations, designers might need to just discourage the use of some candidates and use them only as a last resort. To this extent, the *salience* slot has been added to hold a numeric value representing the appropriateness of a candidate regarding the context on a finer granularity. For instance, Listing 5.5 shows how designers can favour candidates that use the tactile modality over other candidates.

To sum up, like in the WWHT framework, four functions where designed to provide fine grained control of the voting process:

- forbid
- use
- approve
- disapprove

forbid and *use* make non reversible assertions on whether or not it is good to use this candidate. In case a conflict appears – i.e. a candidate is both declared usable and forbidden – different conflict resolution strategies can be envisioned. We chose the strategy to disable candidates for which such conflicts appeared. A candidate should be usable if all of the conditions that make it usable are met. Thus, it seems relevant to prefer disabling candidates if only one of the conditions that make them unusable would be met. *approve* and *disapprove* provide more fine grained tuning by increasing/descreasing the *salience* of the candidate. At the end of the process, candidates are ranked by validity: first, the candidates that are declared valid, then the

```
(defrule preferBigText
    (User (name ?user))
    ?c <- (Candidate (name ?nc) (platform ?ip)
    (mediator ?mediator) (style bigText))
    (usesMediator (platform ?ip)
    (mediator ?mediator)
    (lexComposition ?lc))
    (neededLexicons ?lc display)
    (connect ?ip ?device)
    (speaks ?device display)
    (not (isnear ?device ?user))
    =>
    (approve ?c "User far from screen, need big letters"))
```

Listing 5.6 Rule encouraging the use of big text when user is not near the screen

ones for which no assumption has been done about validity and finally the ones that are forbidden are discarded. Within those three categories, candidates are ranked by decreasing salience.

The slots *validity-reasons*, *pro-reasons* and *cons-reasons* are used to hold *string* explanations about the rules that were applied. It is then easier to trace back which rule was applied to explain the Multimodal Manager's decision in a test case or to the user in real conditions.

The instantiation level of adaptation that involves selecting the morphological attributes of the user interface is addressed via the slot *style* of the candidate. Rules can also be written to select the best style. For instance, Listing 5.6 represents a rule that encourage choosing big text style when the user is not in the proximity of the screen.

It is important to note that any arbitrary contextual information that would have been aligned with the Sphere Ontology can be used in rules, as long as they are defined in the Interaction Ontology. Alignment is impossible if the concept is not sufficiently described in both ontologies. Also, the designer needs a concept name as reference to designate such information in the rules. For that reason, the Interaction Ontology acts as a set of reference terms that can be extended. For instance, in the environmental conditions part of the Interaction Ontology, basic concepts are defined to represent the current level of sound and the current luminosity. It allows for writing rules about generic environmental conditions as if they were generally available. If the sensors from the ecology provides such information and the Ontology Manager is able to align it to these concepts in the Interaction Ontology, then the rule would apply with the sensor value. Otherwise, it would be discarded. The behavioural model is thus adaptative to interaction context and environmental context, and, thanks to the alignment technique[7], the switching from one environment to another is seamless for the designer who writes the rules.

[7] We assume here that such techniques exist. We showed in Section **??** that even though it is the case, the reliability of those techniques is variable. Our model is thus flexible to the extent that third-party ontologies that describe ecology devices are of a sufficient quality to allow alignment

Of course, the allocation and instantiation adaptation that we present here are adapted to the situation when the engine is run, which means at the beginning of the interaction task. We will see in next section how evolution of the interaction task can be achieved during the interaction with the user.

5.5 Continuous adaptation during execution

There are different ways adaptation of a task over time can be seen. Several factors might require evolution of the user interface: [58]

- *Information factor:* Whenever the kernel application status change, the change must be reflected in the user interface. For instance, on a multimedia task such as music control, when the current track in the playlist changes, the new track title has to be reflected by the user interface.
- *User's actions:* The user interface must adapt to the user's actions on the user interface itself. For instance, when the user uses the pointing modality, tool tips can be displayed when the user's pointer hovers an icon.
- *Interaction context:* Because of the natural evolution of the ecology, a candidate might become invalid when for instance a device enters or leaves the space.
- *Time factor:* For some applications, like agendas, the system must pro-actively deliver a message to the user depending on a scheduled event.
- *Spatial factor:* The location of different objects and users in the ecology might change and influence the validity of a candidate.

Those different factors require the Multimodal Manager to be aware of the current status of the Activity sphere. It is needed that the list of valid candidates and their salience continuously evolves during interaction. Section 5.5.1 details how events need to be exchanged with other ATRACO components to adapt during interaction. We can also remark that among these factors, the *user's actions* is the only one that involves only the user and that does not need input from the system. In fact, adaptation to user's actions depends only on the structure of the language that is used to communicate with users, i.e the modality. Such a modality dependent adaptation cannot be handled at the Multimodal Manager level because it does not imply re-evaluating the salience of a candidate. Such kind of modality dependent adaptation is particularly relevant and complex when it comes to use linguistic modalities [8] such as text interaction or natural spoken interaction. Section 5.5.2 introduces an example of such adaptation, the continuous evolution of spoken dialogue.

5.5.1 Revisiting allocation/instantiation over time

Information factor and Time factor are both application dependent factors. They occur as a spontaneous change in the status of the kernel application. In ambient spaces, the kernel application is a composition of web services that expose their status over the network. Thus, the Sphere Manager, which is responsible for the handling the task, is able to monitor those changes by subscribing to the corresponding web services. Every time such a change occur, it will send a message to the Interaction Agent to inform it of the new status. The IA then transmits the new status to the OIO so that it updates its internal representation of the task status and reflects the change.

Other factors – Interaction context and Spatial factor – do not have direct impact on the content of the OIO but they can change the salience of a specific candidate. They might require that the candidate be replaced by a similar candidate with another style – this is called refinement – or by a candidate relying on different interactive devices to interact with the user – this is called re-allocation. An example of refinement required by a spatial factor is the size of a text on "text" modality that must continuously adapt to the user's location. Re-allocation is more frequently used to respond to a change in the set of available interactive devices. For instance, when a candidate makes use of a wireless mouse and this one runs out of battery, the interaction task could be re-allocated onto another screen or using a different modality.

The Multimodal Manager handles allocation and instantiation at the start of the interaction task. For that purpose, it is fed with information gathered by the OM. In order to provide refinement and re-allocation, the candidate's saliences must be re-evaluated as soon as the ecology status get deprecated by a dynamic change. The sphere manager, by managing the ecology of the activity sphere, is the component responsible for being aware of such dynamic changes. It will thus advertise those changes to the OM and the Interaction Agent. There are two sorts of events, each representing the structure of the corresponding factor of change:

- Activity sphere change event : Represents the apparition/removal of an interactive device in/from the ecology
- Location event : Represents a change in the location of objects and users

Every time such an event is received by the Interaction Agent, the state of the Multimodal Manager is updated and the Allocation/Instantiation process is run again. At the end of it, the status of a candidate might have undergone the following changes:

- *Birth/Death* The candidate is removed if it made no sense any more. New candidates can also be created.
- *Validity change* The candidate's validity is changed
- *Salience change* The candidate's salience changes

In case the current candidate becomes outperformed by another candidate, the system is in a non-optimal situation and the conflict needs to be resolved. Different strategies can be envisaged. For instance, designers might prefer sticking to the consistency principle [61] and select the best candidate that uses same devices or the same modalities. Another strategy consists in taking always the best candidate as the current candidate which leads to a much more mobile and adaptive behaviour.

Note that the evolution requirements described until now do not take into account the *User's action* factor. As explained in the introduction, dealing with such kind of context changes requires fine control of the OIO at language level. Such adaptation does not impact the election process but only the dialogue status within one modality. This can be achieved only from within the entity controlling the OIO: the mediator. There is no general solution to that issue and for each modality that is implemented, a different approach must be devised. In next section, we give an example of such kind of adaptation in the complex situation of spoken dialogue.

5.5.2 *Ongoing interaction evolution: The example of spoken dialogue*

Within the scientific area of IEs, Spoken Dialogue System (SDS) technologies offer one of the most natural interfaces. In this context for many tasks such as command-and-control of devices or services, proactive behaviour (warning, information, etc.), and negotiative dialogues, speech is a promising modality. However, to realise spoken dialogue adaptation it is not only necessary to provide advanced voice recognition and speech synthesis capabilities but furthermore to provide a Spoken Dialogue Manager (SDM) residing at the core of an SDS able to manage adaptation. Nowadays one of the most widespread technologies to implement an SDM is the W3C standardized VoiceXML description language [48]. The idea behind this approach is to simplify the development of dialogues that would allow even non-experts to implement speech applications in a similar manner as websites are implemented. VoiceXML has undoubtedly many advantages but the structure of such dialogue systems is limited to system-initiative and mixed-initiative layouts. More complex structures such as negotiative or task-oriented dialogue flows still lack of a sufficient approach that combines the ease-of-use of VoiceXML with the more powerful expressiveness of scientific approaches such as [73, 10] and [41].

In this section, we introduce our approach regarding the third level of adaptation – i.e. evolution – within the spoken dialogue modality. After classifying the situations that require adaptation of the ongoing dialogue, we will present the spoken dialogue manager that we implemented and the ontological model that is associated to it and that affords such flexibility of the dialogue content.

5.5.2.1 A classification of evolution triggers

In [31] we have defined the behaviour of the SDM to be adaptive regarding the following changes that may happen during (spoken) human-computer interaction in the context of IEs and distinguished between three levels of adaptation:

- Environmental changes that relate to *Device Adaptation*: This relates to the continuous modifications of devices and services that are active and accessible within the user's context. Depending on the surrounding and the situation of the user (kitchen, living room, car, etc.), the availability of devices and services may

vary. We assume that users would usually "talk" to devices (i.e., control them by speech) that are within sight and correspond to the current situation of the user. This requires the capability to continuously change grammars, utterances, and system commands to a changing device population and changing user focus within the IE. Furthermore since it seems not to be possible to always be aware of the user's situation (doing housework, relax, prepare a dinner, etc.) it is necessary to provide an option to activate the control of devices or services that even are not within the context of the current location and/or situation. Thus the context of the user may also change depending on the time of day and even on the actual level of trust in nearby entities (guests, newly added technical devices). It is obvious that the spoken commands and/or utterances that can be understood / that are said by an SDM will therefore change continuously as well. In the case of spoken dialogue an important influence on the availability of spoken dialogue is the level of noise. In practice it will hardly be possible to control an environment via speech while music is playing loudly.
- Environmental changes that relate to *Event Adaptation*: Since various tasks within IEs are to be accomplished it is necessary to move the current focus from an on-going dialogue to other (contingently more urgent) dialogues. These dialogues may consist of informative system utterances, alerts, or short yes-no-questions. Afterwards the on-going dialogue would be resumed. We have recognised two kinds of events: external and internal. While external events always need an entity that throws the specific event, internal events can be initiated by the dialogue manager itself. Reasons for initiating an internal event can be various and sundry: fixed priorities, dynamic priorities (i.e., changing over time), semantics (i.e., semantically similar dialogues can extend a dialogue), and depending on the progress done within an on-going dialogue.
- Task-driven changes that relate to *Task Adaptation*: During more complex tasks that might come up during a conversation the task itself may vary. Especially during a spoken human-computer negotiation the requirements and/or constraints may change. The effects of task adaptation may result in task cancellation or in slight changes such as an extension of the initially planned dialogue. Obviously the task adaptation level corresponds to the most complex level of adaption as it relates to automatic learning and/or direct user-system collaboration.

5.5.2.2 The implementation of the SDM

In order to meet the requirements listed in Section 5.5.2.1 the ATRACO SDM is implemented considering the Passive View variation of the Model-View-Presenter (MVP) [52] pattern, itself a model-derivative of the popular Model-View-Controller (MVC) paradigm [53]. The underlying idea of MVP is that an application or system should be divided into three logical parts, the Model, the View and the Presenter. As shown in Figure 5.7 the user only interacts with the View layer. Contrary to MVC the Presenter mediates between Model and View – the Model conveys no functionality, i.e., it is not an application but solely encodes the knowledge that is used by the Presenter. The term Model in this case refers to a *Domain Model*. Therefore especially for systems that directly interface to the user (i.e., user interfaces) MVP is perfectly suitable. To be able to communicate with a user or with other external entities the application needs a knowledge base that describes facts and the relations between such facts. A fact could, for example, be the name of a person or an ID number. A relation could be "has", which could be used for person "A" has ID number "4711".

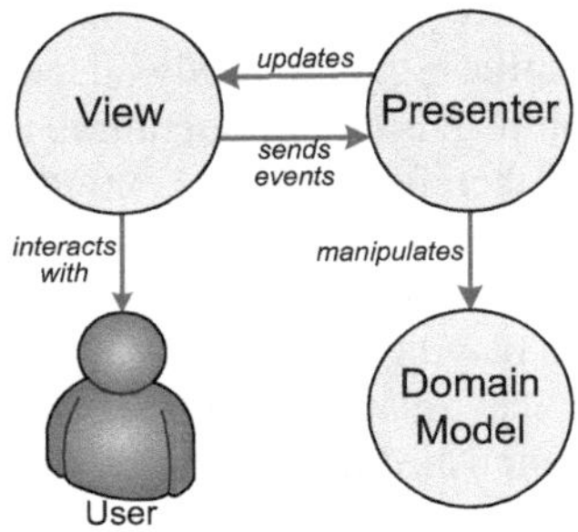

Fig. 5.7 The MVP Passive View pattern as used to implemente the ATRACO SDM.

Without this knowledge the system wouldn't be able to generate useful output, i.e., act as *knowledge source* or to understand input that is provided by external entities, thus acting as *knowledge sink*. The term Domain Model could therefore be specified as the knowledge a system needs in order to be able to interact with the context in a meaningful way. There are many ways to establish such a knowledge base: to name but a few, SQL databases or XML files could be utilised. A more sophisticated option is to make use of ontologies to provide a common understandable knowledge base.

The ATRACO SDM makes use of a specific number of dialogue representations. These representations serve as Domain Models. Each representation provides knowledge about both dialogue flow and state of a specific spoken conversation. Depending on contextual information various sets of spoken dialogues can be activated or deactivated. It is furthermore possible to add new representations for dialogues during runtime and therefore extend the knowledge base, i.e., the Model. For the prototype we use OWL ontologies to implement the Model. In the next paragraphs we provide a detailed look on the capabilities of this structure and the way we use it to describe spoken dialogues.

The underlying knowledge base of the SDM is modelled using OWL ontologies, so called Spoken Dialogue Ontologies (SDOs). We have implemented this tree shaped structure to arrange the data-bearing individuals using a defined set of classes. The root of each knowledge base is DialogueDomain, which has the two subclasses Speech and State. We divide the ontology into these two main branches since we want to distinguish between knowledge that corresponds to the static structure and knowledge that corresponds to the dynamic state of the current dialogue.

Figure 5.8 shows an overview of all classes populating the SDO together with the relations interlinking them.

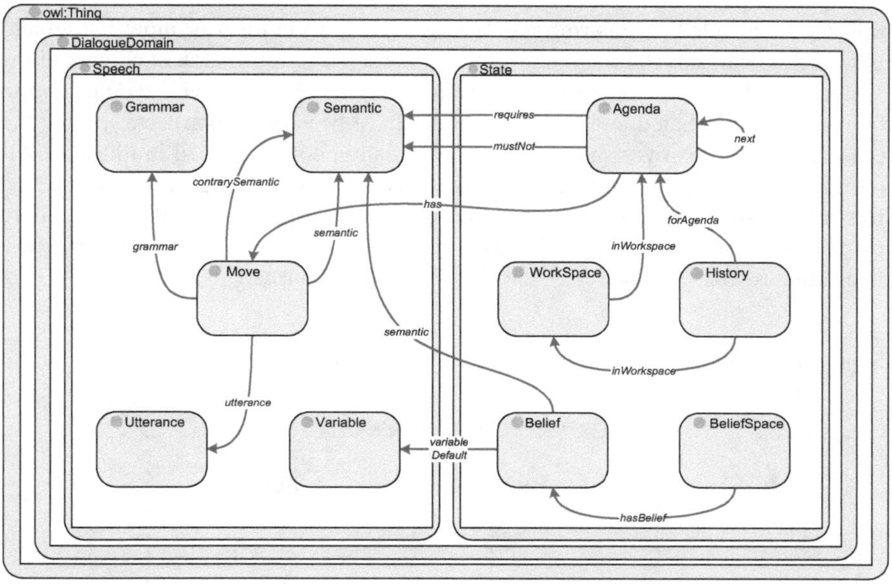

Fig. 5.8 Overview of the classes and main relations of the Spoken Dialogue Ontology.

The main purpose of separating the Domain Model from the actual intelligence – the Presenter – and the View is that the model can be modified, (partly) removed, extended, and exchanged. Furthermore by separating static from dynamic knowledge the data (grammar, utterances, and semantics) that might be needed for various dialogues can be exchanged and reused as well. In the following we present an interactive situation to describe our approach more detailed. After a user returns home there might be several spoken interactive tasks activated in parallel: a "greeting" task, a dinner preparation task, and a light control task. Since one of the main duties an IE should handle is to control specific tasks, it is necessary to provide a (probably varying) set of spoken commands that the IE can interpret and execute (thus device and event adaptation). An example for such a behaviour could be a user telling the system to switch the lights on after entering the apartment. Figure 5.9 shows the three interactive tasks mentioned above that may form the examplary interactive situation.

Since the SDM adapts to the context it needs to be able to receive triggers from the outside world to change its state. The initial phase therefore is triggered by a "user enters room" event. This event might happen only once a day and/or when the user has left the apartment for a specified period depending on the configuration of the environment. In our example the SDM is set up to wait until the user greets the system (Task 1). It further activates a control task that listens to possible lighting control commands the user may utter (Task 3). Initial system studies in the iSpace at the University of Essex [32] revealed that the subjects preferred the Spoken Dialogue System to be as unobtrusive as possible. Thus we have designed the system to behave rather passively and not to proactively initiate a conversation if this can be avoided. A control task such as Task 3 waits for user input by default. However, if the user starts talking to the system by uttering a spoken command the system could take this opportunity to start dialogues that otherwise would have to be proactively initiated and therefore would have been more obtrusive. Figure 5.9 presents two alternatives showing how the situation could proceed in Phase II: Alternative I contains two triggers that might allow the system to perform Task 1; the five-minutes-timer elapsed since the user entered the room or – probably the more usual case – the user greets the system. As mentioned above the reason for such a five-minutes-timer is that the system should act as unobtrusively as possible.

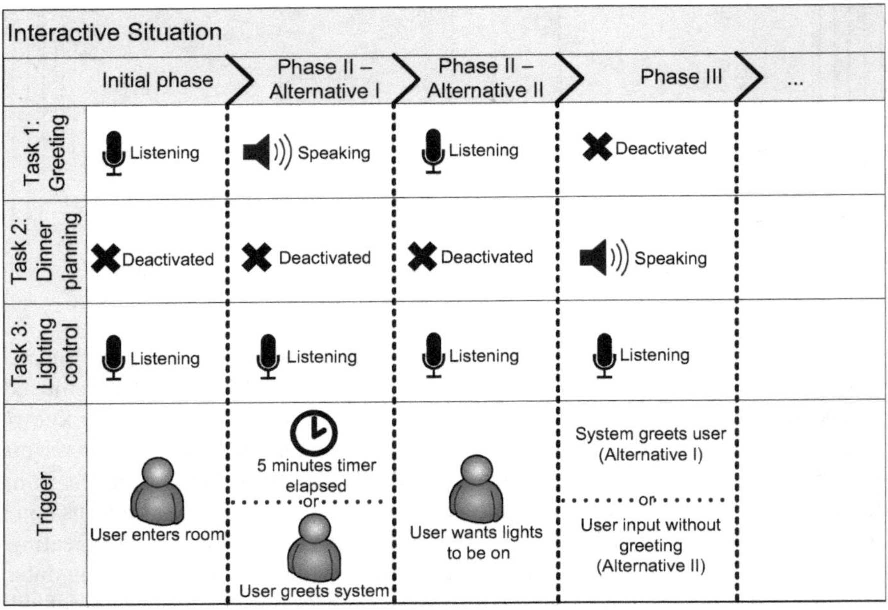

Fig. 5.9 An interactive situation that may occur with two alternatives.

Note that Task 3 is still active since the system is meant for handling more than one interactive task in parallel. If one of the two triggers is actuated the system would greet the user and would add a semantic value such as "userInitiatedConversation" to the knowledge base. This would allow Phase III to start. Table 5.1 shows a possible conversation that might occur using the proposed set of Spoken Dialogue Ontologies.

Alternatively Phase II could be activated by the user telling the system to switch the lights on. This would make Task I obsolete – the system shouldn't greet the user in response to a spoken command. It would be more natural that the system skips the greeting task and activates the proactive Task 2 "Dinner planning" instead. Figure 5.9 shows Phase III constituted by the additionally activated Task 2 and the still-running Task 3. The preceding greeting task has either become obsolete or has already being processed. Since the user can dynamically activate or deactivate the tasks, the SDM may perform, it is possible at any time to terminate a conversation with the system or to start a dialogue the system hasn't been aware of.

Table 5.1 A dialogue snipped that might occur during an interactive situation.

Speaker	Utterance
Suki	Hello Julia!
Julia	Hi Suki!
Suki	Switch the lights on!
Julia	Do you want to start preparing the dinner for your friends now?

The proposed Spoken Dialogue Ontology refers to the informational components and their formal representation as originally described in the Information State theory [41]. It reflects both static and dynamic knowledge and therefore facilitates a generic representation of a model that can be processed by the Presenter layer. Since the SDO is not only interpreted before the dialogue starts but is also involved dynamically in the on-going dialogue our approach allows for a great flexibility and provides many opportunities for adaptation. It is possible to learn new grammars, semantics, or utterances during an on-going dialogue by simply adding new individuals to the SDO. Furthermore by extending or reducing the set of activated SDOs various interactive situations for a changing device and task population can be generated. Thus we provide a fertile ground to cope with the challenges adaptation brings along within Ambient Environments.

This concludes our presentation of the different layers of adaptation regarding user interaction. We explored the evolution through the structure of the user interface, using a model-driven approach to allocate and instantiate user interfaces on existing interaction devices. Though the set of available interaction devices is dynamic, we showed how to evolve the composition during runtime. Finally, we showed that the structural changes are not enough, adaptation is required within an instantiated modality and we demonstrated in this section how such adaptation could be achieved for the most complex modality: the spoken dialogue. Next section will develop our approach regarding the evaluation of this adaptive system.

5.6 Experiments & Results

As a result of the ATRACO project, a prototype of the Interaction Agent was setup in the iSpace at the University of Essex, United Kingdom. We focus in the next sections on the implementation of that prototype and on the validation results we gathered from it.

5.6.1 ATRACO integration

5.6.1.1 Role of the IA in the Activity Sphere

Within the ATRACO architecture, the role of the IA is to interface the system with the user. As such, it needs to provide some functionalities to other components of the system. Table 5.2 shows the features that the Interaction Agent exposes to other components. The main component with which the IA interacts is the SM. The SM will trigger the IA to create interaction tasks on-demand during the execution of the BPEL workflow – defined in Section 1.4. TM and PM will require the IA to disable or hide crucial commands and information when trust or privacy rules are violated – see Sections 7.5.2.2 and 7.4.3.

Table 5.2 The features the IA provides to other ATRACO components

Target component	Features
Sphere Manager	Start/Stop Interaction tasks
	Send user will as predefined events
	Display system status updates
Trust Manager	Inform on where interaction occurs
	Disable a user interface (because of trust breach)
Privacy Manager	Inform on where interaction occurs
	Hide a user interface (because of privacy breach)

5.6.1.2 Controlling remote OIOs

We have highlighted, throughout this chapter, the main role of the IA: to adapt the interaction to the context. A part of this role that we have not focused on is the routing of events. This role, crucial though it may be, doesn't really represent a research topic but rather an architectural and technical challenge. In ATRACO, to keep things simple, we have assimilated events to a pair of strings: one for the event name, and one for the content of the message. Events are divided into two categories: *System Events* represent updates from the system – for instance, the value of the luminosity; *User Will Events* are events generated by the user interface that represent an order from the user to the system – for instance, when the user wants to set the light level, the event $<$ "*light_level*","30%" $>$ will be sent.

The IA is composed of 3 internal modules:

- Multimodal Manager: It is responsible for the allocation, instantiation and evolution of the interaction
- Spoken Dialogue Manager: It is a mediator responsible for handling OIOs for spoken dialogue
- Graphical Dialogue Manager: It is a mediator responsible for handling OIOs for graphical user interfaces

Figure 5.10 shows how system events and user will events are propagated between the User and the System through the IA.

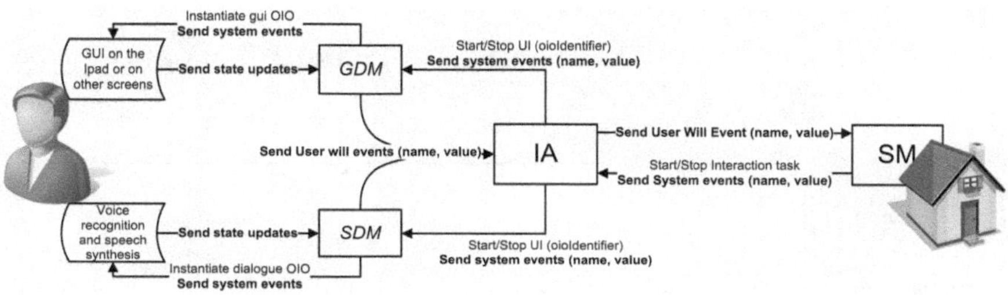

Fig. 5.10 Flow of events within the IA

5.6.2 *Model validation*

The validation of our model for interaction adaptation was made in two steps. After having implemented the Multimodal Manager that reasons over the Interaction Ontology information, we implemented a simulator for interaction. This simulator helped us design a behavioural model for the IA and test it. After this simulation phase, a social evaluation was driven in the iSpace to evaluate the impact of the whole system in real users' life.

5.6.2.1 Simulations

We designed a tool named Simteraction – for SIMulation of inTERACTION. This tool enables designers to write behavioural rules and immediately test them. It offers the possibility to add any interaction component to a simulated system, to give them a location and to evaluate the result of the allocation process in a specific situation.

Figure 5.11 shows a screenshot of the application in a simulation of the iSpace. We can see the different rooms of the apartment, the devices that have been deployed in it, and the user – here, his name is Suki. The application also draws the connections between available devices, thus showing the active Interaction Spaces. On the left side, an interaction task has been started for controlling lights. A list

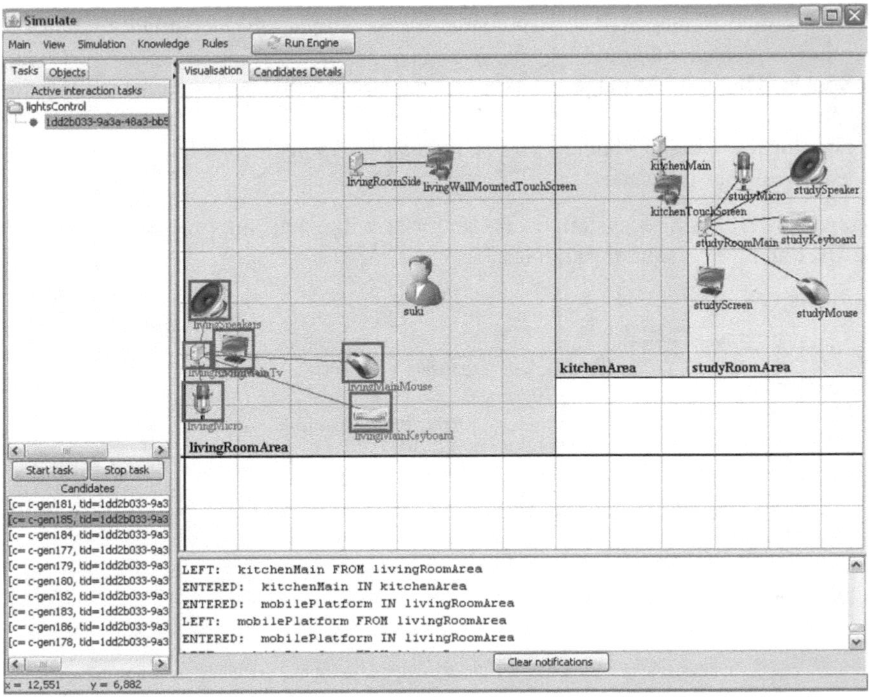

Fig. 5.11 Simulator implemented to test behavioural models – shows the model

of candidates is proposed. By selecting one of them – here the second – the corresponding Interaction Space is highlighted in the main view. More details about a specific candidate can be found in the "Candidates Details" tab – see Figure 5.12. This view is particularly useful for the designer as it gives explanation concerning the salience and validity of each candidate. For instance, in Figure 5.12, we can see that the candidate $c-gen177$ has a salience of $+1$ because it uses the tactile mode.

This simulator was very useful to explore the expressivity of our rule language and of the overall model. It also helped us in imagining and testing the reactions of the system to new devices or interaction languages that do not exist yet – or for which we don't have any implementation. Finally, it has been used – and that was the main reason for its implementation – during the implementation process to design and debug the behavioural model. This first phase of local evaluation by simulation was then followed by a real world application in the iSpace.

5.6.2.2 User evaluation

The social evaluation process and its results are detailed in Chapter 8 and in [32]. Such evaluations gives us an interesting view of how users appreciated the be-

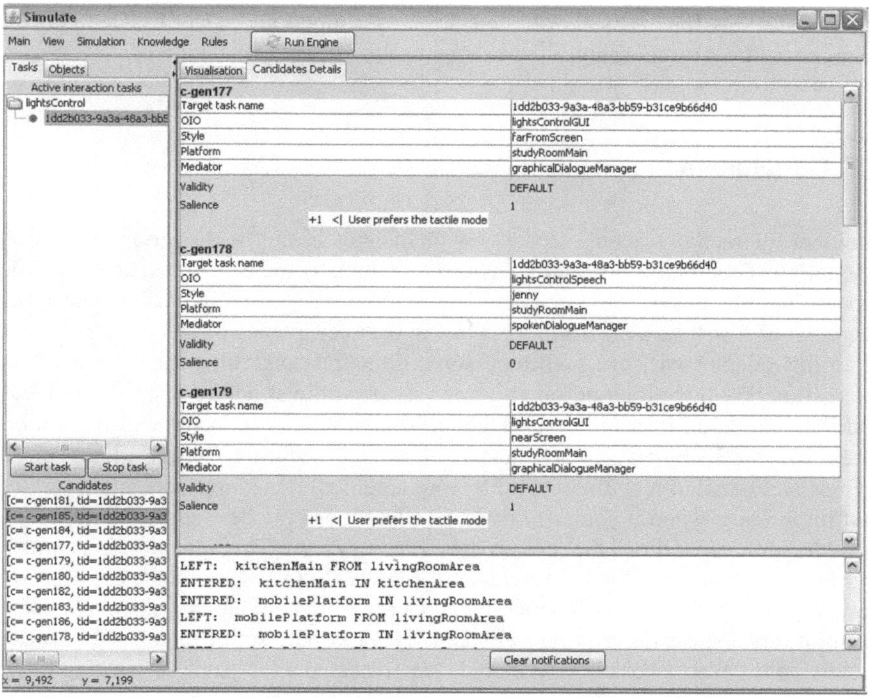

Fig. 5.12 Simulator implemented to test behavioural models – shows the candidates

havioural models we proposed them. However, they do not provide direct feedback on the architecture and the interaction itself, since they are hidden from the user. However, having been able to run such an evaluation is a proof in itself, that the system is consistent.

Moreover, it enabled us to test a first behavioural model that focuses on adapting the user interface to the location of the user. This behavioural model was designed to give users the feeling that the interfaces follow them.

The outcome of this evaluation regarding interaction was mainly positive. Users reacted well on the following behaviour of the interfaces. They however felt sometimes that their personal space was quite invaded, particularly when the system was doing or asking things proactively. It looks like, depending on the cases, the initiation of the dialogue has its importance in the user acceptance. This represents a significant improvement that could be added to the interaction model we propose.

Another interesting outcome was that users felt that their interaction with the system would change with time and depending on their mood. We had not envisaged that the same behaviour of the system could be well-accepted at first and be irritating after some time – or because the user feels in a different mood. Users proposed to have the possibility to change behavioural models of the system. Such a change is theoretically possible, it just requires to have different rule bases available. The main obstacle is for the system to detect the user mood and adapt by changing its behavioural model accordingly.

5.7 Conclusion

Ambient interactive systems create new challenges in the field of user interaction adaptation. Though this adaptation is made complex by the distributed and dynamic aspects of ambient systems, achieving some kind of adaptation in the user interfaces is crucial as it will be a determining factor in user acceptance of these systems.

In this chapter we have presented some directions regarding the adaptation of interaction. We have focused on two main aspects: the structure of user interfaces and the adaptation of a running dialogue. The structural adaptation of user interfaces requires 3 steps: allocation, instantiation and evolution. We introduced those 3 steps and detailed how they could be implemented. Reasoning techniques are of first importance when it comes to dynamically choose the best option in the scope of interaction capabilities that are proposed by a system. This selection is complex and relies on two key points: a structured representation of the context – that we designed as an ontology for interaction, and a reasoning engine that can apply some rules on this context description to make a decision. The set of rules used to make a decision is called a behavioural model. We presented a language for writing those rules and a complete framework for helping designers writing, testing and simulating/evaluating different behavioural models.

After exploring structural adaptation of the user interface, we explored the structural adaptation of the language itself. We focused this analysis on adapting the on-going interaction within one modality, the spoken dialogue. We presented how ontology technologies could be used again, to perform evolution of an on-going dialogue depending on contextual events.

Finally, we showed how we integrated all those concepts in a unique software architecture, the Interaction Agent. We demonstrated how this IA relate to other ATRACO components and how it was perceived by users during the social evaluation.

While our approach proved to be functional, a remaining challenge consists in integrating automatic user interface generation techniques to provide an even more flexible structural adaptation. We imagined a system where components implemented by different designers could be mixed in at runtime to provide just the features that are needed in a unified user interface. Our model was designed in that direction but theoretical aspects of computer assisted generation needs to be solved before our vision can be realized. Furthermore, the realization of such dynamic composition would be even more powerful if it could mix components implementing different modalities. The problem of synergetic multimodality fusion/fission has been investigated but solutions always involve decision taking from a designer. We can only hope that future results will concern its realization as an automated process.

5.8 Further readings

As far as multimodality is concerned in pervasive systems, the book *Multimodality in Mobile Computing and Mobile Devices: Methods for Adaptable Usability* [39] compiles an interesting set of visions ranging from theoretical visions to real-world applications.

Concerning the adaptation of natural spoken dialogue, the book *Spoken Dialogue Systems for Ambient Environments* [42] groups major publications of that field for year 2010.

References

1. André, E.: Handbook of natural language processing, chap. The Generation of Multimedia Presentations, pp. 305–327. Marcel Dekker Inc. (2000)
2. Appert, C., Huot, S., Dragicevic, P., Beaudouin-Lafon, M.: Flowstates : Prototypage d'application interactives avec des flots de donnees et des machines a etats. In: Proceedings of 21eme conference francophone sur l'Interaction Homme-Machine, IHM 2009, pp. 119–128. ACM Press (2009)
3. Athanasopoulos, D., Zarras, A., Issarny, V., Pitoura, E., Vassiliadis, P.: CoWSAMI: Interface-aware context gathering in ambient intelligence environments. Pervasive and Mobile Computing **4**(3), 360–389 (2008)
4. Balme, L., Demeure, A., Barralon, N., Coutaz, J., Calvary, G., et al.: Cameleon-rt: A software architecture reference model for distributed, migratable, and plastic user interfaces. Lecture Notes in Computer Science, Ambient intelligence **3295**, 291–302 (2004)
5. Bass, L., Faneuf, R., Little, R., Mayer, N., Pellegrino, B., Reed, S., Seacord, R., Sheppard, S., Szczur, M.: A Metamodel for the Runtime Architecture of an Interactive System. ACM SIGCHI Bulletin **24**(1), 32–37 (1992)
6. Bellik, Y.: Interfaces multimodales: concepts, modèles et architectures. Ph.D. thesis, Université d'Orsay Paris-Sud (1995)
7. Bellik, Y., Rebai, I., Machrouh, E., Barzaj, Y., Jacquet, C., Pruvost, G., Sansonnet, J.: Multimodal Interaction within Ambient Environments: an Exploratory Study. In: Human-computer Interaction, INTERACT'09: IFIP International Conference on Human-Computer Interaction; Uppsala, Sweden (2009)
8. Bernsen, N.: Modality Theory in support of multimodal interface design. In: Proc. of Intelligent Multi-Media Multi-Modal Systems, pp. 37–44 (1994)
9. Berti, S., Paternó, F.: Migratory multimodal interfaces in multidevice environments. In: Proc. International Conference on Multimodal Interfaces, pp. 92–99. ACM Press (2005)
10. Bohus, D., Rudnicky, A.I.: The ravenclaw dialog management framework: Architecture and systems. Computer Speech & Language **23**, 332–361 (2009)
11. Bordegoni, M., Faconti, G., Maybury, M.T., Rist, T., Ruggieri, S., Trahanias, P., Wilson, M.: A Standard Reference Model for Intelligent Multimedia Presentation Systems. Computer Standards and Interfaces **18**(6), 477–496 (1997)
12. Braffort, A., Choisier, A., Collet, C., Dalle, P., Gianni, F., Lenseigne, B., Segouat, J.: Toward an annotation software for video of Sign Language, including image processing tools and signing space modelling. In: 4th International Conference on Language Resources and Evaluation, held in Lisbon, Portugal (2004)
13. Browne, D., Totterdell, P., Norman, M.: Adaptive User Interfaces. Academic Press (1990)
14. Brusilovsky, P.: Adaptive Hypermedia. User Modeling and User-Adapted Interaction **11**, 87–110 (2001)

15. Calvary, G., Coutaz, J., Thevenin, D., Limbourg, Q., Bouillon, L., Vanderdonckt, J.: A unifying reference framework for multi-target user interfaces. Journal of Interacting With Computers **Vol 15/3**, Elsevier Science B.V (2003)
16. Chikofsky, E., Cross II, J.: Reverse engineering and design recovery: A taxonomy. IEEE software **7(1)**, 13–17 (1990)
17. Coutaz, J.: Meta-User Interfaces for Ambient Spaces. Lecture Notes in Computer Science : Task Models and Diagrams for Users Interface Design **4385**, 1–15 (2007)
18. Coutaz, J., Nigay, L., Salber, D., Blandford, A., May, J., Young, R.: Four easy pieces for assessing the usability of multimodal interaction: the CARE properties. In: Proceedings of INTERACT, vol. 95, pp. 115–120 (1995)
19. Dey, A.K.: Providing Architectural Support for Building Context-Aware Applications. Ph.D. thesis, Georgia Institute of Technology (2000)
20. Dieterich, H., Malinowski, U., Kühme, T., Schneider-Hufschmidt, M.: State of the art in adaptive user interfaces. Human factors in information technology **10**, 13–13 (1993)
21. Dragicevic, P., Fekete, J.: Input device selection and interaction configuration with ICON. People and Computers pp. 543–558 (2001)
22. Duarte, C., Carriço, L.: A conceptual framework for developing adaptive multimodal applications. In: Proceedings of the 11th international conference on Intelligent user interfaces, pp. 132–139. ACM Press (2006)
23. Eisenstein, J., Vanderdonckt, J., Puerta, A.: Applying Model-Based Techniques to the Development of UIs for Mobile Computers. In: in Proc. of the 6th international conference on Intelligent User Interfaces, pp. 69–76. ACM Press Publ., Santa Fe, New Mexico, USA (2001)
24. Florins, M., Trevisan, D., Vanderdonckt, J.: The Continuity Property in Mixed Reality and Multiplatform Systems: a Comparative Study. In: Proceedings of CADUI'04, pp. 13–16. Madeira Island (2004)
25. Forgy, C.: Rete: A fast algorithm for the many pattern/many object pattern match problem* 1. Artificial intelligence **19**(1), 17–37 (1982)
26. Frasincar, F., Houben, G.J.: Hypermedia Presentation Adaptation on the Semantic Web. In: Proceedings of the 2nd International Conference on Adaptive Hypermedia and Adaptive Web-Based Systems, LNCS 2347, pp. 133–142. Malaga, Spain (2002)
27. Friedman, E.: Jess in action: rule-based systems in java. Manning Publications Co. Greenwich, CT, USA (2003)
28. Frohlich, D.M.: The design space of interfaces, multimedia systems, interaction and applications. Proc. of 1st Eurographics Workshop, held in Stockholm, Sweden (1991)
29. Gram, C., Cockton, G.: Design Principles for Interactive Software (IFIP International Federation for Information Processing). Chapman & Hall, Ltd. London, UK, UK (1996)
30. Hagras, H., Goumopoulos, C., Bellik, Y., Minker, W., Meliones, A.: Research Results on Adaptation & Evolution. Tech. rep., 7th Framework program, ATRACO project report - Deliverable D13 (2011)
31. Heinroth, T., Denich, D., Schmitt, A.: Owlspeak - adaptive spoken dialogue within intelligent environments. In: 8th IEEE International Conference on Pervasive Computing and Communications Workshops (PERCOM Workshops), pp. 666 – 671. Mannheim (Germany) (2010). URL `http://ieeexplore.ieee.org/stamp/stamp.jsp?tp=&arnumber=5470518`
32. van Helvert, J., Hagras, H., Kameas, A.: D27 - prototype testing and validation. Restricted deliverable, The ATRACO Project (FP7/2007-2013 grant agreement nï¿½ 216837) (2009)
33. Henricksen, K., Indulska, J., Rakotonirainy, A.: Modeling Context Information in Pervasive Computing Systems. Pervasive Computing **2414**, 79–117 (2002)
34. Horrocks, I., Patel-Schneider, P., Boley, H., Tabet, S., Grosof, B., Dean, M.: SWRL: A semantic web rule language combining OWL and RuleML. W3C Member submission **21** (2004). URL `http://www.w3.org/Submission/2004/03/`
35. Jaimes, A., Sebe, N.: Multimodal human-computer interaction: A survey. Computer Vision and Image Understanding **108**(1-2), 116–134 (2007)
36. Johnston, M., Bangalore, S.: Finite-state multimodal integration and understanding. Natural Language Engineering **11**(02), 159–187 (2005)

37. Kobsa, A., Koenemann, J., Pohl, W.: Personalised hypermedia presentation techniques for improving online customer relationships. The Knowledge Engineering Review **16**(2), 111–155 (2001)
38. Kolski, C., Le Strugeon, E.: A review of intelligent human-machine interfaces in the lights of the arch model. International Journal of Human-Computer Interaction **10**(3), 193–231 (1998)
39. Kurkovsky, S.: Multimodality in Mobile Computing and Mobile Devices: Methods for Adaptable Usability. IGI Global (2010)
40. Landragin, F.: Physical, semantic and pragmatic levels for multimodal fusion and fission (2007). Accessible on HAL server : http://hal.archives-ouvertes.fr
41. Larsson, S., Traum, D.: Information state and dialogue management in the trindi dialoguemove engine. Natural Language Engineering Special Issue pp. 323–340 (2000)
42. Lee, G.G., Mariani, J., Minker, W. (eds.): Spoken Dialogue Systems for Ambient Environments, *Lecture Notes in Computer Science*, vol. 6392. Springer Verlag, Heidelberg (2010)
43. Limbourg, Q., Vanderdonckt, J., Michotte, B., Bouillon, L., López-Jaquero, V.: Usixml: A language supporting multi-path development of user interfaces. Engineering Human Computer Interaction and Interactive Systems pp. 200–220 (2005)
44. Mori, G., Paternò, F., Santoro, C.: Tool support for designing nomadic applications. In: Proceedings of the 8th international conference on Intelligent user interfaces, pp. 141–148. ACM Press (2003)
45. Mori, G., Paternó, F., Santoro, C.: Design and development of multidevice user interfaces through multiple logical descriptions. IEEE Transactions on Software Engineering **30**(8), 507–520 (2004)
46. Navarre, D., Palanque, P., Bastide, R., Schyn, A., Winckler, M., Nedel, L., Freitas, C.: A formal description of multimodal interaction techniques for immersive virtual reality applications. Human-Computer Interaction-INTERACT 2005 pp. 170–183 (2005)
47. Nigay, L., Coutaz, J.: A Generic Platform for Addressing the Multimodal Challenge. In: Proc. of the Conference on Human Factors in Computing Systems, CHI'95 held in Denver, Colorado, USA (1995)
48. Oshry, M., Auburn, R., Baggia, P., Bodell, M., Burke, D., Burnett, D., Candell, E., Carter, J., Mcglashan, S., Lee, A., Porter, B., Rehor, K.: Voice extensible markup language (voicexml) version 2.1. Tech. rep., W3C – Voice Browser Working Group (2007)
49. Padovitz, A., Loke, S., Zaslavsky, A.: The ECORA framework: A hybrid architecture for context-oriented pervasive computing. Pervasive and mobile computing **4**(2), 182–215 (2008)
50. Paganelli, L., Paternó, F.: Automatic Reconstruction of the Underlying Interaction Design of Web Applications. In: Proceedings of SEKE 2002. Ischia, Italy (2002)
51. Paternó, F., Mancini, C., Meniconi, S.: Concurtasktrees: A diagrammatic notation for specifying task models. In: INTERACT '97: Proceedings of the IFIP TC13 Interantional Conference on Human-Computer Interaction, pp. 362–369. Chapman & Hall, Ltd. (1997)
52. Potel, M.: MVP: Model-View-Presenter The Taligent Programming Model for C++ and Java. Tech. rep., Taligent Inc (1996). URL `Availableonhttp://www.wildcrest.com/Potel/Portfolio/mvp.pdf`
53. Reenskaug, T.: Models - views - controllers. Tech. rep., Xerox PARC (1979). URL `http://heim.ifi.uio.no/~trygver/1979/mvc-2/1979-12-MVC.pdf`
54. Riley, G.: Clips: An expert system building tool. In: NASA, Washington, Technology 2001: The Second National Technology Transfer Conference and Exposition, vol. 2 (1991)
55. Rist, T.: Supporting Mobile Users Through Adaptive Information Presentation. Multimodal intelligent information presentation **27**, 113–139 (2005)
56. Rist, T., Andre, E.: Building smart embodied virtual characters. Lecture Notes in Computer Science: Smart Graphics **2733**, 123–130 (2003)
57. Román, M., Hess, C., Cerqueira, R., Ranganathan, A., Campbell, R., Nahrstedt, K.: A middleware infrastructure for active spaces. IEEE Pervasive Computing **1**(4), 74–83 (2002)
58. Rousseau, C.: Présentation multimodale et contextuelle de l'information. Ph.D. thesis, Université d'Orsay Paris-Sud (2006)
59. Rousseau, C., Bellik, Y., Vernier, F., Bazalgette, D.: A framework for the intelligent multimodal presentation of information. Signal Processing **86**(12), 3696–3713 (2006)

60. Samaan, K., Tarpin-Bernard, F.: Task models and interaction models in a multiple user interfaces generation process. In: In Proceedings of TAMODIA 2004, vol. vol. 86, pp. 137–144. ACM Press, Czech Republic (2004)
61. Scapin, D., Bastien, J.: Ergonomic criteria for evaluating the ergonomic quality of interactive systems. Behaviour & Information Technology **16**(4), 220–231 (1997)
62. Serrano, M., Nigay, L., Lawson, J.Y.L., Ramsay, A., Murray-Smith, R., Denef, S.: The open-interface framework: a tool for multimodal interaction. In: CHI '08 extended abstracts on Human factors in computing systems, pp. 3501–3506. ACM, New York, NY, USA (2008). DOI http://doi.acm.org/10.1145/1358628.1358881
63. Stanciulescu, A., Limbourg, Q., Vanderdonckt, J., Michotte, B., Montero, F.: A transformational approach for multimodal web user interfaces based on UsiXML. In: Proceedings of the 7th international conference on Multimodal interfaces, pp. 259–266. ACM New York, NY, USA (2005)
64. Stephanidis, C., Karagiannidis, C., Koumpis, A.: Decision Making in Intelligent User Interfaces. In: Intelligent User Interfaces, IUI'97 held in Orlando, florida, USA, pp. 195–202 (1997)
65. Stephanidis, C., Paramythis, A., Sfyrakis, M., Stergiou, A., Maou, N., Leventis, A., Paparoulis, G., Karagiandidis, C.: Adaptable and adaptive user interfaces for disabled users in avanti project. Lecture Notes In Computer Science. Intelligence in Services and Networks: technology for Ubiquitous Telecom Services **1430**, 153–16 (1998)
66. Stephanidis, C., Savidis, A.: Universal Access in the Information Society: Methods, Tools, and Interaction Technologies. UAIS Journal **1**(1), 40–55 (2001)
67. Tandler, P.: The BEACH application model and software framework for synchronous collaboration in ubiquitous computing environments. The Journal of Systems & Software **69**(3), 267–296 (2004)
68. Tarpin-Bernard, F.: Interaction Homme-Machine Adaptative. Habilitation a diriger des recherches (hdr), Université Claude Bernard de Lyon (2006)
69. Teil, D., Bellik, Y.: The Structure of Multimodal Dialog II, chap. Multimodal Interaction Interface Using Voice and Gesture, chapter 19, pp. pp. 349–366. John Benjamins publishing co. (2000)
70. Thevenin, D., Coutaz, J.: Plasticity of user interfaces: Framework and research agenda. In: Human-computer Interaction, INTERACT'99: IFIP TC. 13 International Conference on Human-Computer Interaction, 30th August-3rd September 1999, Edinburgh, UK, p. 110. IOS Press (1999)
71. Thevenin, D., Coutaz, J.: Adaptation des IHM: taxonomies et archi. logicielle. In: Proceedings of the 14th French-speaking conference on Human-computer interaction, pp. 207–210. ACM New York, NY, USA (2002)
72. Vanderdonckt, J., Grolaux, D., Van Roy, P., Limbourg, Q., Macq, B., Michel, B.: A Design Space For Context-Sensitive User Interfaces. In: In Proceedings of IASSE 2005, held in Toronto, Canada (2005)
73. Young, S., Williams, J., Schatzmann, J., Stuttle, M., Weilhammer, K.: D4.3: Bayes net prototype - the hidden information state dialogue manager. Tech. rep., TALK - Talk and Look: Tools for Ambient Linguistic Knowledge, IST-507802, 6th FP (2006)

Chapter 6
Artificial Intelligence Planning for Ambient Environments

J. Bidot and S. Biundo

Abstract In this chapter, we describe how Artificial Intelligence planning techniques are used in The Adapted and TRusted Ambient eCOlogies (ATRACO) in order to provide Sphere Adaptation. We introduce the Planning Agent (PA) which plays a central role in the realization and the structural adaptation of activity spheres. Based on particular information included in the ontology of the execution environment, the PA delivers workflows that consist of the basic activities to be executed in order to achieve a user's goals. The PA encapsulates a search engine for hybrid planning–the combination of hierarchical task network planning and partial-order causal-link planning. In this chapter, we describe a formal framework and a development platform for hybrid planning, PANDA. This platform allows for the implementation of many search strategies, and we explain how we realize the search engine of the PA by adapting and configuring PANDA specifically for addressing planning problems that are part of the ATRACO service composition. We describe how the PA interacts with the Sphere Manager and the Ontology Manager in order to create planning problems dynamically and generate workflows in the ATRACO-BPEL language. In addition, an excerpt of a planning domain for ATRACO is provided.

6.1 Introduction

Intelligent environments (IEs), such as smart homes, offices, and public spaces, are featured with a large number of devices and services that help users in performing efficiently various kinds of tasks. In the scope of The Adapted and TRusted Ambient

Julien Bidot
Institute of Artificial Intelligence, Ulm University, Ulm, D-89069 Germany, e-mail: `julien.bidot@uni-ulm.de`

Susanne Biundo
Institute of Artificial Intelligence, Ulm University, Ulm, D-89069 Germany, e-mail: `susanne.biundo@uni-ulm.de`

eCOlogies (ATRACO), we use workflows to model how a large number of services should interact with one another as well as with the user in IEs based on available resources, environment characteristics, user's tasks and profile.

Sphere Adaptation (SA) is one of the dimensions of adaptation which are realized within the ATRACO project. In ATRACO, we use the notions of Ambient Ecology to describe the resources of an Ambient Intelligence (AmI) environment and activity spheres (ASs) to depict the specific Ambient Ecology resources, data, and knowledge required to support a user in realizing a particular goal. In our approach, a ubiquitous computing system supports the execution of overlapping or disjoint ASs using the resources provided by the AmI space. An AS which is a temporary entity at run time is set up in order to enable a specific user goal: once the user goal is no longer relevant, the AS is dissolved. As long as this goal persists, AS is adaptive, in the sense that it can be instantiated within different environments containing similar resources and adaptively pursue its goal, whenever it remains possible. An AS is represented by a workflow that consists of activities to be executed and that are described in terms of services required or produced by resources of the Ambient Ecology.

Our approach to service composition consists in two steps: (1) making strategic decisions by considering abstract services; i.e., identifying what basic activities are to be executed and in which order they are to be executed; (2) taking operational decisions; i.e., determining what resources should execute the activities. Describing IEs with ontologies allows us to decompose the problem of orchestrating the services into two parts: in this chapter, we focus on the Planning Agent (PA), an agent integrating Artificial Intelligence (AI) planning techniques, that is responsible for making strategic decisions in order to create workflows with abstract services at design time, while Section 1.4.2 introduces the Sphere Manager, an agent in charge of taking the operational decisions at run time in order to generate executable workflows with concrete services.

In ATRACO, the planning problem can be stated as "discover an execution path of tasks to achieve a user goal given some state of the world." A plan that is solution to the problem is called a task model in the ATRACO terminology. We use a library of planning domains, each of them corresponding to a user goal. AI planning techniques are used for the realization and the structural adaptation of ASs in ATRACO. Based on the available services in the AmI environment, we apply these techniques in order to determine the basic activities to be executed for attaining user goals. The structural adaptation of ASs refers to the persistent achievement of the goals when the type of the available resources changes (as agents and users may come and go, devices and services may appear and disappear over time) and when the cardinality of the available resources varies (as the number of devices or users that participate in the realization of an AS may differ over time).

The formation of a system that realizes adaptive ASs is based on a service-oriented architecture (see Section 1.2.3), which integrates the PA to allow for adaptive planning.

6.2 Related Work

The composition of services has been a hot research topic in the last years, and various AI planning approaches has been used to address this issue [9].

Zhao and Doshi proposed a framework that can handle the uncertainty inherent in Web services [20]. This framework is based on semi Markov decision processes, and it implements symbolic techniques that operate directly on first-order logic based representations of the state space to obtain the compositions. In addition, time constraints of the services are explicitly represented and taken into account. Although this approach is applicable to Web processes that are nested to an arbitrary depth, it is not possible to express complex causal relations between them in the abstraction hierarchy. Since we do not have any information about the uncertainty relative to the services in ATRACO, we use a deterministic AI planning approach to addressing the service composition issue.

Pistore, Traverso, and Bertoli presented an approach to the automated composition of Web services that integrates symbolic model checking [10]. They modeled BPEL4WS Web services with non-deterministic and partially observable behaviors, and they expressed composition requirements with extended goals. Unlike our work, they dealt with AI planning problems with uncertainty.

The work of R-Moreno et al. [11] integrated AI planning and scheduling techniques to automatically generate business process models, avoiding going through all the drawing process, and making sure that the established connections among activities conform to a valid sequence of activities. After the models have been generated, a user can simulate and optimize the process. The authors use Partial-Order Causal-Link planning and constraint propagation techniques to address this problem. Unlike our work, they implemented a system that can deal with explicit resource and time constraints.

EL Falou et al. [3] addressed the problem of automated composition of Web services by using AI planning techniques. They proposed a multiagent framework and an algorithm for guiding the planner of each agent towards the best local plan. Contrary to their work, the search for plans performed by the PA is centralized; i.e. we do not address the issue of merging several local plans generated by a number of planners.

Like the work of Sirin et al. [17], we use Hierarchical Task Network (HTN) planning techniques to model and solve planning problems. However, our implementation is more flexible than theirs, and our planning domain models are purely declarative and do not contain any control structure for guiding search in contrast to theirs. In their application, Web services are specified with OWL-S.

Marquardt and Uhrmacher focus on using AI planning to solve the problem of service composition in smart environments [7]. They compare the runtime performance of four different planning systems using an abstract simulation model, and the evaluation results show that some of these planners are suitable for composing services in time. Their modeling of the problem is different from ours, and none of the evaluated planners implements HTN planning. In addition, the service composition is completely done at design time, which makes replanning from scratch necessary each time a new device appears or disappears in the smart environment.

In the Gaia project, Ranganathan and Campbell presented a paradigm for the operation of pervasive computing environment that is based on AI planning [13]. The first difference with our system lies in the modeling of planning problems, since they use PDDL, the initial world state aggregates the states of all entities (i.e. services, devices, and applications) of the environment along with the context of the environment, and the goal world state is determined given a user goal, a template world state, and a utility function to be maximized, but there are no abstract tasks and expansion methods. In addition, unlike our approach, their planning component is responsible for the binding of concrete services, devices, and applications, which means that scalability problems inevitably appear due to the very large world states for realistic testbeds (no abstraction mechanism) and due to the impossibility to declare abstract tasks and procedures in the planning domains (no HTN planning). Their solving procedure becomes very prohibitive, if it has to restart search for solution plans from scratch several times, each time with a different goal world state. Finally, their planning component is also in charge of executing plans, and it may re-execute an action or replan when an execution failure occurs.

Amigoni et al. proposed an AI planning system, D-HTN, that performs a centralized plan-building activity which is tailored to use the capabilities of the devices currently available in smart environments [1]. In their experimental system, each device is associated with one agent, and another single agent is responsible for building plans. D-HTN implements an HTN planning approach that differs from ours, although their modeling of the problem is similar to ours. D-HTN is not formally grounded and is much more rigid than our search engine. In consequence, D-HTN does not allow for a general backtracking mechanism during the planning process, which is inefficient for addressing complex and large-scale planning problems. An important problem with using D-HTN, also encountered in [17], resides in including search control information in the planning domain models. In contrast to our system, there are no ontologies describing the smart environment, the knowledge about expansion methods is distributed and collected from the present smart devices, and service composition is entirely performed at design time.

There are some similarities between our work and the work of Lundh, Karlsson, and Saffiotti [6]. They address the issue of configuring network robot systems at run time using AI planning techniques, and the role of functional configurations in their work is comparable to the role played by workflows in ATRACO. Their configuration planner uses methods that describe alternative ways to combine functionalities (or other methods) for specific purposes. This technique is inspired by HTN planning. However, there are also differences with our work. First, their planning framework does not deal with causal links, and their probabilistic/possibilistic conditional action planner is based on a forward-chaining planner that searches in the space of world states, while the search engine of the PA explores the space of partial plans at different abstraction levels and addresses deterministic problems with complete information about world states. Second, with our framework, we can explicitly represent and cope with ordering constraints in plans and expansion methods. Third, their action planner is given some search control knowledge in the form of first-order linear temporal logic formulas as input, which is used to prune the search space.

Vuković, Kotsovinos, and Robinson proposed an approach to the composition of Web services based on the context information [19]. The architecture of their system is composed of different layers, and it includes a planning system in the abstract service composition layer. The planner is a forward-chaining planner. Like our system, theirs supports a dynamic on-the-fly adaptation of applications; i.e., it can recover from execution time failures of individual service instances. However, unlike the search engine of the PA, their planning system searches in the space of states and cannot deal with procedural knowledge. In addition, Web service instances are represented as actions in their planning domains, whereas abstract services are encoded as facts in the planning problems the PA addresses. In the same way as the action planner presented in [6], some search control knowledge is used to reduce the search effort. Finally, abstract execution plans are represented in BPEL4WS in their prototype, and the abstract services of these plans are bound to service instances at execution time.

The work of Paluska et al. focuses on automating high-level implementation decisions in pervasive applications [8]. Their system enables a model in which an application programmer can specify the behavior of an adaptive application as a set of open-ended decision points. Each of these decision points may be satisfied by a set of alternative, competing scripts. The set may be extended at run time without needing to modify or remove any existing scripts. Their approach is hierarchical, since the scripts may contain themselves decision points. In the same vein as our work, but without using AI planning techniques and ontologies, their system is able to bind resources at run time.

6.3 Artificial Intelligence Planning

In Artificial Intelligence (AI), the *classical planning problem* consists typically in a set of operators, a single initial world state, and a single goal world state. The instance of an operator is often called a task, and a world state is a set of positive literals. A given world state can evolve; i.e., we can go from this state to another state: A state changes when a task is executed, since every executed task has usually at least one effect on the state. The objective of AI planning is to select and order tasks that allow one to attain the goal state from the initial state. Tasks can be executed in a state, if and only if some preconditions hold in this state. Each task is thus associated with preconditions and effects. A *plan* consists of tasks and ordering and causal relationships between these tasks. In a book, Ghallab, Nau, and Traverso present a survey about AI planning [5].

We consider an illustrative example of such a classical planning problem. A person is living in a flat in the Ulmstreet, and there are services for tuning the ambient temperature and the luminance level in the flat. In the initial state, there are devices that can cool, provide the current ambient temperature and the luminance level, increase and decrease the luminance level in the flat. The person can move with means of transport, can install a radiator in the flat, and can buy a radiator in a hardware store called Buildhouse. In the goal state, we want the ambient temperature and the luminance level in the flat to be tuned according to the person preferences. An intuitive and pragmatic solution would be the following procedure: The person has to go to Buildhouse, buy a radiator there, go back to home, install the radiator, and then devices can start tuning the ambient temperature and the luminance in the flat. We can model this planning problem with five operators: a person moves from one location to another one; a person buys a radiator in a hardware store; a person installs a radiator at a location; devices tune ambient temperature at a location; devices tune luminance level at a location. The operators have preconditions; e.g., devices can tune the ambient temperature at a location, if (i) a person is present at this location; (ii) the location is home; (iii) there is a device that can heat; (iv) there is a device that can cool; (v) there is a device that can sense the current ambient temperature. The operators have also some effects; e.g., when a radiator is installed at a location that is a flat, then this flat can be heated.

A plan that solves the planning problem described above consists of six tasks that are partially ordered (see the acyclic digraph in Figure 6.1, where the nodes represent tasks, and the arrows depict ordering constraints): A person goes from Ulmstreet (their home) to Buildhouse (a hardware store); the person buys a radiator; the person goes from Buildhouse back to Ulmstreet; devices start tuning luminance level and at the same time the person installs a radiator; after this installation, devices (including the radiator) start tuning the ambient temperature in the flat in the Ulmstreet. In Figure 6.1 and Figure 6.2, the first task and the last task are dummy tasks that represent the initial state and the goal state of the planning problem, respectively.

A planning problem is a complex problem to solve, since it is highly combinatorial: A large number of tasks to select and a huge number of conflicts that appear between the tasks. The acyclic digraph on the left of Figure 6.2 shows the causal relationships between the tasks: the nodes represent tasks, and the annotated arrows depict causal links. In ATRACO, an intelligent planner is encapsulated within the Planning Agent (PA) and used to solve planning problems. Section 6.3.1 introduces the basic ingredients of the planning framework on which the intelligent planner is based, and Section 6.3.2 gives a review about planning strategies that guide the planner towards solution plans. [16, 15] give more technical details about the framework and the planning strategies, respectively.

6.3.1 A Formal Framework for Refinement Planning

This section provides the concepts for a generic AI planning approach: Planning by plan refinement. PANDA, an existent development platform, is based on a planning framework that integrates Partial-Order Causal-Link (POCL) planning and Hierar-

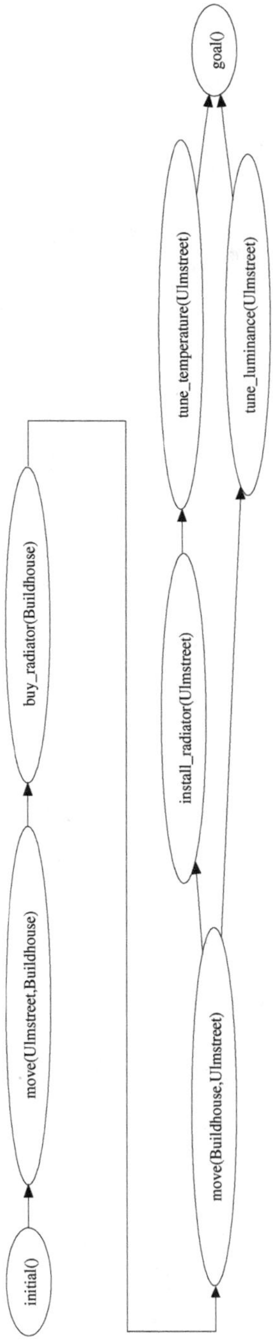

Fig. 6.1 Ordering structure of a plan.

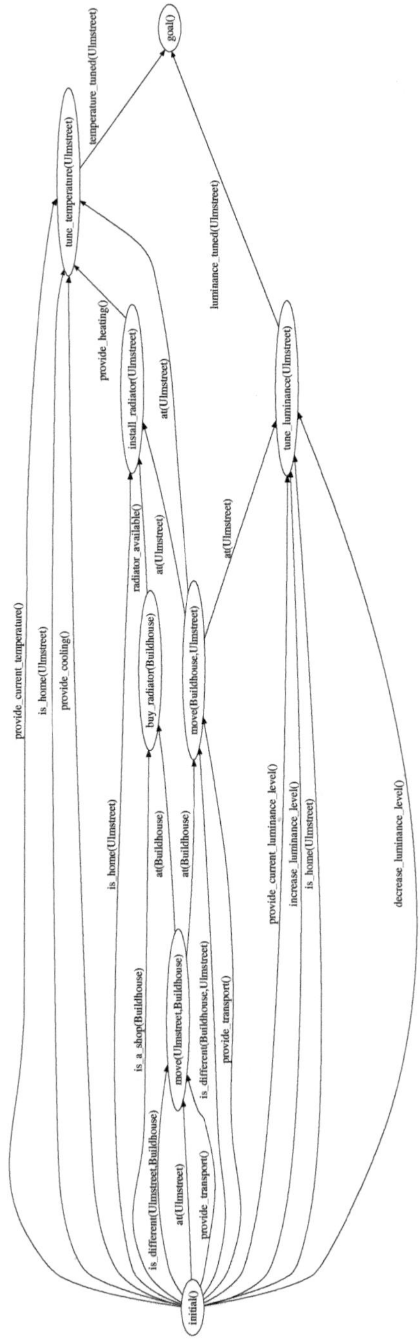

Fig. 6.2 Causal structure of a plan.

chical Task Network (HTN) planning [2]. The planning framework is hybrid, since it is composed of and combines elements coming from both POCL planning and HTN planning. The Planning Agent (PA) of ATRACO encapsulates a search engine which is a particular configuration of PANDA. The formal framework uses an Action Description Language (ADL) like representation of states and basic actions (*primitive tasks*). States and *preconditions* and *effects* of tasks are specified through formulae of a fragment of first-order logic. *Abstract tasks* can be refined by so-called *expansion methods* which provide *task networks* (*partial plans*). The task networks describe how the corresponding abstract task can be solved. Partial plans may contain abstract as well as primitive tasks. With that, hierarchies of tasks and associated methods can be used to encode the various ways to accomplish an abstract task.

A domain model $D = (\mathtt{T},\mathtt{EM})$ for hybrid planning consists of a set of task schemata T and a set of expansion methods EM for implementing the complex tasks in T. Note, that there are in general multiple methods provided for each complex task schema. A *task schema* $t(\overline{\tau}) = (\mathrm{prec}(t(\overline{\tau})), \mathrm{add}(t(\overline{\tau})), \mathrm{del}(t(\overline{\tau})))$ specifies the *preconditions* and the *positive* and *negative effects* of the task. Preconditions and effects are sets of literals, $\bar{\tau} = \tau_1, \ldots, \tau_n$ are the task parameters. A ground instance of a task schema is called an *operation*. A *state* is a finite set of ground atoms, and an operation $t(\bar{c})$ is called *applicable* in a state s, if the positive literals of $\mathrm{prec}(t(\bar{c}))$ are in s and the negative are not. The result of applying the operation in a state s is a state $s' = (s \cup \mathrm{add}(t(\bar{c}))) \setminus \mathrm{del}(t(\bar{c}))$.

In hybrid planning, abstract tasks like primitive tasks show preconditions and effects. The associated state transition semantics is based on axiomatic state refinements that relate task preconditions and effects across various abstraction levels. In contrast to that, in POCL planning, there are no abstract tasks and no expansion methods, but all tasks are described at the same level of abstraction. Though HTN planning allows us to describe tasks at different abstraction levels and decompose them with the help of expansion methods, it does not associate preconditions and effects with the tasks, which precludes any causal reasoning.

A *partial plan* or *task network* is a tuple $P = (TE, \prec, VC, CL)$ with the following sets of *plan components*: TE is a set of *task expressions* (plan steps) $te = l : t(\overline{\tau})$ where l is unique label and $t(\overline{\tau})$ is a (partially) instantiated task. $\prec$ is a set of *ordering constraints* that impose a partial order on the plan steps in TE. VC is a set of *variable constraints*. They include equations and inequations of the form $v \doteq \tau$ and $v \dot{\neq} \tau$ which *codesignate* and *non-codesignate* variables occurring in TE with variables or constants. Moreover, VC contains *co-typing constraints* $v \dot{\in} Z$ and *non-cotyping constraints* $v \dot{\notin} Z$, where Z is a sort symbol, that restrict further codesignations. CL is a set of *causal links* and provides the usual means to establish and maintain causal relationships among the tasks in a partial plan. A causal link $cl = \langle te_i, \phi, te_j \rangle$ indicates, that formula ϕ expressed in first-order logic, which is an effect of task a task te_i, supports (a part of) the precondition of task te_j.

Task networks are also used as pre-defined *implementations* of complex tasks. An *expansion method* $em = (t(\overline{\tau}), (TE_{em}, \prec_{em}, VC_{em}, CL_{em}))$ relates such a complex task schema $t(\overline{\tau})$ to a task network.

A *planning problem* $\pi = (D, s_{\text{init}}, s_{\text{goal}}, P_{\text{init}})$ for hybrid planning consists of a domain model D, an initial world state s_{init}, a goal world state s_{goal}, and an initial task network P_{init}. The solution to a planning problem is obtained by transforming P_{init} stepwise into a partial plan P that meets the following solution criteria: (1) All steps in P are primitive tasks; (2) P is executable in s_{init} and generates a state s_{end} such that $s_{\text{goal}} \subseteq s_{\text{end}}$ holds. Plan P is thereby called *executable* in a state s and *generates* a state s', if all ground linearizations of P, that means all linearizations of all ground instances of the task expressions in TE that are compatible with $\prec$ and VC, are executable in s and generate a state $s'' \supseteq s'$.

On the one hand, the POCL planning process consists in inserting tasks, ordering constraints, and causal links into partial plans until all preconditions are supported by formulae and all causal conflicts disappear. On the other hand, the HTN planning process endeavors to decompose all abstract tasks, until all tasks in the partial plan are primitive.

The XML code in Listing 6.1 defines formally our running planning problem in the planning language of PANDA. Note, that there is no initial task network in this problem.

```
<?xml version="1.0" encoding="UTF-8"?>
<!DOCTYPE problem SYSTEM "problem.dtd">
<problem domainModel="TuneTemperatureDomainModel"
                name="TuneTemperatureProblem">
   <initialStateDescription>
      <fact name="provide_cooling"/>
      <fact name="increase_luminance_level"/>
      <fact name="decrease_luminance_level"/>
      <fact name="provide_current_luminance_level"/>
      <fact name="is_home">
         <constant name="Ulmstreet" sort="location"/>
      </fact>
      <fact name="provide_transport"/>
      <fact name="is_a_shop">
         <constant name="Buildhouse" sort="location"
                                     type="rigid"/>
      </fact>
       <fact name="is_different">
           <constant name="Buildhouse" sort="location"/>
           <constant name="Ulmstreet" sort="location"/>
       </fact>
       <fact name="is_different">
           <constant name="Ulmstreet" sort="location"/>
           <constant name="Buildhouse" sort="location"/>
       </fact>
      <fact name="at">
         <constant name="Ulmstreet" sort="location"/>
      </fact>
      <fact name="provide_current_temperature"/>
   </initialStateDescription>
   <goalStateDescription>
      <fact name="temperature_tuned">
         <constant name="Ulmstreet" sort="location"/>
      </fact>
```

```
        <fact name="luminance_tuned">
            <constant name="Ulmstreet" sort="location"/>
        </fact>
    </goalStateDescription>
</problem>
```

Listing 6.1 Illustrative planning problem.

In this framework, the search space associated with a planning problem depends on several parameters such as the number of operators, the number of expansion methods, the number of preconditions and effects of each operator, the number of parameters of each operator, the number of predicates, and the number of objects.

Using the framework for the implementation of a search engine is particularly advantageous, since (i) it allows us to easily encode and efficiently deal with procedural knowledge given at different abstraction levels (HTN planning) and (ii) it offers a great flexibility thanks to the capacity of reasoning about causal relations between tasks (POCL planning). With these features, we can model and solve a large variety of real-world planning problems.

6.3.2 Planning Strategies

Transforming partial plans into their refinements is done by using so-called *plan modifications*. Given a partial plan $P = (TE, \prec, VC, CL)$ and a domain model D, a plan modification is defined as $\mathtt{m} = (E^{\oplus}, E^{\ominus})$, where $E^{\oplus}$ and $E^{\ominus}$ are disjoint sets of elementary additions and deletions of plan components over P and D, and $E^{\oplus} \cup E^{\ominus} \neq \emptyset$. Consequently, all elements of $E^{\ominus}$ are from TE, $\prec$, VC, or CL, respectively, while $E^{\oplus}$ consists of new plan components. This generic definition makes the changes that a modification imposes on a plan explicit. With this, the available options for a search strategy become comparable qualitatively and quantitatively. $\mathtt{P}$ denotes the set of all plans, while $\mathtt{M}$ constitutes the set of all plan modifications. The application of plan modifications is characterized by the generic plan transformation function $app : \mathtt{M} \times \mathtt{P} \rightarrow \mathtt{P}$ which takes a plan modification $\mathtt{m} = (E^{\oplus}, E^{\ominus})$ and a plan P, and returns a plan P' that is obtained from P by adding all components of $E^{\oplus}$ and removing those of $E^{\ominus}$.

$\mathtt{M}$ is grouped into modification classes $\mathtt{M}_{\mathtt{y}}$. As an example, the class $\mathtt{M}_{\mathtt{AddCL}}$ contains plan modifications $\mathtt{m} = (\{\langle te_i, \phi, te_j \rangle,\ v_1 \doteq \tau_1, \ldots, v_k \doteq \tau_k\}^{\oplus}, \{\}^{\ominus})$ for manipulating a given partial plan $P = (TE, \prec, VC, CL)$ by adding causal links. The plan steps te_i and te_j are in such a modification in TE, and the codesignations represent variable substitutions. They induce a VC'-compatible substitution σ' with $VC' = VC \cup \{v_1 \doteq \tau_1, \ldots, v_k \doteq \tau_k\}$ such that $\sigma'(\phi) \in \sigma'(\mathrm{add}(te_i))$ for positive literals ϕ, $\sigma'(|\phi|) \in \sigma'(\mathrm{del}(te_i))$ for negative literals, and $\sigma'(\phi) \in \sigma'(\mathrm{prec}(te_j))$.

The complete collection of plan modifications for our hybrid planning framework presented above is introduced in [16]. This also covers the decomposition of abstract plan steps and the insertion of new plan steps, ordering constraints, and variable (in-) equations.

For a partial plan P that is a refinement of the initial task network of a given problem, but is not yet a solution, so-called *flaws* are used to make the violations of the criteria defined above explicit. Flaws list those plan components that constitute the deficiencies of the partial plan. The set of all flaws is denoted by $\mathtt{F}$, and subsets $\mathtt{F}_{\mathtt{x}}$ represent classes of flaws. For example, the class $\mathtt{F}_{\mathtt{CausalThreat}}$ contains flaws $\mathtt{f} = \{\langle te_i, \phi, te_j \rangle, te_k\}$ describing causal threats; i.e., such flaws indicate that a task te_k is possibly being ordered between plan steps te_i and te_j, and there exists a variable substitution σ that is consistent with the equations and in-equations imposed by the variable constraints in VC such that $\sigma(\phi) \in \sigma(\mathrm{del}(te_k))$ for positive literals ϕ or $\sigma(|\phi|) \in \sigma(\mathrm{add}(te_k))$ for negative literals. This means, that the presence of te_k in P as it stands will possibly corrupt the executability of at least some ground linearizations of P.

Flaw classes also cover the presence of abstract tasks in the plan, ordering and variable constraints inconsistencies, unsupported preconditions of actions, etc. The complete class definitions can be found in [16]. It can be shown, that these flaw definitions are complete in the sense that for any given planning problem π and plan P that is not flawed, P is a solution to π.

Based on the formal notions of plan modifications and flaws, a generic algorithm and planning strategies can be defined. A strategy specifies *how* and *which* flaws in a partial plan are eliminated through appropriate plan modification steps.

A class of plan modifications $\mathtt{M}_{\mathtt{y}} \subseteq \mathtt{M}$ is called *appropriate* for a class of flaws $\mathtt{F}_{\mathtt{x}} \subseteq \mathtt{F}$, if and only if there exist partial plans P which contain flaws $\mathtt{f} \in \mathtt{F}_{\mathtt{x}}$ and plan modifications $\mathtt{m} \in \mathtt{M}_{\mathtt{y}}$, such that the refined plans $P' = app(\mathtt{m}, P)$ do not contain these flaws any longer.

It is easy to see that the plan modifications perform a strict refinement, that means, that a subsequent application of any modification instances cannot result in the same plan twice; the plan development is inherently acyclic. Given that, any flaw instance cannot be re-introduced once this has been eliminated. This qualifies the appropriateness relation as a valid strategic advice for the plan generation process and motivates its use as the trigger function for plan modifications: the α modification triggering function relates flaw classes with their potentially solving plan modification classes. As an example, causal threat flaws can be solved by expanding abstract actions which are involved in the threat (by overlapping task implementations), by promotion or demotion, or by separating variables through inequality constraints [2]: $\alpha(\mathtt{F}_{\mathtt{CausalThreat}}) = \mathtt{M}_{\mathtt{ExpTask}} \cup \mathtt{M}_{\mathtt{AddOrdCstr}} \cup \mathtt{M}_{\mathtt{AddVarCstr}}$. Please note, that α states nothing about the relationship between the actual flaw and modification instances.

The modification triggering function allows for a simple plan generation process: (1) the flaws of the current plan are collected; if no flaw is detected, the plan is a solution; (2) suitable plan modifications are generated using the modification trigger; if for some flaws no plan modification can be found, the plan is discarded (dead-end); (3) selected plan modification are applied and and generate further refinements of the plan; (4) the next candidate plan is selected and we proceed with (1).

Note, that the tasks of the generated solution plans are partially ordered, which is desirable for realistic applications where activities are often to be performed in parallel.

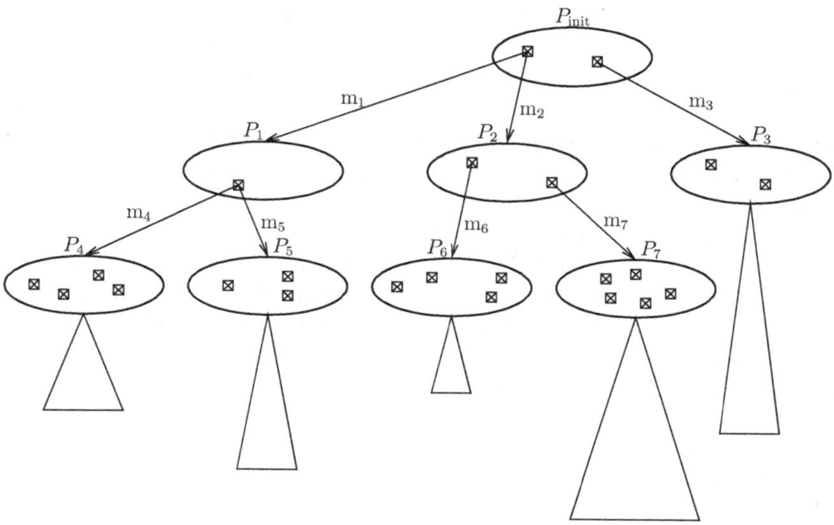

Fig. 6.3 Search space for refinement planning.

The search space for refinement planning can be represented by a graph shown in Figure 6.3. Large elliptic nodes represent partial plans. Small boxes inside nodes correspond to flaws, and arrows symbolize plan modifications. Triangles on the figure mean portions of the search space that are not explicitly shown. For example, partial plan P_2 is created by applying plan modification $\mathtt{m}_2$ to partial plan P_{init} (the initial task network). During the search for finding the plan presented in Figure 6.1 and Figure 6.2, the planning system explores 44 partial plans, including dead-ends.

The above mentioned triggering function completely separates the computation of flaws from the computation of modifications, and in turn both computations are independent from search-related considerations. The system architecture relies on this separation and exploits it in two ways: module invocation and interplay are specified through the α-trigger, while reasoning about search can be performed on the basis of flaws and modifications without taking their actual computation into account. Hence, we map flaw and modification classes directly onto groups of modules which are responsible for their computation.

A *detection module* x is a function that, given a partial plan P and a problem specification π, returns all flaws of type x that are present in the plan: $f_x^{\text{det}} : \mathtt{P} \times \Pi \rightarrow 2^{\mathtt{F_x}}$ where $\mathtt{P}$ is the set of all plans, and Π is the set of all planning problems.

A *modification module* y is a function which computes all plan modifications of type y that are applicable to a given plan P and that are appropriate for given flaws $\mathtt{f}$ with respect to a given domain model D: $f_y^{\text{mod}} : \mathtt{P} \times \mathtt{F_x} \times \mathtt{D} \rightarrow 2^{\mathtt{M_y}}$ for $\mathtt{M_y} \subseteq \alpha(\mathtt{F_x})$ where $\mathtt{D}$ is the set of all domain models.

Please note, that plan modifications carry a reference to the flaw instance they address; i.e., any plan modification is unambiguously linked with its triggering flaw.

While the plan deficiency detectors and the refinement generators provide the basic plan generation functionality, strategy functions can be designed for reasoning about which paths in the refinement space to pursue. To this end, we split up reasoning about search into two compartments. The first compartment is an option evaluation that is performed in the local view of the currently processed plan; it reasons about the detected flaws and proposed refinements in the current plan and assesses the modifications. The second component is responsible for the global view on the refinement space and evaluates the alternative search options.

We begin with the definition of a strategic function that selects all plan modifications that are considered to be worthwhile options, thereby determining the ordered set of successors for the current plan in the plan refinement space. In doing so, the following function also determines the branching behavior of the upcoming refinement-based planning algorithm.

Given a plan P, a set of flaws $\mathtt{f}$, and a set of plan modifications $\mathtt{m}$, a *modification-selection module* is a function $f^{\text{modSel}} : \mathtt{P} \times 2^{\mathtt{F}} \times 2^{\mathtt{M}} \rightarrow 2^{\mathtt{M} \times \mathtt{M}}$ that selects some (or all) of the plan modifications and returns them in a partial order for application to the passed plan.

Strategies discard a plan P, if any flaw remains unaddressed by the associated modification modules. That means, that we reject any plan P for any planning problem π, if for any f_x^{det} and $f_{y_1}^{\text{mod}}, \ldots, f_{y_n}^{\text{mod}}$ with $\mathtt{M}_{\mathtt{y_1}} \cup \ldots \cup \mathtt{M}_{\mathtt{y_n}} = \alpha(\mathtt{F}_{\mathtt{x}})$ the following holds: $\bigcup_{1 \leq i \leq n} f_{y_i}^{\text{mod}}(P, f_x^{\text{det}}(P, \pi), D) = \emptyset$.

The second aspect of search control concerns the selection of those plans that are to be processed next by the detection and modification modules. These unevaluated partial plans, the leaves of the search tree, are usually called the *fringe*. In other words, concrete implementations of the following module are responsible for the general search schema, ranging from uninformed procedures such as depth-first, breadth-first, etc., to informed, heuristic schemata.

A *plan-selection module* is a function that returns a partial order of plans for a given sequence of plans. This function is described as $f^{\text{planSel}} : \mathtt{P}^* \rightarrow 2^{\mathtt{P} \times \mathtt{P}}$.

Based on the building blocks defined so far, we can now assemble a planning system that integrates both Hierarchical Task Network planning and Partial-Order Causal-Link planning. A software artifact that implements the generic refinement planning algorithm (Alg. 1) is making the flaw detection and modification generating modules operational by stepwise collecting plan deficiencies, collecting appropriate modifications, selecting worthwhile modifications, and finally selecting the next plan in the fringe of the search tree. Please note, that the algorithm is formulated independently from the deployed modules, since the options to address existing flaws by appropriate plan modifications are defined via α. The body of the algorithm is basically divided into four sections:

- Termination (2-3): If no more plans in the fringe are due to examination, that means, that the search space is exhausted and does not contain any solution; the algorithm terminates.

Algorithm 1 The generic refinement planning algorithm.

Require: Sets of modules $\mathfrak{Det} = \{f_1^{\text{det}}, \ldots, f_d^{\text{det}}\}$ and $\mathfrak{Mod} = \{f_1^{\text{mod}}, \ldots, f_m^{\text{mod}}\}$
Require: Selection modules f^{modSel} and f^{planSel}

    plan$(P_1 \ldots P_n, \pi)$**:** $\{P_1$ is the plan that is worked on$\}$
    if $n = 0$ **then**
3:     **return** failure
    $P \leftarrow P_1$; Fringe $\leftarrow P_2 \ldots P_n$
    $F \leftarrow \emptyset$
6: **for all** $f_x^{\text{det}} \in \mathfrak{Det}$ **do** {Flaw detection}
        $F \leftarrow F \cup f_x^{\text{det}}(P, \pi)$
    if $F = \emptyset$ **then**
9:     **return** P
    $M \leftarrow \emptyset$
    for $x = 1$ to d **do** {Modification generation}
12:     $F_x = F \cap \mathtt{F_x}$ {Process flaws class-wise as returned by corresponding f_x^{det}}
        for all $\mathtt{f} \in F_x$ **do**
            for all $f_y^{\text{mod}} \in \mathfrak{Mod}$ with $\mathtt{M_y} \subseteq \alpha(\mathtt{F_x})$ **do**
15:                 $M \leftarrow M \cup f_y^{\text{mod}}(P, \mathtt{f}, D)$
            if `f` was un-addressed **then**
                $P_{\text{next}} \leftarrow f^{\text{planSel}}(\text{Fringe})$
18:                 **return** plan$(P_{\text{next}} \circ (\text{Fringe} \setminus P_{\text{next}}), \pi)$
    for all `m` in *linearize*$(f^{\text{modSel}}(P, F, M))$ **do** {Strategic choices}
        Fringe $\leftarrow$ *app*$(\mathtt{m}, P) \circ$ Fringe
21: $P_{\text{next}} \leftarrow$ *first*(*linearize*$(f^{\text{planSel}}(\text{Fringe})))$
    return plan$(P_{\text{next}} \circ (\text{Fringe} \setminus P_{\text{next}}), \pi)$

- Flaw detection (5-9): The results of all deployed detection functions $\mathfrak{Det}$ are collected. If no deficiency can be spotted, the current plan is considered to be a solution to the given planning problem.
- Plan modifications generation (10-18): The applicable plan modification steps are accumulated per flaw class and per class instance from the set of available plan modification generators $\mathfrak{Mod}$ according to the α-defined assignments. If any flaw is found unaddressed by its associated plan modification generation functions, the current plan is discarded and the algorithm is called recursively with a newly selected current candidate plan.
- Strategy (19-22): All plan modifications that pass the strategic plan modification selection function are applied to the current plan and thereby constitute the set of its refinements, that is, the set of its successor plans. This fringe extension is established by the strategic decisions in $f^{\text{modSel}}$and inserted at the beginning of the fringe. The plan selection function f^{planSel}finally chooses the next current plan and the procedure is called recursively.

The algorithm uses a function *linearize* to compute linear sequences of plans, respectively plan modifications that are consistent with the partial orders obtained from the appropriate strategic selection function.

The hybrid planning framework introduced in the previous section and the generic refinement planning algorithm (Alg. 1) are very adequate to building planning systems for real-world applications in IEs, since: (i) we can encode formally the procedural knowledge about everyday activities of people thanks to abstract tasks and expansion methods (owing to HTN planning); and (ii) the algorithm offers the means necessary to resolve causal conflicts between tasks that share resources of the IE (thanks to POCL planning). In addition, the integration of POCL and HTN planning techniques allows for the causal reasoning at different abstraction levels in the task hierarchy due to decomposition axioms, which is very powerful.

A large number of search strategies can be realized in the proposed refinement-planning framework by sequencing the respective selection modules and using the returned partially ordered sets of modifications, respectively plans, to modulate preceding decisions: If the primary strategy does not prefer one option over the other, the secondary strategy is followed and so on, until finally a random preference is assumed.

Some strategies are *unflexible* in the sense that they represent a fixed preference schema on the flaw type they want to get eliminated and then select appropriate modification methods. A traditional form of modification selection is either to prefer or to disfavor categorically specific classes of plan modifications; e.g., we may prefer the decomposition of tasks to task insertions.

With our refinement planning framework, it is also possible to design modification-selection strategies that are capable of operating on a more general level than unflexible strategies by exploiting flaw/modification information: They are neither flaw-dependent, as they do not primarily rely on a flaw type preference schema, nor modification-dependent, since they do not have to be biased in favor of specific modification types. A representative is the modification-selection strategy Least Committing First (LCF) which selects those modifications that address flaws for which the smallest number of alternative plan modifications has been found:

$$\begin{aligned}&\mathtt{m}_i < \mathtt{m}_j \in f^{\text{modSel}}_{\text{LCF}}(P, \{\mathtt{f}_1,\ldots,\mathtt{f}_m\}, \{\mathtt{m}_1,\ldots,\mathtt{m}_n\})\\&\quad \text{if } 1 \le x_i, x_j \le m,\ 1 \le i,j \le n,\ \mathtt{m}_i \in \mathit{modFor}(\mathtt{f}_{x_i}, P),\ \mathtt{m}_j \in \mathit{modFor}(\mathtt{f}_{x_j}, P)\\&\qquad \text{and } |\mathit{modFor}(\mathtt{f}_{x_i}, P)| < |\mathit{modFor}(\mathtt{f}_{x_j}, P)|.\end{aligned}$$

It can easily be seen that this is a *flexible* strategy, since it does not depend on the actual types of issued flaws and modifications: It just compares answer set sizes in order to keep the branching in the search space low. In [15], more details about flexible strategies are presented.

In the next section, we will explain (i) how the planning techniques offered by the formal framework can participate in the composition of services in IE, and (ii) how we have shaped a search engine for the PA, based on this formal framework, that is particularly suited for solving planning problems during the composition of services.

6.4 Planning in Intelligent Environments

In ATRACO, each Planning Agent (PA) encapsulates an AI planning system. This system is a search engine for hybrid planning that relies on the formal framework presented in Section 6.3.1. This search engine is actually a specific configuration of the existent development platform PANDA. For specifying a PANDA planning problem, we need two sources: the planning domain and the planning problem.

The planning domain contains the various operators and the alternatives (by means of expansion methods) to implement each abstract operator. Furthermore, predicates and types of objects of the application domain are represented.

The planning problem describes the problem instance to be solved. The description is made up of three parts: The initial world state, the specification of the goal world state, and the initial task network. In addition, we declare the objects of the problem at hand; e.g., the abstract services that are available in the room where a particular user is.

Searching for solutions to a planning problem is complex, since the search space associated with such a problem may be huge (e.g., see Figure 6.3). In order to cope with this issue, we need planning strategies that guide the search towards solutions.

In ATRACO, we address the deployment of ASs using a service-oriented architecture approach in which resources in the Ambient Intelligence environment provide independent, loosely coupled services, see Section 1.3. A new PA is created, each time a new AS is instantiated; when an AS is dissolved, the associated PA is destructed. The PA is part of the advanced service composition mechanism of ATRACO.

In the literature, many research efforts addressing the Web service composition problem via AI planning have been reported (see Section 6.2). In terms of Web services, the initial world state and the goal world state are specified in the requirement of Web services requesters. The set of operators correspond to the set of available Web services, unlike our approach where abstract services are represented by facts. For example, McDermott [9] presented a Web service composition method based on PDDL and introduced a type of knowledge called value of an action that persists and is not treated as a truth literal.

In our approach, each user goal corresponds to a particular planning domain. From the abstract services offered currently by the AmI environment and known by Ontology Manager (OM), we generate the planning problem π to be solved by the PA. The planning problem depends on the user goal to be attained.

The OM is responsible for the creation of the Sphere Ontology which will include references to all the IE resources relevant to the AS that have been discovered.

The PA is in charge of finding a solution plan P to π. This plan describes abstract tasks that require abstract services in order to support the user goals. From the set of applied plan refinements that have led to P, the PA generates a workflow expressed in the ATRACO-BPEL language.

ATRACO-BPEL abstract service workflows comprise a set of activities and abstract services associated with these activities. These activities are structured with special constructs and correspond to the task of the plan generated by the PA. Actually, an abstract service workflow represents the abstract process model of an AS.

A flow diagram for the PA is shown in Figure 6.4. The PA communicates with the Sphere Manager (SM) and OM for the creation of abstract service workflows.

The SM is in charge of (i) binding services provided by IE resources to the abstract services of the abstract service workflows, and then (ii) executing and controlling the execution of the executable service workflows that result from the binding. During the execution of these workflows, the SM can handle execution errors and rebind abstract services if necessary. If the rebinding fails, the SM calls the PA again for replanning.

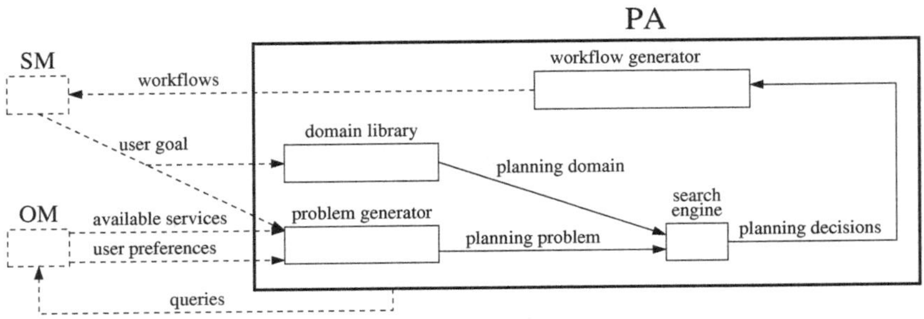

Fig. 6.4 Flow diagram of the Planning Agent.

The life cycle of the PA is described as follows:

1. The SM instantiates a new PA for user goal *goal*.
2. The PA receives *goal* from the SM. The planning domain that matches *goal* is selected, and a new planning problem $\pi = (D, s_{\text{init}}, s_{\text{goal}}, P_{\text{init}})$ is then defined dynamically; i.e., s_{init}, s_{goal}, and P_{init} are defined.
3. A backtracking search starts in the space of partial plans in order to find a plan P that solves π.
4. From the set of plan modifications that have been applied during search and that have led to the solution plan, a workflow W, expressed in the ATRACO-BPEL language, is generated by the PA.
5. The PA transmits W to the SM.
6. The SM makes dynamic service binding for W (see Section 1.4.2) and executes W:

   - During the execution of W, if a concrete service that is bound to an abstract service of W is no longer available, SM searches for a new concrete service to be bound to this abstract service. If no concrete service is found, the binding fails and replanning takes place:
     a. a new planning problem π' is defined by querying OM (π' differs from π either in the initial state or in the goal state);
     b. we proceed with step 3 by replacing π with π'.

- When the execution of W is finished, the corresponding PA is destructed.

The initial generation of W in step 3 relies on the generic refinement planning algorithm (see Algorithm 1) and supports the creation of ASs. The replanning procedure described in step 7 provides the structural adaptation of ASs.

6.4.1 Generation of Planning Problems

The search engine of the PA is based on the formal framework for refinement-based planning presented in 6.3.1. In this framework, states and *preconditions* and *effects* of tasks are specified through formulae of a fragment of first-order logic. A world state is thus represented by a list of facts. Contrary to other approaches presented in the literature (e.g., [9, 17]), we do not create planning domains dynamically, but instead the PA creates planning problems dynamically: (1) Depending on the user goal to be achieved, the currently available abstract services are modeled either by the facts the initial world state or by the facts of the goal world state (e.g., `ambientLightLevel(service0)` in the domain "entertainment"); (2) if the initial world state s_{init} is not empty, then the user goal is modeled by tasks of the initial task network P_{init}. For determining s_{init} and s_{goal} of the planning problem at hand, the PA queries the OM to get the list of available abstract services that are relevant to the current planning domain; e.g., the PA asks the OM whether there are some abstract services available in the IE for sensing and controlling the ambient temperature in the room where a particular user is. Section 3.3 introduces ontologies and ontology managers.

In the planning formalism presented in 6.3.1, an ATRACO planning problem is described as follows:

- the user goal indicates the planning domain to be selected and is possibly modeled by the tasks of the initial task network P_{init};
- the currently available abstract services that can support the current user's goals are modeled by the facts of the world state s_{init} or s_{goal}.

We have designed a planning domain for AS "prepare an unexpected dinner," and an excerpt of which is included in Appendix.

6.4.2 Backtracking Search for Solution Plans

Some of the ATRACO planning domains we have designed contain a large number of expansion methods, the initial task network contains some tasks, and the goal state of each ATRACO planning problem π is then empty ($s_{\text{goal}} = \emptyset$). In this context, the insertion of tasks during backtracking search is not a necessity: in practice, we can find concrete plans without applying plan modifications of class $\mathtt{M}_{\mathtt{InsTask}}$. The configuration of PANDA reduces quite a bit the size of the search space to be explored. In addition, search is not going to be trapped into partial plans that contain arbitrary long sequences of reversible tasks. Such reversible tasks and causal links are inserted recursively in POCL planning in order to support open preconditions, but the insertion creates cycles. Avoiding the creation of these cycles during search makes the backtracking search more efficient.

For addressing the planning problems, we use the following unflexible plan-selection strategy:

$$P_i < P_j \in f^{\text{planSel}}_{\text{MorePlanSteps}}(P_1, \ldots, P_n) \quad \text{if } |TE_i| > |TE_j|$$

that prefers partial plans with more plan steps. We apply this plan-selection strategy, since each solution plan should contain as many tasks (i.e. services) as possible to support a user's goals.

The PA uses a flexible modification-selection strategy to solve ATRACO planning problems: Least Committing First (see Section 6.3.2).

In theory, searching for solutions to a planning problem is complex, since the search space associated with such a problem may be huge. The complexity of the backtracking search done by the PA is limited in practice for ATRACO, since (1) the planning domains and the planning problems are designed at a high level: the number abstract services is small; i.e, the size of world states is small, (2) the number of preconditions and effects per tasks is small, and (3) the number of causal interactions between the task networks of different abstract tasks is small. For our practical application, the size of search space is small enough to be explored in a few seconds using a Pentium 4 processor 3GHz. For example, for a planning domain with 5 primitive tasks, 3 abstract tasks, 8 expansion methods, 8 predicates, and 2 preconditions and 2 effects per task, a corresponding planning problem consisting of one single abstract task and 6 objects is solved after about 8s, after having explored 154 partial plans including dead-ends. Note, however, that this planning domain involves a large number of causal interactions between tasks.

Moreover, we can still expect a significant speedup of the planning process by integrating some inference mechanisms. For example, there is not yet any special provision to deal with duplicate partial plans and cycles in the search space such as the approach presented in [18]. In addition, there is a recent paper that presents a landmark technique for HTN planning [4]. This technique filers out some parts of a planning domain that are irrelevant for the planning problem at hand.

6.4.3 Generation of Workflows

In ATRACO, the SM manages and executes workflows with services. A workflow consists of abstract and primitive tasks that are partially ordered. We have decided to use a variant of the language BPEL (also known as WS-BPEL, Web Service Business Process Execution Language) for the description of the ATRACO workflows. This XML language is called ATRACO-BPEL. The PA is in charge of generating workflows expressed in ATRACO-BPEL at design time, and then the PA transmits these workflows directly to the SM (see step 5 of the life cycle of the PA above).

There are three kinds of basic BPEL activities called respectively `<bpel:invoke>`, `<bpel:receive>`, and `<bpel:reply>`, and they are associated with primitive tasks of the planning domain. The structured activities of the BPEL-like language, such as `<bpel:while>` and `<bpel:repeatUntil>` used to express the repeated execution of activities are associated with expansion methods of the planning domain. The structured activities of type `<bpel:pick>` are also associated with expansion methods.

The BPEL constructs `<bpel:partnerLink>` and `<ATRACO:role>` used to express the abstract services required and provided by each basic activity are directly inserted into a workflow depending on what primitive tasks are present in the corresponding solution plan. These constructs are used to associate activities with abstract services.

The BPEL construct `<bpel:sequence>` used to describe precedence constraints between workflow activities corresponds to an abstract task and the expansion method that has been applied to decompose it and that describes a network of tasks that are totally ordered with ordering constraints. The construct `<bpel:link>` is used to synchronize workflow activities that belong to different structured activities. This construct also corresponds to any ordering constraint in the solution plan that has been explicitly added during the planning process; e.g., the insertion of such an ordering constraint is sometimes necessary to remove causal threats.

The BPEL construct `<bpel:flow>` used to describe workflow activities that execute in parallel corresponds to an abstract task and the expansion method that has been applied to decompose it and that describes a network of tasks that are not ordered.

For the generation of abstract service workflows in our BPEL-like language, the workflow generator of the PA looks for and analyze the set of plan modifications $\mathtt{M}_{\mathtt{success}}$ that have led to solution plans and that belong to two classes: $\mathtt{M}_{\mathtt{ExpTask}}$ (decomposition of abstract tasks) and $\mathtt{M}_{\mathtt{AddOrdCstr}}$ (addition of ordering constraints). The formal definition of these plan modification classes is given in [16]. Each plan modification of $\mathtt{M}_{\mathtt{ExpTask}}$ is interpreted as a BPEL structured activity such as `<bpel:flow>` composed possibly of other structure activities and basic activities (`<bpel:invoke>`, `<bpel:receive>`, and `<bpel:reply>`). The analysis of the plan modifications of $\mathtt{M}_{\mathtt{AddOrdCstr}}$ leads to the identification of `<bpel:sequence>` or `<bpel:link>` constructs. Since we have decided to deactivate the modification module $f^{\mathrm{mod}}_{\mathtt{InsTask}}$ of PANDA in the PA, there are no plan modifications of class $\mathtt{M}_{\mathtt{InsTask}}$ in $\mathtt{M}_{\mathtt{success}}$.

The search space explored during the planning process is represented by a tree. The cost of generating workflows after the planning process is linear in the number of successful plan modifications $\mathtt{M}_{\mathtt{success}}$, since all the information we need is stored inside one single path: the cost depends on the depth of the tree (i.e., the number of applied plan modifications) from the initial task network P_{init} to the solution plan.

The XML code in Listing 6.2 defines a simple ATRACO-BPEL workflow with abstract services. The workflow consists of three activities that are not ordered (i.e., they are to be executed in parallel): musicControl, lightsControl, and photoViewerControl.

```
<bpel:process name="entertainment"
              targetNamespace="http://eclipse.org/bpel/sample
                 "
  xmlns:tns="http://eclipse.org/bpel/sample"
  xmlns:bpel="http://docs.oasis-open.org/wsbpel/2.0/
                                  process/executable"
  xmlns:ATRACO="http://daisy.cti.gr/ATRACO_BPEL">
  <!-- ============================================ -->
  <!-- PARTNER LINKS                                 -->
  <!-- List of services participating in this BPEL   -->
  <!-- process                                       -->
  <!-- ============================================ -->
  <bpel:partnerLinks>
    <bpel:partnerLink name="ListenMusic"
          partnerLinkType="Trigger"
          partnerRole="ATRACO:ListenMusic"/>
    <bpel:partnerLink name="SetupLight"
          partnerLinkType="Continuous"
          myRole="ATRACO:LightControl"
          partnerRole="ATRACO:TriggerLight"/>
    <bpel:partnerLink name="ShowPhotos"
          partnerLinkType="Trigger"
          partnerRole="ATRACO:ShowPhotos"/>
  </bpel:partnerLinks>
  <!-- ============================================ -->
  <!-- ORCHESTRATION LOGIC                           -->
  <!-- Set of activities coordinating the flow of    -->
  <!-- messages across the                           -->
  <!-- services integrated within this business process -->
  <!-- ============================================ -->
  <bpel:sequence name="main">
    <bpel:flow name="Flow">
      <bpel:reply name="musicControl"
                  partnerLink="ListenMusic"
                  operation="music"/>
      <bpel:invoke name="lightsControl"
                   partnerLink="SetupLight"
                   operation="light"/>
      <bpel:reply name="photoViewerControl"
                  partnerLink="ShowPhotos"
                  operation="photo"/>
    </bpel:flow>
  </bpel:sequence>
  <!-- ============================================ -->
  <!-- ROLES                                         -->
  <!-- ============================================ -->
  <ATRACO:roles>
    <ATRACO:role name="ListenMusic" type="output"
                 agentMonitored="no"
```

```
                  interactionType="direct">
        <ATRACO:device type="Music"/>
      </ATRACO:role>
      <ATRACO:role name="ShowPhotos" type="output"
                  agentMonitored="no"
                  interactionType="direct">
        <ATRACO:device type="ShowPhoto"/>
      </ATRACO:role>
      <ATRACO:role name="LightControl" type="input"
                  agentMonitored="no" interactionType="no">
        <ATRACO:service type="AmbientLightLevel"
                        triggerValue="On" resetValue="Off"
                        quantity="1" specialRules=""/>
      </ATRACO:role>
      <ATRACO:role name="TriggerLight" type="output"
                  agentMonitored="yes"
                  interactionType="direct">
        <ATRACO:device type="OutputLightLevel"/>
      </ATRACO:role>
    </ATRACO:roles>
  </bpel:process>
```

Listing 6.2 Illustrative workflow with abstract services

6.5 ATRACO Contribution

The Planning Agent (PA) proposes to the Sphere Manager (SM) the service of generating workflows with abstract services that describe what basic activities are to be executed and in which order they are to be executed for achieving user goals. The PA uses some services offered by the Ontology Manager (OM) in order to define the initial world state of the planning problem at hand.

The SM is in charge of (i) binding services provided by IE resources to the abstract services of the abstract service workflows, and then (ii) executing and controlling the execution of the executable service workflows that result from the binding. The OM creates the Sphere Ontology which will include references to all the IE resources relevant to the Activity Sphere that have been discovered. More details about SM and OM are given in Section 1.3.3.

In the prototype system, The PA integrates some components of PANDA, and also includes a generator of ATRACO-BPEL workflows. The workflows are stored in files to be parsed and used by SM, which creates and executes workflows with concrete services.

6.6 Conclusion

Sphere Adaptation (SA) is a major aspect of adaptation and evolution with AmI environments. In the ATRACO project, we address SA in the context of a distributed, ontology-based framework with a service-oriented architecture approach. The objective is to support users in their home for various everyday activities.

In this chapter, we elaborated the concept of SA within ATRACO. We described how Artificial Intelligence planning techniques contribute to SA for the realization and the structural adaptation of activity spheres (ASs).

An existing formal framework for refinement-based hybrid planning that integrates hierarchical task network planning and partial-order causal link planning was presented. In addition, we detailed an existent development platform for hybrid planning, PANDA, which is based on the formal framework. We explained that the large flexibility, the high reasoning capacities and the important expressive power offered by this formal framework and the development platform are well suited to participating in the composition of services in Intelligent Environments. We introduced the Planning Agent (PA) that encapsulates a search engine based on PANDA and which takes part in the advanced service composition of ATRACO and is responsible for the generation of workflows with abstract services at design time. The workflows are part of ASs and consist of the basic activities to be executed in order to achieve user goals in different and changing environments. At run time, the activities are bound to concrete services dynamically by the Sphere Manager (SM). This approach for the composition of services is indeed powerful, since it can cope with complex and large-scale environments. The adaptations of services and devices are indeed possible due to the dynamic binding of services at run time. The flexibility offered by abstract service workflows prevents the running system from replanning each time a service appears or disappears in the AmI environment.

This chapter also explained how we configure and adapt PANDA to shape the search engine of the PA: particular modules and planning strategies are selected in order to deal efficiently with ATRACO planning problems. The problems are tractable and solved in a couple of seconds at design time, since the number of abstract services is limited and there are few causal interactions between tasks. Finally, we detailed how the PA generates workflows with abstract services in the ATRACO-BPEL language in cooperation with SM and the Ontology Manager, based on particular information included in the ontology of the execution environment.

6.7 Further readings

A journal paper of Ramos, Augusto, and Shapiro introduces some general ideas about exploiting Artificial Intelligence (AI) for addressing the issues of Ambient Intelligence [12].

In a book, Ghallab, Nau, and Traverso present a survey about AI planning [5]. They explain a broad range of planning approaches. In particular, a book chapter is dedicated to hierarchical task network planning.

The PhD thesis of Schattenberg is focused on hybrid planning and scheduling, and it describes in detail the PANDA experimental platform [14].

The most prominent journals that focus on AI are Artificial Intelligence (AI Journal) and Journal of Artificial Intelligence Research (JAIR). The International Joint Conference on Artificial Intelligence (IJCAI), the National Conference on Artificial Intelligence (AAAI), and the European Conference on Artificial Intelligence (ECAI) are renowned general AI conferences. The International Conference on Artificial Intelligence Planning and Scheduling (AIPS) and the International Conference on Automated Planning and Scheduling (ICAPS) are devoted specifically to AI planning and scheduling.

Appendix

This appendix consists of an excerpt of the planning domain for the activity sphere "prepare an unexpected dinner."

```
<?xml version="1.0" encoding="UTF-8"?>
<!DOCTYPE domainModel SYSTEM "model.dtd">
<domainModel name="Kitchen_threads">
   <sortDefinition name="hotplate" type="concrete"/>
   <sortDefinition name="oven" type="concrete"/>
   <sortDefinition name="microwave" type="concrete"/>
   <sortDefinition name="device" type="abstract">
      <documentation>Locations with additional functionality
          and only suitable for a number of containers.</
          documentation>
      <subSortStatement subsort="hotplate"/>
      <subSortStatement subsort="oven"/>
      <subSortStatement subsort="microwave"/>
   </sortDefinition>

   <!-- ... -->

   <sortDefinition name="resource" type="abstract">
      <documentation>Parent sort to all objects of the
          kitchen subjected to planning.</documentation>
      <subSortStatement subsort="food"/>
      <subSortStatement subsort="equipment"/>
   </sortDefinition>


   <!-- ... -->

   <relationDeclaration name="has_context" type="flexible">
      <sortExpression name="resource"/>
      <sortExpression name="context"/>
```

```
    </relationDeclaration>

    <!-- ... -->

    <decompositionAxiom name="clean_or_no_context_axiom">
       <varDeclaration name="coca_res" sort="resource"/>
       <varDeclaration name="coca_cxt" sort="context"/>
       <leftHandSide>
          <atomic name="clean_or_no_context">
             <variable name="coca_res"/>
             <variable name="coca_cxt"/>
          </atomic>
       </leftHandSide>
       <rightHandSide>
          <atomic name="clean">
             <variable name="coca_res"/>
          </atomic>
          <not>
             <atomic name="clean_or_no_context">
                <variable name="coca_res"/>
                <variable name="coca_cxt"/>
             </atomic>
          </not>
       </rightHandSide>
    </decompositionAxiom>

    <!-- ... -->

    <taskDeclaration name="move_container" type="primitive">
       <documentation>Move a container from one non_storage
           location to another.</documentation>
       <varDeclaration name="move_container_obj" sort="
           container"/>
       <varDeclaration name="move_container_from" sort="
           non_storage"/>
       <varDeclaration name="move_container_to" sort="
           non_storage"/>
       <and>
          <atomic name="free">
             <variable name="move_container_to"/>
          </atomic>
          <atomic name="ready">
             <variable name="move_container_obj"/>
          </atomic>
          <atomic name="at">
             <variable name="move_container_obj"/>
             <variable name="move_container_from"/>
          </atomic>
       </and>
       <and>
          <atomic name="free">
             <variable name="move_container_from"/>
          </atomic>
          <atomic name="at">
```

```
            <variable name="move_container_obj"/>
            <variable name="move_container_to"/>
         </atomic>
         <not>
            <atomic name="at">
               <variable name="move_container_obj"/>
               <variable name="move_container_from"/>
            </atomic>
         </not>
         <not>
            <atomic name="free">
               <variable name="move_container_to"/>
            </atomic>
         </not>
      </and>
   </taskDeclaration>

   <!-- ... -->

   <taskDeclaration name="procedure_fry" type="complex">
      <varDeclaration name="p_fry_food" sort="food"/>
      <varDeclaration name="p_fry_container" sort="
          fry_in_here"/>
      <varDeclaration name="p_fry_tool" sort="tool"/>
      <varDeclaration name="p_fry_cxt" sort="context"/>
      <and>
         <atomic name="ready">
            <variable name="p_fry_container"/>
         </atomic>
         <atomic name="ready">
            <variable name="p_fry_tool"/>
         </atomic>
      </and>
      <and>
         <atomic name="has_context">
            <variable name="p_fry_container"/>
            <variable name="p_fry_cxt"/>
         </atomic>
         <atomic name="has_context">
            <variable name="p_fry_food"/>
            <variable name="p_fry_cxt"/>
         </atomic>
         <atomic name="has_context">
            <variable name="p_fry_tool"/>
            <variable name="p_fry_cxt"/>
         </atomic>
      </and>
   </taskDeclaration>

   <!-- ... -->

   <methodDeclaration name="method_procedure_fry" taskRef="
       procedure_fry">
```

```
        <varDeclaration name="method_procedure_fry.p_fry_food"
            sort="food"/>
        <varDeclaration name="method_procedure_fry.
            p_fry_container" sort="fry_in_here"/>
        <varDeclaration name="method_procedure_fry.p_fry_tool"
            sort="tool"/>
        <varDeclaration name="method_procedure_fry.p_fry_cxt"
            sort="context"/>
        <taskNode name="mpfmc_into_pan" taskRef="move_food">
           <varDeclaration name="mpfmc_into_pan.move_food_obj"
               sort="movable"/>
           <varDeclaration name="mpfmc_into_pan.move_food_from"
               sort="container"/>
           <varDeclaration name="mpfmc_into_pan.move_food_to"
               sort="container"/>
        </taskNode>
        <taskNode name="mpfmc_context_container" taskRef="
            set_context">
           <varDeclaration name="mpfmc_context_container.
               scxt_obj" sort="resource"/>
           <varDeclaration name="mpfmc_context_container.
               scxt_cxt" sort="context"/>
        </taskNode>
        <taskNode name="mpfmc_context_food" taskRef="
            set_context">
           <varDeclaration name="mpfmc_context_food.scxt_obj"
               sort="resource"/>
           <varDeclaration name="mpfmc_context_food.scxt_cxt"
               sort="context"/>
        </taskNode>
        <taskNode name="mpfmc_context_tool" taskRef="
            set_context">
           <varDeclaration name="mpfmc_context_tool.scxt_obj"
               sort="resource"/>
           <varDeclaration name="mpfmc_context_tool.scxt_cxt"
               sort="context"/>
        </taskNode>
        <taskNode name="mpfmc_flip" taskRef="use_tool">
           <varDeclaration name="mpfmc_flip.use_tool_tool" sort
               ="tool"/>
           <varDeclaration name="mpfmc_flip.use_tool_on" sort="
               container"/>
        </taskNode>
        <taskNode name="mpfmc_onto_hotplate" taskRef="
            move_container">
           <varDeclaration name="mpfmc_onto_hotplate.
               move_container_obj" sort="container"/>
           <varDeclaration name="mpfmc_onto_hotplate.
               move_container_from" sort="non_storage"/>
           <varDeclaration name="mpfmc_onto_hotplate.
               move_container_to" sort="non_storage"/>
        </taskNode>
        <orderingConstraint predecessor="mpfmc_context_tool"
            successor="mpfmc_flip"/>
```

```
        <orderingConstraint predecessor="
            mpfmc_context_container" successor="mpfmc_into_pan
            "/>
        <orderingConstraint predecessor="mpfmc_context_food"
            successor="mpfmc_into_pan"/>
        <orderingConstraint predecessor="mpfmc_into_pan"
            successor="mpfmc_flip"/>
        <orderingConstraint predecessor="mpfmc_onto_hotplate"
            successor="mpfmc_into_pan"/>
        <valueRestriction type="eq" variable="
            method_procedure_fry.p_fry_container">
          <variable name="mpfmc_into_pan.move_food_to"/>
        </valueRestriction>
        <valueRestriction type="eq" variable="
            mpfmc_context_container.scxt_obj">
          <variable name="method_procedure_fry.p_fry_container
              "/>
        </valueRestriction>
        <valueRestriction type="eq" variable="
            method_procedure_fry.p_fry_cxt">
          <variable name="mpfmc_context_container.scxt_cxt"/>
        </valueRestriction>
        <valueRestriction type="eq" variable="
            method_procedure_fry.p_fry_food">
          <variable name="mpfmc_into_pan.move_food_obj"/>
        </valueRestriction>
        <valueRestriction type="eq" variable="
            mpfmc_context_container.scxt_obj">
          <variable name="method_procedure_fry.p_fry_container
              "/>
        </valueRestriction>
        <valueRestriction type="eq" variable="
            mpfmc_context_food.scxt_cxt">
          <variable name="method_procedure_fry.p_fry_cxt"/>
        </valueRestriction>
        <valueRestriction type="eq" variable="
            mpfmc_context_food.scxt_obj">
          <variable name="method_procedure_fry.p_fry_food"/>
        </valueRestriction>
        <valueRestriction type="eq" variable="
            method_procedure_fry.p_fry_cxt">
          <variable name="mpfmc_context_tool.scxt_cxt"/>
        </valueRestriction>
        <valueRestriction type="eq" variable="
            method_procedure_fry.p_fry_tool">
          <variable name="mpfmc_flip.use_tool_tool"/>
        </valueRestriction>
    </methodDeclaration>

    <!-- ... -->

</domainModel>
```

Listing 6.3 Excerpt of the planning domain for AS "Prepare an unexpected dinner."

References

1. Amigoni, F., Gatti, N., Pinciroli, C., Roveri, M.: What planner for ambient intelligence applications? IEEE Transactions on Systems, Man, and Cybernetics–Part A: Systems and Humans **35**(1), 7–21 (2005)
2. Biundo, S., Schattenberg, B.: From abstract crisis to concrete relief–a preliminary report on combining state abstraction and HTN planning. In: Proceedings of the 6th European Conference on Planning (ECP'01), pp. 157–168. Toledo, Spain (2001). Preprint
3. El Falou, M., Bouzid, M., Mouaddib, A.I., Vidal, T.: A distributed planning approach for web services composition. In: Proceedings of the 2010 IEEE International Conference on Web Services (ICWS), pp. 337–344. IEEE Computer Society, Miami, Florida, USA (2010)
4. Elkawkagy, M., Schattenberg, B., Biundo, S.: Landmarks in hierarchical planning. In: H. Coelho, R. Studer, M. Wooldridge (eds.) ECAI, *Frontiers in Artificial Intelligence and Applications*, vol. 215, pp. 229–234. IOS Press (2010)
5. Ghallab, M., Nau, D.S., Traverso, P.: Automated Planning: Theory and Practice. Morgan Kaufmann (2004)
6. Lundh, R., Karlsson, L., Saffiotti, A.: Autonomous functional configuration of a network robot system. Robotics and Autonomous Systems **56**(10), 819–830 (2008)
7. Marquardt, F., Uhrmacher, A.M.: Evaluating AI planning for service composition in smart environments. In: M. Wiberg, A.B. Zaslavsky (eds.) Proceedings of the 7th International Conference on Mobile and Ubiquitous Multimedia (MUM 2008), pp. 48–55. ACM, Umeå, Sweden (2008)
8. Mazzola Paluska, J., Pham, H., Saif, U., Chau, G., Terman, C., Ward, S.: Structured decomposition of adaptive applications. In: PerCom, pp. 1–10. IEEE Computer Society (2008)
9. McDermott, D.V.: Estimated-regression planning for interactions with web services. In: M. Ghallab, J. Hertzberg, P. Traverso (eds.) Proceedings of the Sixth International Conference on Artificial Intelligence Planning and Scheduling (AIPS), pp. 204–211. AAAI, Toulouse, France (2002)
10. Pistore, M., Traverso, P., Bertoli, P.: Automated composition of web services by planning in asynchronous domains. In: S. Biundo, K.L. Myers, K. Rajan (eds.) Proceedings of the Fifteenth International Conference on Automated Planning and Scheduling (ICAPS), pp. 2–11. AAAI, Monterey, California, United States of America (2005)
11. R-Moreno, M.D., Borrajo, D., Cesta, A., Oddi, A.: Integrating planning and scheduling in workflow domains. Expert Systems with Applications **33**(2), 389–406 (2007)
12. Ramos, C., Augusto, J.C., Shapiro, D.: Ambient intelligence–the next step for artificial intelligence. IEEE Intelligent Systems **23**(2), 15–18 (2008)
13. Ranganathan, A., Campbell, R.H.: Autonomic pervasive computing based on planning. In: ICAC, pp. 80–87. IEEE Computer Society (2004)
14. Schattenberg, B.: Hybrid planning and scheduling. Ph.D. thesis, University of Ulm, Institute of Artificial Intelligence, Ulm, Germany (2009). URN: urn:nbn:de:bsz:289-vts-68953
15. Schattenberg, B., Bidot, J., Biundo, S.: On the construction and evaluation of flexible plan-refinement strategies. In: J. Hertzberg, M. Beetz, R. Englert (eds.) Proceedings of the 30th German Conference on Artificial Intelligence (KI), *Lecture Notes in Artificial Intelligence*, vol. 4667, pp. 367–381. Springer, Osnabrück, Germany (2007)
16. Schattenberg, B., Weigl, A., Biundo, S.: Hybrid planning using flexible strategies. In: U. Furbach (ed.) Proceedings of the 28th German Conference on Artificial Intelligence (KI), *Lecture Notes in Artificial Intelligence*, vol. 3698, pp. 258–272. Springer, Koblenz, Germany (2005)
17. Sirin, E., Parsia, B., Vu, D., Hendler, J., Nau, D.: HTN planning for web service composition using SHOP2. Web Semantics: Science, Services and Agents on the World Wide Web **1**(4), 377–396 (2004)
18. Smith, D.E., Peot, M.A.: Suspending recursion in causal-link planning. In: B. Drabble (ed.) Proceedings of the Third International Conference on Artificial Intelligence Planning and Scheduling (AIPS), pp. 182–190. AAAI, Edinburgh, Scotland (1996)

19. Vuković, M., Kotsovinos, E., Robinson, P.: An architecture for rapid, on-demand, service composition. Service-Oriented Computing and Applications **1**, 197–212 (2007). DOI http://dx.doi.org/10.1007/s11761-007-0016-x
20. Zhao, H., Doshi, P.: Haley: A hierarchical framework for logical composition of web services. In: Proceedings of the 2007 IEEE International Conference on Web Services (ICWS), pp. 312–319. IEEE Computer Society, Salt Lake City, USA (2007)

Chapter 7
Privacy & Trust in Ambient Intelligence Environments

B. Könings, B. Wiedersheim, and M. Weber

Abstract Privacy and trust are critical factors for the acceptance and success of next generation ambient intelligence environments. Those environments often act autonomously to support a user's activity based on context information gathered from ubiquitous sensors. The autonomous nature, their accessibility to large amounts of personal information, and the fact that actuators and sensors are invisibly embedded in such environments, raise several privacy issues for participants. Those issues need to be addressed by adequate mechanisms for privacy protection and trust establishment. In this chapter, we provide an overview of existing privacy enhancing technologies (PETs) in the area of ambient intelligence environments and present the ATRACO approach to achieve privacy within those environments. Further, we will discuss how computational trust mechanisms and social trust aspects can be utilized to support privacy protection and the establishment of trust between system components and between the system and participants. After describing the integration of these mechanisms in the overall system architecture of ATRACO, we conclude by giving an outlook on future directions in this area.

Bastian Könings
Institute of Media Informatics, Ulm University, 89081 Ulm, Germany,
e-mail: bastian.koenings@uni-ulm.de

Björn Wiedersheim
Institute of Media Informatics, Ulm University, 89081 Ulm, Germany,
e-mail: bjoern.wiedersheim@uni-ulm.de

Michael Weber
Institute of Media Informatics, Ulm University, 89081 Ulm, Germany,
e-mail: michael.weber@uni-ulm.de

7.1 Introduction

The field of *Ambient Intelligence* (AmI), or ubiquitous computing, has been the focus of much research in the last decades. New technological solutions emerging from this research area offer great opportunities for a large number of new applications ranging from assisted or daily living systems, entertainment systems, to intelligent transportation systems. Such systems will be invisibly embedded into our everyday environments through a pervasive transparent infrastructure consisting of a multitude of sensors, actuators, processors and networks. The interplay of those components allows the system to support, interact with and adapt to individuals in a seamless and unobtrusive way.

However, in order to allow such a high flexibility, support and adaptation, AmI systems require a large amount of information, such as real-time data gathered from ubiquitous sensors, personal user information, and the ability to intervene in the user's physical environment. These requirements raise several issues regarding privacy and trust which need to be addressed in order to satisfy user needs and acceptance. Mechanisms to achieve these goals include trust establishment and *privacy enhancing technologies* (PETs). The system itself should be trustworthy to the user and the user should be able to control the system in any privacy relevant situation.

In this chapter, we first discuss the state of the art in privacy enhancing technologies and trust establishment mechanisms in the context of ambient intelligence environments. We then discuss the different forms of privacy and trust in our context in Section 7.3. Sections 7.4 and 7.5 present how these forms of privacy and trust are achieved and integrated in the overall ATRACO architecture. After the conclusion in Section 7.6, we suggest some future directions in this area and propose further readings in Section 7.7.

7.2 Related Work

AmI environments have been the focus of several research in the last decades [60, 18, 5]. However, privacy and trust issues have often been neglected and research addressing those issues started to emerge only in the last few years [11, 38]. We will give an overview of existing work in privacy protection and trust establishment mechanisms in general, and for AmI environments in particular after discussing common definitions for privacy and trust.

7.2.1 Definitions of Privacy

As there is no universally accepted definition of privacy, the formulation of an adequate definition is one of the most intractable problems in privacy research. Besides hundreds of existing definitions one of the oldest and most cited still remains influential: Samuel Warren and Louis Brandeis's 1890 declaration of privacy as the *"right of an individual to be let alone"* [59]. This traditional way of understanding privacy can also be described as *"a state in which one is not observed or disturbed by others"* [58].

However, most of today's common definitions and understandings of privacy in the context of information technology are based on Westin's definition from 1967. Westin defined privacy as *"the claim of individuals, groups or institutions to determine for themselves when, how, and to what extent information about them is communicated to others"* [61]. This definition is often referred to as *Information Privacy*.

One definition including both former privacy aspects was made by Philosopher Sissela Bok [12] who defined privacy as *"the condition of being protected from unwanted access by others – either physical access, personal information, or attention."*

Also Robert Ellis Smith [53] covers the former privacy aspects in his definition of privacy as *"the desire by each of us for physical space where we can be free of interruption, intrusion, embarrassment, or accountability and the attempt to control the time and manner of disclosures of personal information about ourselves."*

Based on these definitions we will discuss privacy classifications in the context of AmI environments in Section 7.3.1.

7.2.2 Definitions of Trust

Although trust is a common concept, on which we rely in our everyday life, it still remains hard to define. Like privacy, the term trust has a huge definitional diversity, which often leads to more confusion than to a better understanding of the term. As a consequence, some researchers argue that their *"purposes may be better served ... if they focus on specific components of trust rather than the generalized case"* [42].

Nevertheless, there is a set of trust properties, which are commonly recognized among most trust researchers in the context of information technologies. They agree that trust is *subjective*, *asymmetric*, *context- and situation-dependent*, *dynamic and non-monotonic*, and *not transitive* [41, 1].

Some of these properties are reflected by one of the most adopted definitions of trust by sociologist Diego Gambetta [24]: *"trust is a particular level of the subjective probability with which an agent assesses that another agent or group of agents will perform a particular action, both before he can monitor such action ... and in a context in which it affects his own action."*

Depending on the environment in which trust is specified, it can be considered as a composition of different attributes like *reliability*, *dependability*, *honesty*, *truthfulness*, *security*, *competence*, and *timeliness*. Utilizing these attributes, Grandison and Sloman [27] define trust as *"the firm belief in the competence of an entity to act dependably, securely, and reliably within a specified context"*.

Chopra and Wallace [17] tried to find a common denominator of various definitions and argued that an integrated definition of trust consists of at least three elements. A *trustee* to whom trust is directed, *confidence* that the trust will be upheld, and a *willingness* to act on that confidence. Based on these elements they define trust as *"the willingness to rely on a specific other, based on confidence that one's trust will lead to positive outcomes"*.

Further, trustworthiness of an entity can be interpreted as a level of trust, that could be established over time by monitored information. Therefore, *"trust can be seen as a complex predictor of the entity's future behavior based on past evidence"* [50].

Although the given definitions do not explicitly use the term risk in their formulation, the inclusion of risk is implied by most definitions in terms of uncertainty and negative outcomes if trust assessment fails. Thus, most researchers agree that trust is inherently related to risk but often fail to understand the precise relation between the two notions [54, 40].

7.2.3 Privacy in Smart Environments

In general Privacy protection mechanisms can be divided into the four main categories of *regulatory strategies*, *policy matching*, *prevention & control* and *detection*. Mechanisms belonging to the last three categories are often referred to as *privacy enhancing technologies* (PETs). Even though some of the PETs discussed below were not primarily designed for AmI environments, they provide important approaches which can be further extended to create new AmI privacy solutions.

7.2.3.1 Regulatory Strategies

Regulatory strategies refer to governmental rules on the use of personal information. Because individual privacy claims often differ in several ways, it is important to find general principles for privacy protection that fit to the most common requirements. Those principles can be used to expand voluntary agreements or regulations enforced by law, but can also serve as important input for the design process of AmI systems and the subsequent treatment of personal information in such systems.

The most famous principles relating to information privacy are listed in the *Organization for Economic Cooperation and Development* (OECD) guidelines on the protection of privacy and transborder flows of personal data [44]. The guidelines state eight principles for the protection of privacy together with numerous other and similar sets of privacy protecting rules. The four core principles are:

1. **Collection Limitation Principle:** There should be limits to the collection of personal data and any such data should be obtained by lawful and fair means and, where appropriate, with the knowledge or consent of the data subject.

2. **Data Quality Principle:** Personal data should be relevant to the purposes for which they are to be used and, to the extent necessary for those purposes, should be accurate, complete and kept up-to-date.
3. **Purpose Specification Principle:** The purposes for which personal data are collected should be specified not later than at the time of data collection and the subsequent use limited to the fulfillment of those purposes or such others as are not incompatible with those purposes and as are specified on each occasion of change of purpose.
4. **Use Limitation Principle:** Personal data should not be disclosed, made available or otherwise used for purposes other than those specified in accordance with the Purpose Specification principle except:
    a. with the consent of the data subject; or
    b. by the authority of law.

These four principles contain the essence of privacy protection whereas the remaining four describe procedural aspects. The fundamental role of these principles is reflected by the fact that they were adopted for many international privacy regulations, including the EU Data Protection Directive 95/46/EC [23] and the Canadian PIPEDA law [14].

However, the OECD guidelines as well as similar existing regulations often are in conflict with common AmI characteristics. For example, the principle of limitation and purpose specification of data collection are conflicting with the active, pervasive and continuous collection of data in AmI environments. Privacy enhancing technologies which try to enforce existing guidelines are therefore often a trade-off between privacy, and benefit or usability of the AmI system.

7.2.3.2 Policy Matching

Policy matching techniques try to minimize privacy risks by specifying and matching user privacy policies with involved service provider policies. The most common policy languages of that kind are the *Platform for Privacy Preferences* (P3P) [19], the *Enterprise Privacy Authorization Language* (EPAL) [4] and the *eXtensible Access Control Markup Language* (XACML) [26], which are XML-based languages to support the definition and enforcement of privacy policies and obligations. The main drawback of privacy policy systems is that they generally cannot enforce privacy and instead rely on trustworthiness and regulatory pressures to ensure policy compliant behavior.

7.2.3.3 Prevention & Control

Prevention and control mechanisms seek to prevent misuse of personal information by adapting information in several dimensions or by avoiding privacy-critical operations or access. Such mechanisms include *pseudonymization* [16], *obfuscation* [62] or *access control* [43].

Pseudonymization and obfuscation techniques gained popularity in the area of location privacy, which was one of the first privacy concerns raised by ubiquitous computing [60]. As mobile devices with integrated GPS-receivers are becoming more and more ubiquitous, location privacy has attracted considerable research in the last decades [21, 35]. Beresford and Stajano [7] employ mix zones and anonymity sets to analyze location privacy. They propose to use changing pseudonyms instead of unique identifiers to prevent tracking of users. Gruteser and Grunwald [28] applied the concept of k-anonymity to location privacy. Instead of pseudonymously reporting exact location information, a person reports a region containing k-1 other persons. This way k-anonymity provides privacy protection by guaranteeing that each released record of an individual cannot be distinguished from records of at least k-1 other individuals. The larger the value of k, the greater the implied privacy since no individual can be identified through linking attacks with probability exceeding 1/k. Duckham and Kulik [20] propose the use of obfuscation, that is decreasing the accuracy of exchanged personal information to the minimum, needed by a service provider in order to provide the required service. A simple solution is giving a more general location, e.g., only the street or city instead of detailed coordinates.

Also several architectures for privacy protection and control in ubiquitous environments have been proposed in the last decades. Hong and Landay [29] proposed Confab, a privacy-aware architecture for ubiquitous computing. Confab uses the concept of information spaces by Jiang et al. [30] in combination with in- and out-filters to manage the flow of context information about a person. Langheinrich [36] discussed how privacy guidelines, similar to the OECD guidelines, can be adapted to ubiquitous computing scenarios. He pointed out six principles, which he followed to design the privacy awareness system PawS [37] in order to address third party data collection in smart environments.

7.2.3.4 Detection

Detection mechanisms try to identify and penalize privacy violators. A system for privacy violation detection continuously monitors access to personal data and detects misuse or abnormal behavior.

PRIVDAM [8] is a data mining based architecture for privacy violation detection and monitoring. The system attempts to detect possible privacy violations based on network characteristics and seeks to prevent them in the future.

An et al. [3] propose to use Bayesian network-based methods to detect insider privacy intrusion in database systems. The advantage of Bayesian networks is that they can effectively deal with uncertainties involved in the activities of database operations.

Ortmann et al. [45] discuss the idea of modeling information flow graphs to identify directly and indirectly available data in a pervasive system and link them to the sources of the raw input data. The graphs can be used to determine which information or data sources need to be disabled to hide forbidden information and adapt the modeled system to given privacy requirements.

7.2.4 Trust in Smart Environments

As introduced in Section 7.2.2, the concept of trust may be used in interactions for estimating the future behavior of an interacting entity. Trust allows people to deal with uncertainty and assess risk in uncertain interactions or situations.

In social interactions, trust can be established based on personal assessments, experience or reputation. In the context of AmI environments and IT-systems in general, the establishment of trust is often based on the concept of *computational trust*. Existing mechanisms and models for computational trust can be classified into *credential-based* and *reputation-based* mechanisms [13, 34, 48].

7.2.4.1 Credential-based Trust

Credential-based computational trust refers to cryptographic solutions for establishing trust by obtaining and verifying credentials of an entity. These credentials are usually issued in form of digital certificates by a trusted third party or *certificate authority* (CA). This approach is commonly used to authenticate identities or memberships in Internet applications, often based on *public key infrastructures* (PKIs) [47, 25]. A PKI is a hierarchically arrangement of digitally signed certificates for assuring bindings of public keys to specific identities.

Based on this notion of trust, Blaze et al. [9] introduced the term *trust management*, which involves the formulation of policies and credentials, determining satisfaction of credentials to policies and deferring trust to third parties. They proposed a simple language for specifying trusted actions and trust relationships, which was used in their trust management system *PolicyMaker* [10]. Li et al. [39] combined the strengths of role-based access control with similar trust management systems in their *RT framework* to support delegation of role definition and role activations.

Lund et al. [40] view trust management as a particular form of risk management with the same weaknesses but with a higher complexity and dynamic. In order to account for evolution in risk and trust assessment, they distinguish between three main scenarios: *maintenance*, *before-after*, and *continuous-evolution*. They focus on understanding the impact of subjective trust on factual risks in each of these scenarios.

7.2.4.2 Reputation-based Trust

Reputation-based trust establishment uses the history of an entity's past behavior or recommendations and experiences of third entities to compute a certain trust level [31]. Well known applications for reputation-based systems are electronic markets like eBay[1] or Amazon[2]. The main challenge here is to find adequate trust metrics and models to support this kind of computation. One of the first attempts at formalizing such trust metrics was made by Marsh [41]. His model was based on linear equations and was further extended by Abdul-Rahman and Hailes [1] to address trust in virtual communities.

Other approaches to compute trust and reputation values are based on the use of probability theory [49], fuzzy logic [15], or the use of entropy [51]. Similar models are adopted by distributed mechanisms for computing reputation values in peer-to-peer networks [2, 32, 63, 56].

7.2.5 The Relation of Privacy and Trust

Most researchers agree that privacy and trust are two concepts which cannot be addressed without respecting each other and often even depend on each other.

From the privacy point of view, trust in entities can be utilized as a policy restriction on which privacy decisions are based. For example, a person might want to reveal his mobile number only to a trusted entity or to an entity of a certain level of trust. In this context, trust is needed to deal with the risk, that privacy preferences will not be respected by the other entity. For example, a person registering at a web shop needs to trust that the shop owner will not forward any sensitive information to third parties.

On the other hand, the verified use of well integrated privacy mechanisms can support trust establishment. If a person knows that an online shop applies adequate privacy mechanisms to protect sensitive user data, his trust in this shop will be higher than without these guaranteed privacy protection mechanisms.

Some argue that privacy and trust are conflicting [50] and need to be balanced by ensuring minimal trade of privacy for the required trust. This statement implies that establishment of trust in an entity always comes with a degradation of its privacy, which might be true if trust is solely established by gathering information about an entity, as it is the case in most reputation based systems. Some research has been done which tries to find solutions for this problem. Pavlov et al. [46] present three protocols based on probabilistic schemes that allow partial privacy for feedback providers in dezentralized reputation systems. Steinbrecher [57] proposed privacy-respecting design options for centralized reputation systems. However, if trust establishment between two parties is based on assessing former interactions between those parties, privacy is not necessarily affected.

---

[1] http://www.ebay.com

[2] http://www.amazon.com

7.3 Classification of Privacy & Trust in ATRACO

In order to cope with all relevant aspects of privacy and trust in the context of AmI environments we classify the terms into *information privacy*, *territorial privacy*, *technical trust* and *social trust*. These classifications will be discussed in Section 7.3.1 and 7.3.2, respectively.

7.3.1 Privacy Classification

In ATRACO we distinguish between two forms of privacy, namely *information privacy* and *territorial privacy* as depicted in Figure 7.1. Information privacy refers to the protection of personal information (or personal data) whereas territorial privacy refers to the protection of private spaces (or territories). Some privacy classifications [22] also list *bodily privacy* and *communication privacy* as separate privacy categories, whereas we consider them to be subcategories of information privacy and territorial privacy.

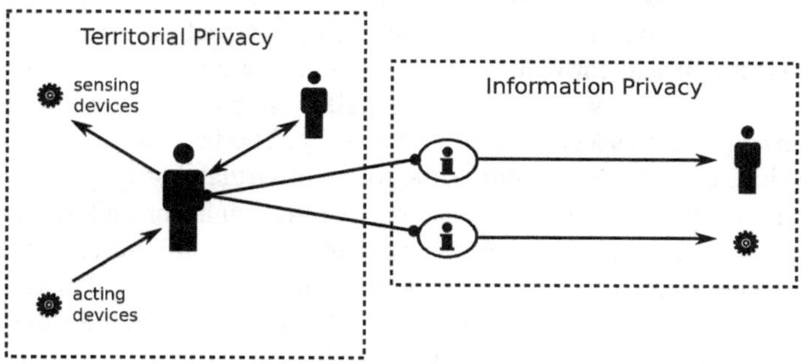

Fig. 7.1 Privacy classification in ATRACO: information privacy and territorial privacy.

7.3.1.1 Information Privacy

Information privacy refers to the definition of Westin [61] given in Section 7.2.1. Thus, persons want to control "when, how, and to what extent information about them is communicated to others". In the context of AmI environments, this means that any kind of sensitive information needs to be protected from unintentionally leaving the borders of the surrounded AmI environment and needs to be protected from unauthorized access of other entities. Thus, personal information must remain at any time under full control of the user. Such information includes *personal data* (e.g., pictures, e-mails, web history), *personal properties* (e.g., age, size, weight), *personal preferences* (e.g., favourite music, movies, fashion) or *personal behavior* (e.g. doing what, when, how often).

Personal information can either be static (e.g., mails, pictures) or dynamic (e.g., location, activity) and can be collected directly (e.g., database access, physically sensed) or indirectly (e.g., analytically derived, compositionally derived).

7.3.1.2 Territorial Privacy

While protecting information privacy is sufficient in the context of most common IT domains, it is only one aspect of privacy protection in the context of AmI systems. The pervasive nature of such systems equipped with multiple sensors, actuators and computational devices, constitute a new facet of privacy protection, that we call *territorial privacy*. Territorial privacy aims to satisfy a more traditional expectation of privacy such as being in *"a state in which one is not observed or disturbed by others"* [58].

In an AmI environment a user might be continuously observed by cameras, microphones or other sensors. Therefore, it is required that the user is able to control whether or not an entity is allowed to observe him in a certain way. We call the different ways of observation *observation channels*. For example, an entity receiving a live video stream of a user is connected by a visual observation channel. A detailed discussion and formalization of this observation model is given in [33]. Further, the control of how other entities are allowed to disturb a user in his private space is an important aspect of territorial privacy. The level of disturbance can be either physical, e.g., if a person enters the room, or virtual. Virtual disturbances, for instance, are undesired outputs of visual or acoustic signals or undesired initiations of interactions. The goal of territorial privacy is to control all present observers and disturbers. An example of a person with surrounding observers and disturbers is depicted in Figure 7.2. Desired observers and disturbers are separated from undesired ones by the territorial privacy boundary.

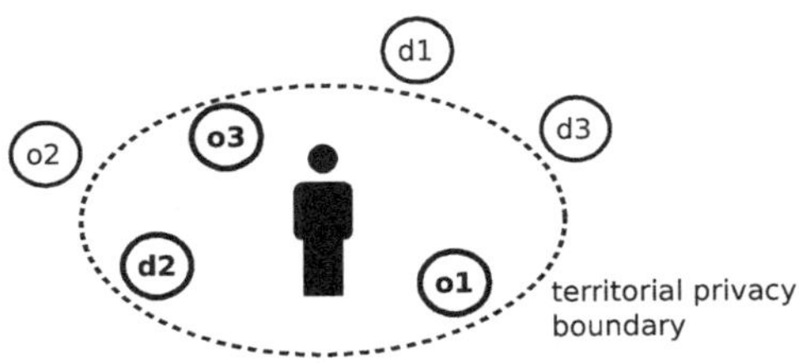

Fig. 7.2 A territorial privacy boundary with desired and undesired observers and disturbers.

One of the most important factors influencing territorial privacy decisions is the user's current context. The most obvious context information is the user's current location. But also other contextual constraints like time, activity, or user's mood need to be respected in order to address territorial privacy needs.

7.3.2 Trust Classification

In order to realize a trustworthy AmI system, different forms of trust need to be addressed. We distinguish between *technical trust* in system components and devices, and *social trust* in other persons (see Figure 7.3). The two forms will be discussed in the following sections.

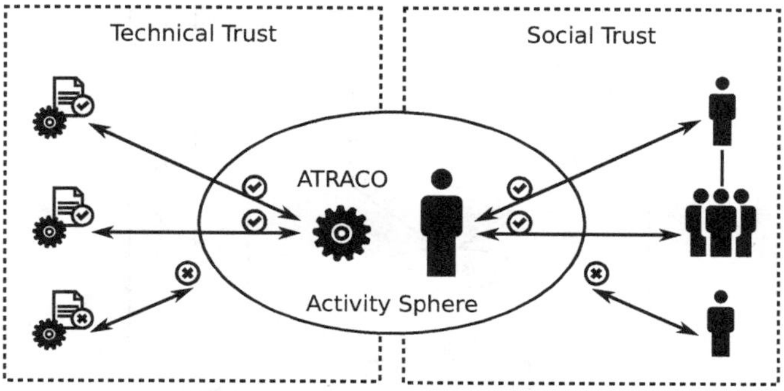

Fig. 7.3 Trust classification in ATRACO: technical trust and social trust.

7.3.2.1 Technical Trust

An AmI system consists of several interacting software and hardware components of different types. Technical trust refers to trust in these components with respect to their intended functionality, reliability, and safety. These aspects are particularly important in AmI environments with high flexibility and adaptivity of involved components. This form of trust can be established by credential-based trust mechanisms and PKIs. For example, a component can provide a certificate issued by the manufacturer's CA. If the manufacturer CA is in a valid PKI path to a trusted root CA, the device will be trusted in the current AmI environment. Including manufacturers in a trusted PKI might depend on several factors, like quality guidelines or performance and reliability assessments. This approach allows an initial binary trust assignment, which can support the decision if a component should be used or not in a given context.

Further, the use of digital signed certificates allows verification of several component properties, which could be utilized for trust establishment. Thus, a certificate might not only certify the device manufacturer, but also the place of origin, application fields, or other device specific properties. This way different credentials can be rated to achieve a more fine granular trust level assessment.

To further enhance trust assessment in terms of dynamic establishment and adaptations, we propose to combine credential-based mechanisms with reputation-based mechanisms, as depicted in Figure 7.4. For example, a new component of an un-

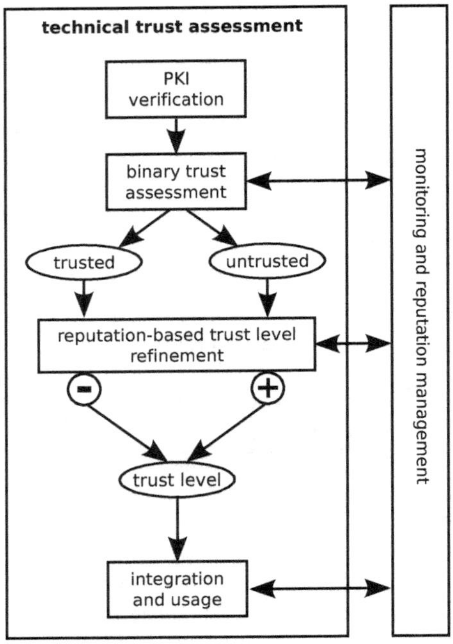

Fig. 7.4 Technical trust assessment process combining credential-based and reputation-based trust establishment.

known manufacturer will be classified as untrusted when it initially appears in an AmI environment. The operations and interactions of the component will be continuously monitored to dynamically assess the device's proper functioning and reliability. The results of this monitoring phase will be fed to a central database in order to collaboratively collect and receive monitored information about a specific component. On the one hand, the creation of such a reputation database will ease the trust assessment of components, which are not certified by a trusted CA. On the other hand, trust of properly certified components could further be refined, for example decreased if the reputation database reports unreliable behavior for the given component.

7.3.2.2 Social Trust

Social trust refers to the way human beings perceive and understand the notion of trust in social interactions and social relationships. As AmI environments are not always single-user oriented, social trust needs to be respected whenever a system component is involved in multi-user situations. In this context, we focus on the establishment of trust groups, which represent a group of individuals with the same level of assigned trust. The concept of trust groups addresses the problem of defining specific trust levels. Approaches for computing such specific levels of trust need to map real life trust to a given computational metric. The problem is to find metrics that are flexible enough to model the dynamic and context dependent trust perception of real persons. Trust groups avoid the need of specifying those metrics and specific trust levels by providing an adequate abstraction layer for user-oriented trust management.

We propose to have several predefined trust groups, which could be further refined and rearranged by a user. Those groups refer to common social groups, namely *family*, *friends*, and *colleagues*. By default a trust group will not have assigned a specific trust level. Trust will be implied by specifying specific access policies either for personal information that utilize trust for privacy protection, or for control instances of a personal AmI environment. For example, a home AmI system which controls the heating or light levels should be accessible only by family members.

7.4 Privacy in ATRACO

The ATRACO approach aims to realize the concepts of an AmI system by addressing heterogeneity of artifacts, system transparency, discovery & management of various artifacts, and autonomous behavior of learning agents. ATRACO uses a *Service-Oriented Architecture* (SOA) at the resource level to support numerous devices and sensors, and at the system level to support AmI applications. Several agents complement the SOA infrastructure by providing high level adaptation to a user's task. ATRACO agents support adaptive planning, task realization and enhanced human computer interaction. The concept of ontologies is used to address the semantic heterogeneity that arises in AmI environments. The instantiation of agents to support a specific user task is based on the concept of *Activity Spheres* (AS). An AS is the utilization of knowledge, services and other resources required to realize an individual user goal within ATRACO. A more detailed discussion of the ATRACO core concepts and architecture is given in Section 1.3.

Our basic approach to privacy enforcement in ATRACO is based on policy matching and access control mechanisms. All privacy relevant functionalities and components are encapsulated within the *Privacy Manager* (PM). The integration of the PM within the overall architecture and its interaction with other system components is discussed in Section 7.4.3.

7.4.1 Privacy Requirements

Most of the discussed PETs in Section 7.2.3 only focus on a small part of privacy protection in the context of AmI environments and in most cases address only information privacy aspects. As a consequence, we need to find solutions which are widely applicable by combining policy matching techniques with adequate policy enforcement mechanisms. In addition, a comprehensive privacy solution needs to respect and address territorial privacy aspects as well.

Another fundamental requirement for privacy in AmI environments is the dependence on contextual information, e.g., location, activity, or presence of other persons. Furthermore, users prefer to rely on context instead of setting access control policies on specific content [52]. This aspect must be taken into consideration when developing policies and enforcement techniques as it leads to the need of dynamic adaptation of privacy policies or enforcement. In addition, privacy protection needs to respect given user preferences and trust in involved devices or other persons.

Previous studies [29, 6] have shown that the deployment of adequate feedback and control mechanisms is an essential requirement for the user acceptance of privacy systems. This fact needs to be respected in the design process of privacy solutions as well.

Taking all previous aspects into consideration we identified the following privacy requirements in ATRACO:

- Privacy protection must depend on user privacy preferences, trust, and context.
- User privacy preferences must be represented as privacy policy ontologies.
- Privacy protection must be deployed in an activity sphere when sensitive data is requested, or when the user wants to reduce observations or disturbances.
- A user should have the option to be aware of any event relating to privacy.

7.4.2 Privacy Policy Ontologies

In general a *Privacy Policy Ontology* (PPO) describes how, under which conditions, and in which context an entity is allowed to handle personal information or is allowed to participate in a user's activity. In the case of information privacy, policies need to be specified on the user side as privacy preferences, and on the receiver side as privacy obligations in order to allow a matching between the two parties. The policies of receivers (e.g., persons, web services, but also system components) can either be specified as policies in PPO representation or in a different policy language, like P3P [19]. For the latter case, a policy wrapper translates such policy languages into PPO representation.

We distinguish between *Information Privacy Policies* (IPPs) and *Territorial Privacy Policies* (TPPs). The key attributes of an IPP are *information item* (the information to apply the policy to), *purpose* (what is allowed to be done with the information), *recipient* (who is allowed to use the information), *retention* (how long can the information be used/stored), and *context* (further access constraints).

A TPP consists of the key attributes *territory* (the space to apply the policy to, explicitly identified by a location, or implicitly by an activity), *participant* (the entity which is allowed to observe or disturb the user), *observation type* (the allowed observation channel, declared by sensor types), *disturbance type* (the allowed type of intervention), and *context* (further access constraints).

For both IPP and TPP a default *deny-all* rule is assumed. Therefore, any access to information or territories must be explicitly granted by either a IPP or TPP, respectively.

In the ATRACO prototype we implemented policies as part of the *User Profile Ontology* (UPO). Each policy is an individual of one of the two class types: `InformationPrivacyPolicy` or `TerritorialPrivacyPolicy`. Policies of the first type may permit access to specific private information, indicated by the object property `canUse`, for different entities via the `permitsAccess` object property. If access to private information should be granted only for specific actions, a policy can specify the `allowsAction` object property. In addition, time constraints can be specified with the data properties `hasStartDate` and `hasEndDate` to limit the usage of private information to a predefined time interval. Figure 7.5 shows a simplified example of an IPP, granting Bob access to view a picture for a time period of one year.

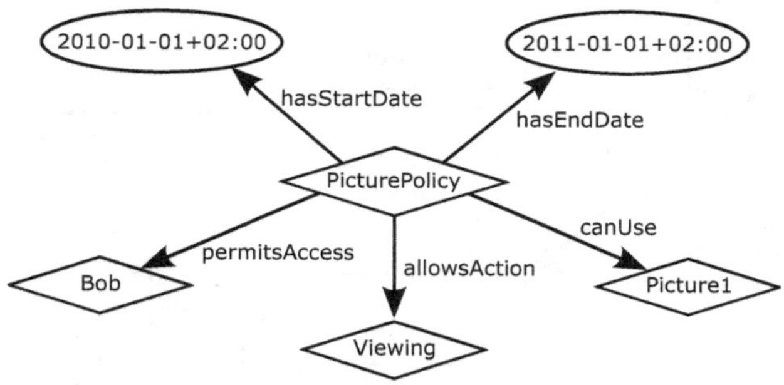

Fig. 7.5 Example of a Privacy Policy Ontology.

7.4.3 Privacy Manager

In ATRACO, all privacy components are encapsulated in the Privacy Manager which consists of the *Policy Processing Engine* (PPE), the *Information Privacy Controller* (IPC), the *Territorial Privacy Controller* (TPC) and the *Feedback & Control Component* (FCC). Figure 7.6 provides an overview of the main components of the ATRACO privacy architecture and its interactions with the *Sphere Manager* (SM), *Ontology Manager* (OM), *Interaction Agents* (IAs), and *Trust Manager* (TM).

The SM is responsible for initializing or dissolving an Activity Sphere (AS) with its associated resources for a specific user goal. The SM acts as an event service to other components of the AS. It monitors the execution of the task workflow and adapts the composition of resources in case of any conflicts, such as failing devices or other exceptions. For a detailed description of the SM see Section 1.3.

The OM is responsible for managing the sphere ontology and responds to ontology queries of other components. Ontologies provide a central knowledge base for properties, state and context information as well as user preferences. The most privacy relevant ontology is the User Profile Ontology (UPO), which contains user preferences, user properties and other personal related information. The management of ontologies is discussed in Section 3.3.

Several IAs offer multimodal services for user interaction. In order to provide smooth interaction, IAs use distributed multimodal interaction widgets which adapt to the user's context. Therefore, the user interaction can be distributed among available modalities and devices at runtime. For more information about the ATRACO IAs see Section 5.3. The interaction between PM and IA is managed by the *Interaction Agent Controller* (IAC) which is part of the PM as well as of the TM, and provides shared functionalities for controlling and event handling of the IA. The TM will be discussed in detail in Section 7.5.2.

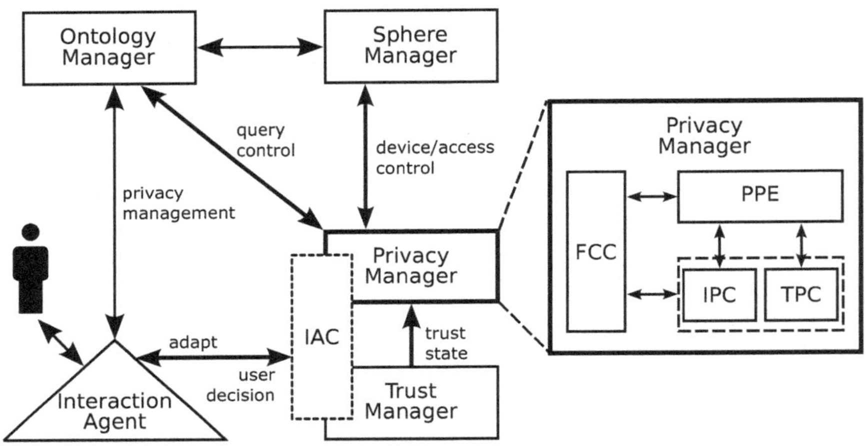

Fig. 7.6 Integration of the Privacy Manager.

The Privacy Manager serves as a privacy interface for other components in the ATRACO architecture. Whenever a privacy relevant event occurs, such as a query of personal information, a territory access request, or context change, the PM requests related policies and contextual information from the OM. The Policy Processing Engine (PPE) uses this information to perform policy reasoning and provides the results to the corresponding privacy controller, which will perform further privacy enhancing mechanisms based on the results. If any conflict occurs during this reasoning process, the PPE can trigger the IA via the FCC and IAC in order to ask the user for a conflict resolution, such as modifying policies or adding policy exceptions. Furthermore, the FCC provides a user interface for controlling and modifying privacy settings that are stored as part of the User Profile Ontology. The interplay and role of privacy controllers (IPC and TPC) is described below, along with a discussion of potential privacy affecting events.

Queries of Personal Information: The main knowledge base for personal information is the User Profile Ontology (UPO). In order to ensure information privacy, the OM triggers the PM to validate queries on ontologies, in particular on the UPO. The privacy validation will be handled by the PPE, involving the requester ID, associated policies, and current context. Based on the validation result the Information Privacy Controller (IPC) will grant or deny access to the query results. Optionally, the IPC can perform obfuscation and pseudonymization techniques to further enhance privacy or react to policy conflicts.
A special case occurs whenever IAs request personal information. The IPC then checks if any other persons are located inside the user's sphere by querying the OM. If this is the case, the IPC uses the IA's ontology to determine the presentation modality for the information and whether it is accessible by other persons. If the presentation is accessible by other persons, for example a large screen display, the PPE checks the related policies as described above. On a deny decision, the IPC may either retain the personal information or adapt the interaction modalities of the IA via the IAC to ensure that unauthorized access to personal information is prevented.

Territory Access Requests: The Territorial Privacy Controller (TPC) performs territorial access control in order to protect territorial privacy whenever external entities attempt to join the user's sphere. Access requests are forwarded by the SM. An access request may ask either for an observation or disturbance. Observation access is requested when an entity attempts to receive sensed user data, for example via a camera or microphone. A disturbance access is requested whenever an entity attempts to pass the physical borders (e.g. enters a room) or wants to actively intervene in the user's territory, for example, by acoustic or visual output, or by initiating interactions. Access decisions are based on the associated policies, which are validated by the PPE. In addition, access decisions can respect the current trust state of a device which is provided by the TM. This will be discussed further in Section 7.5.2.1.

Contextual Changes: Contextual changes, such as an activity change, a location change or a newly arrived person, are reflected as events by the SM. Whenever such an event occurs, the IPC and TPC perform appropriate privacy measures. In the case of a new person, the IPC checks whether personal information is currently presented by any interaction modality the new person would have access to. This information is obtained either from the interaction ontology or directly from the interaction agent. If so, the IPC may again adapt the interaction modalities via the IAC based on the policy reasoning results of the PPE, as described above. In case of activity or location changes, the TPC checks if any device or remote entity is currently observing or disturbing the user in his new territory. This process is analogous to the IPC's operation, in that the PPE validates the associated privacy policies. Based on the results, the TPC will exclude the undesired observing and disturbing entities by switching off relevant sensors or active devices, respectively.

7.4.4 Prototype Evaluation

A prototype of the overall ATRACO architecture has been deployed and evaluated at the University of Essex in its iSpace laboratory, a real world testbed for home technology. A first evaluation of the ATRACO privacy components was performed as part of a user test during the project's second year. The next section briefly describes the privacy relevant scenario and results.

7.4.4.1 Scenario

A participant, Alice, sits on the sofa at home. The system asks Alice if she wants to look at holiday pictures on the TV. Alice accepts and the slideshow starts. Some of the pictures have been marked as *private*, i.e., only accessible to the owner. After a couple of minutes a guest arrives at the front door. The system informs Alice that "*Bob has arrived*" and asks "*would you like him to enter?*". The participant responds by saying "*Yes*". The door opens and as Bob enters the private pictures are seamlessly excluded from the slideshow.

7.4.4.2 User Reactions

The service of automatically hiding private information in presence of other persons was generally accepted by the participants. However, participants were more concerned about the ability to control the system. In situations where the user felt out of control, reactions to the system were mostly negative. In some cases, participants were comfortable with the system initiating interactions, such as when a guest arrives at the door. However, in other cases, it was regarded as an invasion of personal space, for example, when participants were asked if they would like to look at some pictures. Perceptions of control may also be influenced by the channel of interaction. Voice interactions appeared to imply the existence of a separate social presence which could lead to a feeling that the personal space has been invaded. Also mood changes could have an impact on the way a user would prefer to interact with the system. These early results show that privacy concerns of participants in AmI environments often involve territorial privacy aspects, such as the fear of losing control, the perception of privacy invasion, or particular forms of interaction. These aspects are covered by our privacy architecture and will be further integrated in the prototype's interaction behavior for further evaluation.

7.5 Trust in ATRACO

Trust in ATRACO is comprised of technical trust and social trust, as described in Section 7.3.2. This section discusses the main requirements for building these two forms of trust in an AmI environment and then describes how this is achieved in the context of ATRACO by the Trust Manager component.

7.5.1 Trust Requirements

The consideration of trust in the context of AmI environments is an essential requirement for the user acceptance in such systems. The high dynamic of available devices as well as their autonomous nature make the establishment of technical trust for those devices indispensable. A user relies on the technical functionalities of devices in the environment and must therefore have the assurance that the devices are trustworthy. In ATRACO we refer to a device as trustworthy in terms of reliability, compatibility, intended functioning, and functioning without unwanted or even malicious behavior. As a first step towards such trustworthiness, all available devices need to be authenticated when they connect to an ATRACO activity sphere. In addition, the technical reliability and compatibility with the ATRACO environment needs to be proven and certified by an ATRACO certificate authority which issues a digital certificate to a valid device.

As certification allows only binary trust assessment, the process should be extended by reputation mechanisms. This allows finer trust assessment on the one hand, and the establishment of trust for devices which do not own a certificate of an ATRACO CA, on the other hand. The latest trust state of a device needs to be respected by ATRACO when devices are used in an activity sphere. The trust state should further influence device selection when more than one device of a specific category is available. Assessed trust states of devices should be visible to an activity sphere's user in an intuitive way. Further, the user should be able to influence selection or usage of devices when trust states are not sufficient for automatic selection.

In addition to technical trust considerations, we need to consider social trust assessment in ATRACO whenever an activity sphere includes more than one person. This is essential to keep the control of an activity sphere completely to the owner. Thus access to user interfaces which can influence the state of an activity sphere need to be controlled. To achieve such access control, a user needs to be supported in managing social trust groups and access control policies in an intuitive manner.

In summary, the following basic design requirements for realizing technical trust and social trust in ATRACO were defined:

- Device binding and usage must depend on technical trust decisions.
- Technical trust should be provided in form of digital certificates and signatures.
- Monitoring and reputation mechanisms should be integrated to refine binary trust assignments.
- A user should be aware of the trust states of active and involved components.
- social trust must be respected in multi-user scenarios.
- A user should be able to manage and control access policies and trust groups.

7.5.2 Trust Manager

The Trust Manager (TM) is the central component for controlling and managing trust decisions in ATRACO. The TM consists of the *PKI Verifier*, the *Technical Trust Controller* (TTC), the *Social Trust Controller* (STC), the *Monitoring and Reputation Component* (MRC) and a *Feedback & Control Component* (FCC). Figure 7.7 provides an overview of the TM integration in the overall ATRACO architecture and its interactions with the Sphere Manager (SM), Ontology Manager (OM), Interaction Agent (IA), and Privacy Manager (PM). The interaction between TM and IA is managed by the shared Interaction Agent Controller (IAC). The interplay of these different components will be discussed for technical trust decisions and social trust decisions, respectively.

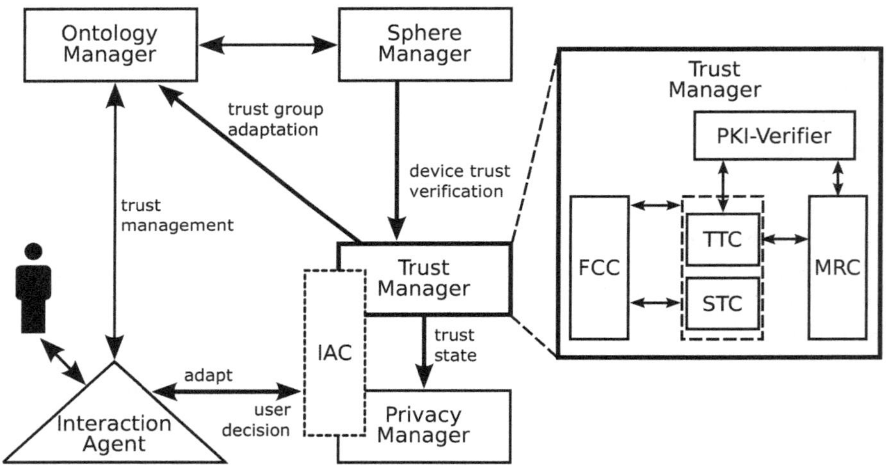

Fig. 7.7 Integration of the Trust Manager.

7.5.2.1 Technical Trust Decisions

Technical trust decisions are made by the TM whenever a new activity sphere starts or a new component appears in an already running activity sphere. Upon the initialization phase of an activity sphere, the SM requests a trust verification for relevant components from the TM. Based on the computed trust states, the SM decides which of the components can be used to create the new activity sphere. Thus an activity sphere will always be created with the most trusted components. The SM will request the same trust verification if a new component appears in an already existing activity sphere.

The verification requests of the SM are processed by the TTC. First, the PKI-Verifier verifies the provided certificate of the given component. A certificate confirms the validity of several properties of the component, such as unique identifier, manufacturer, type, or serial number. In this way, the component is uniquely identifiable and can get assigned a reputation-based trust level by the MRC, even if the certificate was signed by a CA, which is untrusted or unknown in the context of ATRACO. The MRC has a local database of monitored components, which stores statistical information needed to infer reputation metrics. This information includes the number of failures or incorrect behavior as well as the duration of reliable functioning. The database can be further synchronized with an online repository to enable collaborative reputation gathering.

In the case of a new device which is both rated as untrusted by the PKI-Verifier and unable to get a trust level assigned by the MRC due to insufficient reputation information, the TTC will use the FCC to involve the user. The user may decide to use and monitor the new component in his current activity sphere, or to deny the usage of the device. The FCC will receive this user decision from the Interaction Agent Controller (IAC) which in turn will trigger the IA.

The current trust state of a device can further support access control decisions of the PM for protecting territorial privacy.

7.5.2.2 Social Trust Decisions

A decision about social trust is needed whenever multi-user situations occur during the lifetime of an activity sphere. These decisions primarily support privacy protection and mechanisms for access control of user interfaces needed to control activity sphere components. As mentioned in Section 7.3.2 we adopted the concept of trust groups to support the user in managing social trust preferences and, based on that, to create appropriate access policies. Figure 7.8 shows a possible arrangement of trust groups and access control policies, which are stored and provided as ontologies by the OM. In order to ease the management of access control policies, controls are organized in different control domains, such as *environment controls* or *entertainment controls*. In a similar way, sensitive information items can be organized in different categories, for instance *personal information* or *personal pictures*.

The main task of the STC is locking or unlocking controls depending on social trust settings. For instance, if Charlie (as depicted in Figure 7.8) is in physical range of the heat control, the STC will lock it because Charlie is denied access to environmental controls. For this scenario to work, the STC receives the location events of persons who are present from the SM. In addition, the IA registers each user interface belonging to a specific control domain via the IAC with the TM. Thus the STC always knows the location of a control and can lock or unlock it depending on each person's location.

Another task of the STC is to find new members of existing trust groups or to suggest new trust groups based on explicit user interactions. For instance, in Figure 7.8 Bob does not have access to personal information. If the user of the activity

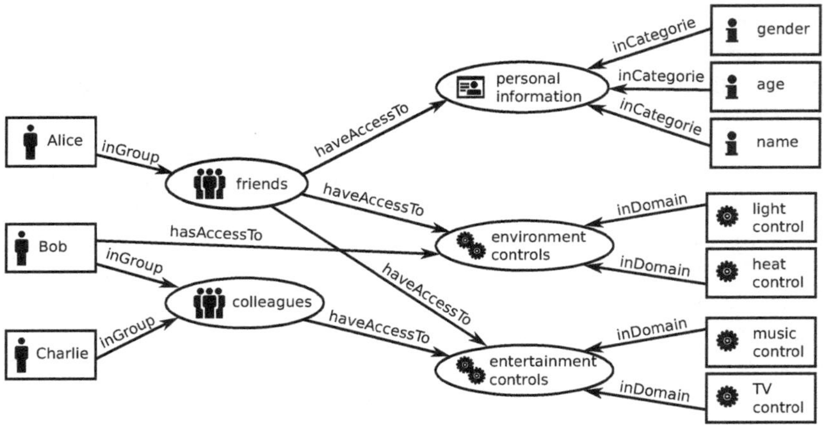

Fig. 7.8 Representation of trust groups and associated access control policies.

sphere is viewing personal information on a screen when Bob enters the room, the information will be protected by the PM and thus disappear from the screen. However, if the user explicitly makes the information reappear in Bob's presence, the PM will create a new policy which grants Bob access to the information with the context constraint that the user is present. Whenever a new policy is created in such a way, the STC compares access policies of individuals with those of trust groups. If the policies match to a specific threshold, the STC will suggest adding the person to that group. In our example, Bob's individual access policies will match those of trust group *friends*. As a consequence, the STC will suggest making Bob a member of the group friends.

7.6 Conclusions & Future Work

In this chapter we have discussed privacy and trust in the context of ambient intelligence environments. We identified information privacy and territorial privacy as the main privacy aspects in those environments. While information privacy has been the dominating aspect of privacy research in recent years, next generation AmI environments raise new demand for the more traditional aspect of territorial privacy. The privacy components presented here were designed to achieve both of these aspects by respecting user preferences and contextual information.

We classified trust into aspects of technical trust and social trust, which need to be addressed accordingly. The combination of credential-based and reputation-based trust establishment mechanisms facilitates dynamic trust level assessment for technical components. Social trust is addressed by the concept of trust groups with support of implicit adaptation and creation of group arrangements.

The proposed privacy and trust concepts were integrated in ATRACO in the form of the Privacy Manager and Trust Manager components, respectively. The interplay of these components with the overall ATRACO architecture allows the creation of trustworthy and privacy preserving activity spheres.

There are still several research challenges and open issues, which need to be solved in the future in order to achieve a comprehensive protection of privacy and trust establishment in AmI environments. Our conceptual solution proposes the protection of territorial privacy by disabling observers and disturbers, e.g., an active camera. However, even if this is possible in a user-controlled environment such as the home, it drastically increases the complexity of the AmI system and raises new problems. For example, a system which depends on active cameras to support a user's activity, will fail if cameras are suddenly disabled. Solutions need to be found that address this issue by allowing a dynamic trade-off between privacy and functionality. Further, in environments which are not under full control of the user, e.g., in public environments, disabling of devices might not work at all. Therefore, trustworthy privacy mechanisms need to be found that a user can rely on. The establishment of trust in those mechanisms will be a key requirement for their success.

The dynamic creation and adaptation of privacy policies and trust is another important area for future work. In ATRACO, policies and trust groups can be adapted by explicit user interactions. However, an implicit adaptation depending on several contextual information and constraints can further improve the reliability of adaptation and provide a more fine-grained access control.

As AmI environments are becoming more and more reality in our everyday lives, it is important to consider privacy and trust issues from the very beginning. If we have to trade our privacy for the benefit we get out of those environments, their successful establishment will likely fail.

7.7 Further readings

Marc Langheinrich presents a good overview of the privacy problem in ubiquitous computing and existing approaches for protecting privacy in the book *Ubiquitous Computing Fundamentals* [38].

A comprehensive discussion of privacy problems in the context of legislation and a new taxonomy of privacy is provided by Daniel J. Solove in his book *Understanding Privacy* [55].

Privacy International, the Electronic Privacy Information Center (EPIC) and the Center for Media and Communications Studies (CMCS) provide an overview of international privacy development in different states in their annual report *Privacy and Human Rights*[3].

[3] http://www.privacyinternational.org/

Used icons in graphics from http://www.picol.org and http://www.pictoico.org

References

1. Abdul-Rahman, A., Hailes, S.: Supporting trust in virtual communities. In: System Sciences, 2000. Proceedings of the 33rd Annual Hawaii International Conference on, p. 9 pp. vol.1 (2000). DOI 10.1109/HICSS.2000.926814. URL `10.1109/HICSS.2000.926814`
2. Aberer, K., Despotovic, Z.: Managing trust in a peer-2-peer information system. In: Proc. of the 10th Intl. Conference on Information and Knowledge Management, pp. 310–317. ACM, Atlanta, Georgia, USA (2001). DOI 10.1145/502585.502638. URL `http://portal.acm.org/citation.cfm?id=502638`
3. An, X., Jutla, D., Cercone, N.: A bayesian network approach to detecting privacy intrusion. In: Proceedings of the 2006 IEEE/WIC/ACM International Conference on Web Intelligence and Intelligent Agent Technology, pp. 73–76. IEEE Computer Society (2006)
4. Ashley, P., Hada, S., Karjoth, G., Powers, C., Schunter, M.: Enterprise privacy authorization language (EPAL 1.2). Tech. rep., W3C (2003). URL `http://www.w3.org/Submission/2003/SUBM-EPAL-20031110/`
5. Aztiria, A., Izaguirre, A., Augusto, J.C.: Learning patterns in ambient intelligence environments: a survey. Artificial Intelligence Review **34**(1), 35–51 (2010). DOI 10.1007/s10462-010-9160-3
6. Bellotti, V., Sellen, A.: Design for privacy in ubiquitous computing environments. In: Proceedings of the third conference on European Conference on Computer-Supported Cooperative Work, pp. 77–92. Kluwer Academic Publishers, Milan, Italy (1993)
7. Beresford, A., Stajano, F.: Location privacy in pervasive computing. Pervasive Computing, IEEE **2**(1), 46–55 (2003). DOI 10.1109/MPRV.2003.1186725
8. Bhattacharya, J., Dass, R., Kapoor, V., Gupta, S.: Utilizing network features for privacy violation detection. In: Proc. of the 1st Intl. Conference on Communication System Software and Middleware, pp. 1–10 (2006). DOI 10.1109/COMSWA.2006.1665184
9. Blaze, M., Feigenbaum, J., Lacy, J.: Decentralized trust management. Tech. rep., Certer for Discrete Mathematics \& Theoretical Computer Science (1996)
10. Blaze, M., Feigenbaum, J., Strauss, M.: Compliance checking in the PolicyMaker trust management system. In: Financial Cryptography, *Lecture Notes in Computer Science*, vol. 1465, pp. 1439–1456. Springer (1998)
11. Bohn, J., Coroama, V., Langheinrich, M., Mattern, F., Rohs, M.: Social, economic, and ethical implications of ambient intelligence and ubiquitous computing. In: Ambient Intelligence, p. 5–29. Springer, Berlin (2005)
12. Bok, S.: Secrets: On the Ethics of Concealment and Revelation. Vintage (1989)
13. Bonatti, P., Duma, C., Olmedilla, D., Shahmehri, N.: An integration of reputation-based and policy-based trust management. In: In Proc. of the Semantic Web Policy Workshop. Galway, Ireland (2005)
14. Canadian law: Personal information protection and electronic documents act (PIPEDA) (2000). URL `http://laws.justice.gc.ca/eng/P-8.6/`
15. Carbo, J., Molina, J.M., Davila, J.: Trust management through fuzzy reputation. International Journal of Cooperative Information Systems **12**(1), 135–155 (2003)
16. Chaum, D.L.: Untraceable electronic mail, return addresses, and digital pseudonyms. Communications of the ACM **24**(2), 84–90 (1981). DOI 10.1145/358549.358563
17. Chopra, K., Wallace, W.: Trust in electronic environments. In: Proc. of the 36th Annual Hawaii Intl. Conference on System Sciences, p. 10 pp. (2003). DOI 10.1109/HICSS.2003.1174902
18. Cook, D.J., Augusto, J.C., Jakkula, V.R.: Ambient intelligence: Technologies, applications, and opportunities. Pervasive and Mobile Computing **5**(4), 277–298 (2009)
19. Cranor, L., Dobbs, B., Egelman, S., Hogben, G., Humphrey, J., Langheinrich, M.: The platform for privacy preferences 1.1 (P3P1.1) specification. Tech. rep., W3C (2006). URL `http://www.w3.org/TR/2006/NOTE-P3P11-20061113/`
20. Duckham, M., Kulik, L.: A formal model of obfuscation and negotiation for location privacy. Pervasive Computing pp. 152–170 (2005)

21. Duckham, M., Kulik, L.: Location privacy and location-aware computing. Dynamic & Mobile GIS: Investigating Change in Space and Time pp. 34–51 (2006)
22. EPIC, Privacy International: Privacy and Human Rights Report 2006: An International Survey of Privacy Laws and Developments, 2006 edn. Electronic Privacy Information Center (2007)
23. EU: 95/46/EC-Data protection directive. Official Journal of the European Communities (1995)
24. Gambetta, D.: Can we trust trust? In: Trust: Making and Breaking Cooperative Relations, electronic edition edn., pp. 213–237. Blackwell Publishers, Department of Sociology, University of Oxford (2000)
25. Gerck, E.: Overview of certification systems: X. 509, PKIX, CA, PGP & SKIP. The Bell **1**(3), 8 (2000)
26. Godik, S., Anderson, A., Parducci, B., Humenn, P., Vajjhala, S.: OASIS eXtensible access control 2 markup language (XACML) 3. Tech. rep., OASIS (2002)
27. Grandison, T., Sloman, M.: A survey of trust in internet applications. IEEE Communications Surveys and Tutorials **3**(4) (2000)
28. Gruteser, M., Grunwald, D.: Anonymous usage of Location-Based services through spatial and temporal cloaking. In: Proceedings of the 1st Intl. Conference on Mobile Systems, Applications and Services, pp. 31–42. ACM, San Francisco, CA (2003)
29. Hong, J.I., Landay, J.A.: An architecture for privacy-sensitive ubiquitous computing. In: Proc. of the 2nd Intl. Conference on Mobile Systems, Applications, and Services, pp. 177–189. ACM, Boston, USA (2004)
30. Jiang, X., Hong, J., Landay, J.: Approximate information flows: Socially-Based modeling of privacy in ubiquitous computing. In: Proc. of the 4th Intl. Conference on Ubiquitous Computing, pp. 176–193. Springer (2002)
31. Jusang, A., Ismail, R., Boyd, C.: A survey of trust and reputation systems for online service provision. Decision Support Systems **43**(2), 618–644 (2007)
32. Kamvar, S.D., Schlosser, M.T., Garcia-Molina, H.: Eigenrep: Reputation management in p2p networks. In: 12th Intl. World Wide Web Conference. Budapest, Hungary (2003)
33. Könings, B., Schaub, F., Kargl, F., Weber, M.: Towards territorial privacy in smart environments. In: Intelligent Information Privacy Management Symposium (Privacy 2010). Stanford University, USA (2010)
34. Krukow, K., Nielsen, M., Sassone, V.: Trust models in ubiquitous computing. Philosophical Transactions of the Royal Society A **366**, 3781–3793 (2008)
35. Krumm, J.: A survey of computational location privacy. Personal and Ubiquitous Computing **13**(6), 391–399 (2008). DOI 10.1007/s00779-008-0212-5
36. Langheinrich, M.: Privacy by design - principles of Privacy-Aware ubiquitous systems. In: Proc. of the 3rd Intl. Conference on Ubiquitous Computing, pp. 273–291. Springer, London, UK (2001)
37. Langheinrich, M.: A privacy awareness system for ubiquitous computing environments. In: Proc. of the 4th Intl. Conference on Ubiquitous Computing, pp. 237–245. Springer, London, UK (2002)
38. Langheinrich, M.: Privacy in ubiquitous computing. In: J. Krumm (ed.) Ubiquitous Computing Fundamentals, 1 edn. Chapman & Hall/CRC (2009)
39. Li, N., Mitchell, J., Winsborough, W.: Design of a role-based trust-management framework. In: Security and Privacy, 2002. Proceedings. 2002 IEEE Symposium on, pp. 114–130 (2002). DOI 10.1109/SECPRI.2002.1004366. URL `10.1109/SECPRI.2002.1004366`
40. Lund, M.S., Solhaug, B., Stolen, K.: Evolution in relation to risk and trust management. Computer **43**(5), 49–55 (2010)
41. Marsh, S.: Formalising trust as a computational concept. Ph.D. thesis, University of Stirling (1994)
42. McKnight, D.H., Chervany, N.: Trust and distrust definitions: One bite at a time. Trust in Cyber-societies pp. 27–54 (2001)
43. Ni, Q., Bertino, E., Lobo, J., Calo, S.B.: Privacy-Aware Role-Based access control. Security & Privacy, IEEE **7**(4), 35–43 (2009). DOI 10.1109/MSP.2009.102
44. OECD: Guidelines on the Protection of Privacy and Transborder Flows of Personal Data. OECD Publishing (1980)

45. Ortmann, S., Langendörfer, P., Maaser, M.: Enhancing privacy by applying information flow modelling in pervasive systems. In: On the Move to Meaningful Internet Systems 2007: OTM 2007 Workshops, OTM'07, pp. 794–803. Springer-Verlag (2007)
46. Pavlov, E., Rosenschein, J.S., Topol, Z.: Supporting privacy in decentralized additive reputation systems. Trust Management **2995**, 108–119 (2004)
47. Perlman, R.: An overview of PKI trust models. IEEE network **13**(6), 38–43 (1999)
48. Sabater, J., Sierra, C.: Review on computational trust and reputation models. Artificial Intelligence Review **24**(1), 33–60 (2005). URL `http://www.springerlink.com/content/rw03811201223550/`
49. Schillo, M., Funk, P., Rovatsos, M.: Using trust for detecting deceitful agents in artificial societies. Applied Artificial Intelligence **14**(8), 825–848 (2000)
50. Seigneur, J., Jensen, C.D.: Trading privacy for trust. In: C. Jensen, S. Poslad, T. Dimitrakos (eds.) Trust Management, *Lecture Notes in Computer Science*, vol. 2995, pp. 93–107. Springer Berlin / Heidelberg (2004)
51. Sierra, C., Debenham, J.: An information-based model for trust. In: Proc. of the 4th Intl. joint Conference on Autonomous Agents and Multiagent Systems, pp. 497–504. ACM, New York, NY, USA (2005)
52. Smetters, D.K., Good, N.: How users use access control. In: Proceedings of the 5th Symposium on Usable Privacy and Security, pp. 1–12. ACM, Mountain View, CA (2009)
53. Smith, R.E.: Ben Franklin's web site: privacy and curiosity from Plymouth Rock to the internet. Privacy Journal (2000)
54. Solhaug, B., Elgesem, D., Stolen, K.: Why trust is not proportional to risk. In: Proc. of the 2nd Intl. Conference on Availability, Reliability and Security, pp. 11–18 (2007). DOI 10.1109/ARES.2007.161
55. Solove, D.J.: Understanding Privacy. Harvard University Press (2008)
56. Song, S., Hwang, K., Zhou, R., Kwok, Y.K.: Trusted P2P transactions with fuzzy reputation aggregation. IEEE Internet Computing **9**(6), 24–34 (2005)
57. Steinbrecher, S.: Design options for privacy-respecting reputation systems within centralised internet communities. Security and Privacy in Dynamic Environments **201**, 123–134 (2006)
58. The Oxford English Dictionary: "Privacy" Definition, 2 edn. Oxford University Press, USA (2005)
59. Warren, S., Brandeis, L.: Right to privacy. Harvard Law Review **4**, 193–220 (1890)
60. Weiser, M.: Some computer science issues in ubiquitous computing. Communications of the ACM **36**(7), 75–84 (1993). DOI 10.1145/159544.159617
61. Westin, A.F.: Privacy and Freedom. NY: Atheneum (1967)
62. Wishart, R., Henricksen, K., Indulska, J.: Context privacy and obfuscation supported by dynamic context source discovery and processing in a context management system. In: Proc. of the 4th Intl. Conference on Ubiquitous Intelligence and Computing, vol. 4611, pp. 929–940. Hong Kong, China (2007)
63. Xiong, L., Liu, L.: PeerTrust: supporting reputation-based trust for peer-to-peer electronic communities. Knowledge and Data Engineering, IEEE Transactions on **16**(7), 843–857 (2004)

Chapter 8
From scenarios to "free-play": Evaluating the user's experience of ambient technologies in the home

J. van Helvert and C. Wagner

Abstract The Ambient Intelligence (AmI) vision of interactive home environments with embedded technologies that learn from our behaviour and provide services in anticipation of our needs has been with us since the 1990's. While the technical knowledge and capability to realise the physical aspects of the vision is now within our grasp, user involvement in developing and refining the concepts underlying this new intimate relationship between humans and their technologies appears so far to have been limited. This may, in part, be due to the very nature of the research and innovation process, in that technical competence is often far ahead of the potential users ability to envisage how such services could be usefully and affordably incorporated in their everyday lives. It is understandable therefore that user involvement in the early stages of development may be seen as hindering innovation. Instead, one approach has been to capture the user perspective in visionary scenarios, often written by the researchers themselves. These are useful for elaborating the vision and driving further technical innovation, however, they often assume an established relationship between system and user and thus avoid more mundane issues such as how the user might practically incorporate AmI technologies in his or her everyday life today or in the near future. If we are to make the transition from future vision to present reality we must at some point move away from the visionary scenario and engage with users in the process of evolving our existing home environments to incorporate practical and grounded AmI solutions. This chapter looks first at the notion of user experience and options for the location of AmI evaluation studies in general. It moves on to describe some of the key features of the Adaptive and Trusted Ambient Ecologies (ATRACO) concept and prototype from the user perspective and proceeds to discuss related research. The second part of the chapter describes a preliminary evaluation of ATRACO followed by a description of the design for the participant oriented final study that draws on Dervin's Sense-Making

Joy van Helvert
University of Essex in Colchester CO4 3SQ, United Kingdom, e-mail: jvanhe@essex.ac.uk

Christian Wagner
University of Essex in Colchester CO4 3SQ, United Kingdom, e-mail: chwagn@essex.ac.uk

approach. The authors conclude that the user experience evaluation has significantly contributed to the development of the ATRACO concepts in a way that ensures they are relevant to everyday users.

8.1 Introduction

Our home environments have changed in the last 100 years. Take for example an average British Victorian semi-detached home; while its character and solidity remain unchanged, the patterns of everyday life conducted within it have been revolutionised by wave upon wave of technological innovation. If the house could talk it would tell, for example, of the advent of domestic electricity bringing light and home appliances, in turn liberating women from many domestic chores. It would tell of the transformation of its fire places as sources of heat to central pieces of indoor decoration, and the availability of hot water from the tap leading to the conversion of a bedroom into an internal bathroom and the incorporation of the scullery (separate room for washing dishes and cutlery) into the kitchen. It would tell of the arrival of television and its influence on the time and attention of household members and its effect on the fabric of the family life. It would note the appearance of the telephone, connecting friends and relatives instantly from across the neighbourhood to across the world; and towards the end of the 1990s, it would tell of the rise of the internet and wireless networks, leading to computers in children's bedrooms and the conversion of the loft space to allow its occupants to work from home. Right now the house is witnessing a revolution in personal relationships through on-line social networking and mobile connectivity, and new patterns of living resulting from the convergence of devices, for example the streaming of media from the computer to the living room TV. Given this continual process of technological transformation, the unavoidable and exciting question is: what inventions, what technology and what stories are in the pipeline for the next twenty years? How will the dynamic relationship between space, lifestyle and technology continue to develop and importantly, how can we steer technological innovation towards maximum benefit for the user?

Researchers in Ambient Intelligence (AmI) make it their aim to investigate and explore the future potential of technology in and around the home by envisaging spaces that will be sensitive and responsive to human presence and behaviour through the use of networked sensors and microchips embedded in the fabric of the built environment. Based on detailed private profiles we will carry with us, it is predicted these spaces will endeavour to know us personally, they will learn from our patterns of activity, our likes and dislikes, our aims and desires, and seek to support our needs [2, 17, 23].

Their predictions and visions stem from futuristic narrative scenarios, originating in 1990s, that were generated by technologists who could see the potential of embedded networked devices within the home. Their influence has been far-reaching, galvanising research across disciplines such as 'electrical engineering, computer science, industrial design, user interfaces and the cognitive sciences' [1].

Narrative scenarios are recognized as a valuable tool and a necessary part of the research process. Based on storytelling, they have the 'power to create in our minds an image of a world... so that we almost feel we are there... [They] bring with them a wealth of context, mostly unwritten, from our shared culture... [and provide] a frame of reference in which the reader can evaluate the story as a whole, and make sense of its individual statements... Stories provide an internal logic – of a sequence of events in time; of causality – which is valuable to engineering as it permits, indeed encourages validation' [3]. Scenarios are used extensively in design and engineering to communicate new ideas, stimulate further research, guide user evaluations and communicate with the public.

Used as a visionary tool in the field of AmI, they are also intended to align resources, innovation focus and commercial interest in order to bring the dream into reality. Therefore, it is appropriate to ask why, despite the technology being within our grasp, do the scenarios for ambient intelligence in the home remain just out of reach [2, 17, 16]?

One reason may be that the scenarios written by technical visionaries project a landscape of future daily life that ordinary people do not identify with and find difficult to evaluate. Many scenarios assume an established relationship between the system and user and avoid mundane issues such as how the user might practically begin to incorporate the technologies into his or her current daily life routines. Also, in contrast to the straightforward life enhancement offered, for example, by the transition from gas to electric lighting in the 1890's (where some home owners were so eager that they converted their houses long before the power supply was available in their area [11]), the apparent benefits of ambient technologies in the home are multiple, subtle and complex, at times addressing needs that we may have not yet fully articulated to ourselves. Thus, scenarios that paint pictures of frequently all-knowing systems, intervening in every aspect of daily life, can engender impressions of powerlessness and subjugation. In short, it seems there is a disconnection between the way potential users envisage their future home life and the technologists vision; and while potential users are not captivated by a particular technological advance, private investment, research and product development in that sector remains minimal. This, in turn, can inhibit public interest and prevent the kind of self-fuelling adoption we have seen with social networking.

Interestingly Alvin Toffler [20] identified this type of disconnection at a societal level and termed it "Future Shock". He explains the phenomenon in relation to the more widely-used term "Culture Shock":

> Culture shock is the effect that immersion in a strange culture has on the unprepared visitor... It is what happens when the familiar psychological cues that help an individual to function in a society are suddenly withdrawn and replaced by new ones that are strange or incomprehensible... Future shock is a time phenomenon, a product of the greatly accelerated rate of change in society. It arises from the superimposition of a new culture on an old one.It is culture shock in one's own society. But its impact is far worse... most people are grotesquely unprepared to cope with it.

If, as Toffler suggests, the man or woman in the street is ill prepared to cope with the new world as envisaged in the AmI scenarios, then perhaps, there is a point at which we need to abandon the predictions and visions and acknowledge users as stakeholders that are directly engaged in the co-creation and evaluation of these new highly personalised environments. After all, as Rodden and Benford [18] point out, the new smart home will evolve from our existing homes and the process will necessarily involve ordinary people. If this is the case, we need new approaches to evaluation that engage users and incorporate understanding of their existing living patterns.

The concern of this chapter is to illustrate how the design of a user-centred evaluation of the Adaptive and Trusted Ambient Ecologies (ATRACO) prototype attempts to address difficulties with scripted scenario based approaches that manage and confine user response, by adopting an iterative process of co-creation that acknowledges existing patterns of living and integrates the participant's voice alongside that of the researcher and/or technologist.

Specifically, this chapter looks first at the notion of user experience and, intrinsically related, at the options for the setting/location of AmI evaluation studies in general. It moves on to describe some of the key features of ATRACO from the user perspective in the context of user experience and proceeds to discuss related research. The second part of the chapter describes the preliminary evaluation of ATRACO and its limitations followed by a review of the resulting, modified design for the final, participant oriented study. Conclusions are drawn to close the chapter.

8.2 User Experience (UX) and study settings

Prototypes and demonstrators[1] are the logical outputs from the technical AmI research process. They provide not only the opportunity to test the technical feasibility of the concepts, designs, interoperation of components and usability of the interface, but also the actual user experience : how the user feels about the system, it's acceptability, it's ease of use and the level of comfort or satisfaction it induces [19]. User experience (UX) has become a strong theme in academic approaches to design and evaluation; it goes beyond the efficient accomplishment of a single task with a single system to consider multiple tasks and/or systems in a broader flow of interaction emphasising the users' perceptions of fulfilment [14, 19, 12, 21, 9]. It is particularly relevant in the context of AmI system evaluation where the number of possible choices presented to the individual user have increased 'to a level that no longer allows evaluation of each individual option' [1]. Add this to the high levels of personalisation and adaptability/learning capabilities of AmI systems and it is clear that traditional structured scenario or use case driven approaches that script or predetermine the user's path through the system are no longer adequate.

Law et al [13] suggest that as yet there is no consensus within the academic community on the nature and scope of UX , although 'most. . . agree that it is dynamic, context-dependent and subjective.' Hassenzahl in [9] defines experience as:

[1] Prototypes and demonstrators in terms of AmI implementations such as intelligent homes or ambient intelligent subsystems such as intelligent kitchens, intelligent offices etc.

> ... an episode, a chunk of time that one went through – with sights and sounds, feelings and thoughts, motives and actions; they are closely knitted together, stored in memory, labelled, relived and communicated to others. An experience is a story, emerging from the dialogue of a person with his or her world through action.

A "user" experience might be described similarly but the dialogue is between the person and a system or device designed to serve a particular set of user needs. The story that emerges is multidimensional, dependent not only on the usability and efficiency of the system but also a range of contextual factors such as past experience and attitudes towards technology, cultural and personal beliefs/values and even prevailing mood on the day and time of evaluation. It is difficult to reduce the story to its constituent parts without losing essential information related to their interconnectedness [9].

The importance of context in UX evaluation means the setting of the study is likely to have an impact on the outcome. With regard to AmI in the home, naturalistic household settings allow participants to align their experience with their own everyday life patterns and thus give valuable accounts of what would or would not work for them if they were living with the system on a daily basis. This could mean using the technology in their own homes or alternatively in a simulated home environment such as a "Smart Space" or "Living Lab".

Genuine in-home field studies can be costly and problematic where systems are designed to be embedded in the fabric of the building. Researcher observations in the home can also raise privacy issues and multiple instances of a system located in different participants' homes can generate additional variations in the data. Alternatively, the Living Lab or Smart Space provides a single controlled environment with embedded technologies such as networking and sensors that looks and feels like a home. Such environments are research friendly while the familiarity of the surroundings allows participants to relax and interact with the system in a relatively natural way. The two phases of prototype evaluation discussed here, take place in the setting of a Living Lab.

8.3 The User Experience within ATRACO

In order to further discuss approaches to evaluation is it necessary to provide some further detail about ATRACO and how the user might experience it. As a Future and Emerging Technology project, the focus of the research has been to develop an underlying technical framework for the development of a symbiotic relationship between the user and her/his home devices and services. It can be described as an ambient ecology consisting of people, context-aware artefacts and digital commodities (e.g. services and content) that can be grouped together in what is referred to as an Activity Sphere (AS), where each AS supports a particular type of user activity; for example, cooking, relaxing, watching TV, etc. The components within each AS are related with each other, and the purpose of a specific AS is to learn and adapt to support a user's activity in a meaningful way; from simple co-operation to 'smart', or anticipatory behaviour.

From the perspective of the user, the ambient ecology is the space surrounding and including her or him. It encompasses digital services and physical objects (TV, computer, washing machine, music centre, home security etc.) as well as devices such as sensors (temperature, light, location etc), touch screens and displays. Within a specific AS, all members of the AS are interrelated via the medium of ATRACO which interprets the conditions and activities in the space, and provides appropriate services to the user according to their needs. As a simple example, when the user wants to relax, ATRACO can pool all available relaxation options as part of "relaxation AS" (TV, music, games...), put them on standby and activate them according to the users preferences (e.g. favourite TV station, favourite relaxing music, most played game and so on). The user can initiate any of the options via an integrated interface that recognises several modalities of interaction such as physical controls (e.g. dimmers, switches), voice and touch screen commands. Also, once an option is selected, ATRACO makes appropriate changes to the environment to support the activity, such as adjusting the lighting or opening/closing the curtains. Importantly however, functionality within ATRACO is not scripted; in other words, while functionality is made available, it is the user who decides what aspects to engage with in the context of her/his current activity. For example, the lights or curtains are only adjusted as part of a specific AS if ATRACO has previously learnt the user's preference to adopt this specific state (e.g. closed curtain during relaxation).

As such, the AS is a key concept of service delivery, it groups devices and services dynamically according to the type of activity the user wants to perform. "Relax" is one example of an AS, encompassing the appliances and services described above. "Work" is another example; it could encompass a completely different set of appliances and services or it could share some of the same. A sphere can be associated with a particular space or geographical location; for example relaxation might be centred around the couch or TV area of the living space, but it can also be adapted to other locations. When a sphere is mobile, it adapts its services to the appliances/devices available in the particular space. For example, the bedroom might not be equipped with a TV but may provide a radio which can be incorporated into the 'Relax' AS when executed in the bedroom. Similarly, adaptation occurs if a device in the ecology fails: ATRACO aims to continue to support the user activity by transferring the roles adopted by a particular device to another device with compatible attributes. For example if the user is listening to a particular radio program when the Hi-Fi fails, the system will locate the nearest available device able to emit sound, such as the TV, switch it on and reroute the music to it so that the user can continue the entertainment experience.

It is clear that the underlying framework of ATRACO would potentially allow for a complex matrix of possibilities of interaction between the user, the services and the devices that populate her/his space. At present this vision is in the process of being partially realised in prototype form; some aspects will however remain conceptual. Any evaluation will be required to address the acceptability of this abstract view as well as the constituent parts embodied in the prototype. While there are a wide range of assessment approaches under the heading of UX , it is necessary to consider what is relevant to the particular ATRACO context. .

8.4 Related Work: Assessing the UX of AmI

Considering approaches that have been used to asses AmI systems, several have focused on evaluating UX in terms of quality metrics. For example Wang et al [22] consider user group experience in an AmI or Smart Museum setting and aim to control context by classifying groups according to the relationships between the individuals within them (homogeneous groups are assumed to have equal relationships, heterogeneous groups are assumed to have unequal/hierarchical relationships and loosely coupled groups are assumed to be strangers/no relationship). They propose a framework for the evaluation of group experiences using metrics such as user rating (on a scale from one to ten) and duration of user attention in relation to group classifications. They claim the framework can also be deployed in the evaluation of home settings such as the "optimisation of family TV viewing". While this is a useful approach for systems that perform with a tightly defined set of services for users in a group, it can be argued that it reduces real-world complexity to an extent that cannot adequately represent the continually evolving and multi-dimensional user-system relationship as it exists in a real world setting and as is partially incorporated by ATRACO.

In a contrasting approach, Mourouzis et al [15] combine contextual and psychological perspectives in a heuristic framework for evaluating UX in terms of the extent to which an interactive product is user-oriented. They claim their approach is suitable for diverse user groups such as those with different cultural backgrounds or with different levels of physical ability. Their framework focuses on three groups of metrics:

1. The characteristics of the user (gender, physical and cognitive abilities, language, culture etc.).
2. The context of system use (tasks, social and environmental conditions).
3. The user's behavioural situation, perceived usefulness and perceived ease of use).

In addition, it incorporates the notion of progressive levels of usage from discovering the product, through to exploring it at a high level to using it in depth for specialised purposes. Although comprehensive in its recognition of many of salient aspects of user experience , including a focus on both the internal state of the participant and the context of use, the researchers' choice of metrics/questions still limits and controls the participant's response, i.e. the full scope of the participant's perspective is lost. However, the concept of exploring how the user progresses or becomes familiar with the system is highly relevant to the evaluation of AmI environments and in particular to the learning and anticipatory elements of ATRACO. It shifts the focus of inquiry from the participant's state after completing the experience to what happens in the process of moving through it.

Hole and Williams [10] are also concerned with what happens throughout the process of user-system interaction. They use an Emotion Sampling Device (ESD) which can be accessed by mobile phone or PDA to gather the participants' emotional responses each time they hit a positive or negative event during the period of interaction with the system. Participants register their feelings by answering a set of questions which are then analysed to identify the emotions experienced. According to the authors, Emotion Sampling aims to provide insight into the 'hidden reasons for users' responses'; their 'love/hate/tolerate' moments which the authors claim can provide useful indications for improving specific aspects of a product or service. This approach has a significant degree of participant orientation in that there are no restrictions on the way the system is used and the participant is empowered with identifying her/his own moments of emotional upheaval or salience. However, in the context of the aims of this chapter, it can be argued that the approach is limited as it is the researchers who define and categorise the actual emotion, not the participant. Additionally, the approach does not consider user characteristics or context of use. It would therefore be used most appropriately in conjunction with other techniques.

Ethnography is a research strategy used predominantly in Sociology and Anthropology that has more recently been applied to aspects of user inquiry in both academic and commercial technology research. It is concerned with a holistic understanding of the way groups of people live; their material, cultural and spiritual practices. Koskela and Vaananen-Vainio-Mattila [12] use an ethnographic approach to design and evaluate three prototype interfaces (for laptop, remote control and mobile phone) to provide access to a range of standard smart devices (automated curtains, status aware plant pots etc.). The process starts with a highly user-oriented requirements elicitation phase involving contextual inquiry, home interviews that identify people's living patterns and focus groups within which participants are allowed to choose options for the development of the User Interfaces (UIs). Outcomes from the focus groups are also used to shape the evaluation study. The smart devices with their prototype UIs are installed in an ordinary apartment. Following this, two participants, selected for their neutral approach to new technology, are asked to live in the space and use the interfaces as part of their normal patterns of living. The authors study the participants' device usage over a six month period by monitoring their interactions with the UIs and conducting contextual interviews as well as participatory walkthroughs. The study shows that incremental smart additions to existing home devices are welcomed by the participants; an outcome supporting the argument that progress towards the AmI home will be evolutionary. It also reveals additional requirements particularly with respect to tailoring UIs to different types of activity patterns. Overall, this study demonstrates a high degree of user-orientation, including elements of co-creation and the careful preparation of the prototypes to fit into everyday life patterns, leading to a focussed and productive evaluation. However, the number of participants was small and the six months duration of the experiments, while highly valuable, would be difficult to reproduce as a general approach as significant resources are required to sustain long term studies. Nevertheless, we feel that the inquiry into existing living patterns is of particular relevance to ATRACO as the concept of Activity Spheres is intended to support natural behaviour in the home.

These examples illustrate some of the diversity of approaches to UX evaluation. Of course, each study depends on numerous local conditions such as the stated aims of the study (more requirements – usability scores), the time and resources available, whether it is a research exercise or refining of a near market product, etc. While none of the approaches detailed above are wholly suitable to the evaluation of ATRACO, they provide insights to help shape a user-oriented study within the ATRACO context. Specifically, the inclusion of user background/characteristics, existing living pattern inquiry, a focus on what happens throughout the experience as opposed to end point satisfaction, and empowering the user to identify her/his own moments of emotional upheaval or salience were identified as highly relevant.

8.5 Own approach to user evaluation

ATRACO is an EU funded project focussed predominantly on developing the underlying standards and frameworks to deliver an intelligent home environment based on the concept of the ASs (as described above). The components of the prototype are being developed in the various European partner institutions and are subsequently integrated and installed in the University of Essex's Living Lab, referred to as the "iSpace". From the outset of the project, prototype development and evaluation has been considered as an iterative and bi-directional process. At the time of writing, a preliminary prototype with limited functionality has been developed, deployed and evaluated and the outcomes are being used to shape both the final prototype and the design of the corresponding evaluation study which is scheduled to take place in early 2011. In this part of the chapter we will describe the iSpace evaluation environment, the context of the ATRACO evaluation, and proceed to give details of the preliminary evaluation and its outcomes. Following on from this, we will present the outline of the final study.

8.5.1 The iSpace

Simulated living spaces or "Living Labs" are becoming more commonplace around the globe and the research carried out within them ranges from the longer term visionary experimentation to the refining of commercial products prior to launch [7]. Fowler et al define a Living Lab as

> ... an environment that is designed to support innovation through co-creation and evaluation of products and services being used in realistic but familiar contexts.

The University of Essex iSpace (see Figs. 8.1 and 8.2) is a simulated home environment comprising a two bedroom apartment with specially designed walls and ceilings able to conceal the extensive networking infrastructure. There are numerous sensors and networked artefacts throughout, including location tracking systems and pressure sensitive furniture. The core space is an open-plan kitchen, sitting and dining area which generates a relaxed modern home ambience. It includes appliances such as two large plasma TV's, a music system, light sensors, remotely operated curtains and a networked picture frame. The main bedroom is also equipped with a touch screen enabled flat screen, and the smaller bedroom doubles as an office containing a desk and computer alongside a fold-away bed.

Fig. 8.1 The iSpace at the University of Essex, view of the main living room area.

Importantly, the iSpace provides ATRACO with an appropriate test-bed for the integration of the prototypes and their subsequent user evaluations. It enables participants to interact with and experience ATRACO in a relaxed and natural setting that they can begin to imagine being their own home. This in turn helps enable and stimulate co-creative interview responses and focuses group discussions to provide a rich data resource for user experience evaluation.

Fig. 8.2 The iSpace at the University of Essex, view of the open plan kitchen area.

8.5.2 The ATRACO evaluation context

As each partner institution in the ATRACO project is responsible for developing a different component of the prototype system, it was important to ensure the broad design of both rounds of evaluation were produced collaboratively involving all the relevant stakeholders. At the beginning of each iteration, the components had not yet been integrated and given that ATRACO offers multiple possibilities for supporting daily living, there was a certain amount of flexibility in the nature and scope of the prototype design. This meant that the development of both prototypes and the activities required to evaluate them were/are interdependent to a certain degree, with the broad design for each emerging through informal rounds of collaboration between the partners.

In the lifecycle of a commercial product, user evaluation would generally be conducted once the core technology was robust and the user interface reasonably polished, allowing participants to get a feel for a device or service that may be incremental to technologies they are already familiar with and could envisage owning or subscribing to. In contrast, as a research project, ATRACO's aims are longer term and more focused on the exploration of visionary concepts, i.e. less focussed on an immediate product. This presents a challenge for evaluation design as participants with limited technical expertise might not be familiar with the concepts or technologies deployed and could potentially find it difficult to imagine how they would be relevant to their lives. Furthermore, terms of reference for the project meant that prototype development was focussed primarily on the underlying technical and conceptual frameworks allowing interoperability between devices – the user interface was only a secondary concern. Participants would therefore experience ATRACO through an unrefined front end, again making it potentially more difficult to envisage its benefits. Finally, for various reasons and in particular time constraints, a longitudinal study involving participants living in the iSpace was/is not an option for either the preliminary or final evaluation. This meant that it was necessary to find alternative approaches to evaluate the learning and user experience aspects of ATRACO.

8.5.3 Preliminary Evaluation

As a future and emerging technologies research project with no pre-existing user base, the user requirements for ATRACO were driven by a number of visionary narrative scenarios compiled by members of the project during its initial phase. In the first round of evaluation, these scenarios were simplified into five vignettes (short interaction scenes) illustrating a range of possibilities for the deployment of ATRACO in the home. Due to its early stage of development, the prototype was only able to support the specific interactions defined in the scenarios. 8.3 shows the layout of the iSpace and the devices available to ATRACO.

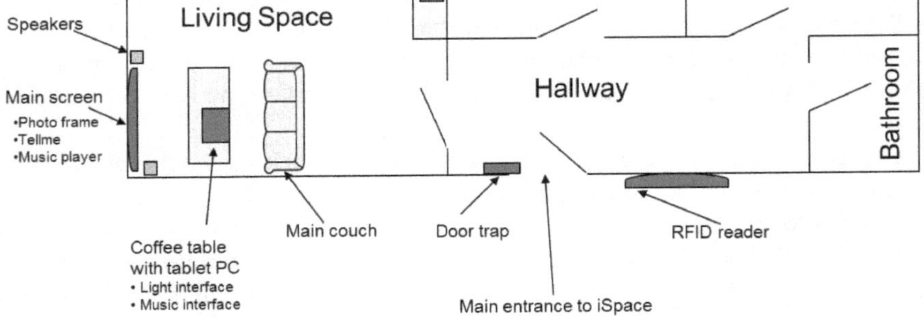

Fig. 8.3 The iSpace layout for the preliminary study.

Each vignette prescribed either a single interaction or a flow of interactions between the participant and the system. In this first stage of the evaluation, individual participants were introduced to the iSpace and asked to complete the five vignettes imagining that the iSpace was their home. A short interview was conducted after each vignette in which the participant was asked questions about the appropriateness of the system response in relation to specific interactions within the vignette. This was followed by some broader questions about their overall experience. The approach attempted to build a representative image of what the participant experienced throughout each vignette as well as their overall satisfaction. The specific vignettes investigated were the following:

- **Arriving home after work:** The participant starts outside the iSpace. She/he holds an RFID tag to the reader, the door unlocks and the participant enters the hallway and then walks into the living space. The voice interaction system says 'please adjust the lights' as the participant walks towards the main couch. The lights are off (there is sufficient light from the hallway to see but the light level is not pleasant for everyday use). The participant sits on the main couch and adjusts the lights using a touch screen interface on a tablet PC located on the coffee table. The system asks 'would you like to view photos?' If the participant replies with a 'yes', a rolling photo display appears on the main screen. The system proceeds to ask 'would you like some music?' On a positive response, the music starts to play. Vignette ends.
- **Guest and privacy:** The participant sits on the sofa looking at holiday photographs on the TV. The tablet PC on the coffee table displays a touch screen interface allowing the participant to control the photo viewing using 'next' and 'previous' options. Some of the photographs have been marked as 'public', i.e. anyone can see them, others are 'private' and as such marked only for the 'consumption' of the owner. The researcher is outside the iSpace (playing the part of an arriving guest). After a couple of minutes the researcher places her/his RFID tag against the reader outside the front door. The system announces to the participant that 'X has arrived' (where X is the researchers name) and asks 'would you like her/him to enter?' If the participant responds by saying 'yes', the door opens and the researcher enters the iSpace – as she/he does so, the picture set is modified to display only 'public' pictures; the private pictures are removed from the slideshow. Vignette ends.
- **Photo frame adaptation:** The participant sits on the sofa viewing holiday photos displayed in the photo frame on the main screen. The tablet PC on the coffee table displays a touch screen interface allowing the participant to control the photo viewing using 'next' and 'previous' options. The researcher is in the second bedroom/office of the iSpace. After a few minutes the researcher presses a button that simulates a failure of the photo frame device on the main screen. The system autonomously adapts to the failure by moving the photo display to the side screen. Vignette ends.
- **Follow me:** The touch screen music interface is displayed on the side screen. Holding a tag, the participant walks from the doorway to the side screen and starts playing some music by touching the appropriate symbol(s) on the user interface.

She/he then move across the room to sit on the sofa in front of the main screen. After a few seconds the interface autonomously moves from the side screen to the tablet PC on the coffee table in front of the participant. Vignette ends.

- **Light adaptation:** The lights are on (to the level specified by the user during the first vignette) and the lighting control interface is visible on the tablet PC on the coffee table. The participant has been asked to readjust the lights if there is any change in light level (requirement). The participant sits on the sofa with a magazine. She/he adjusts the lights so they are comfortable for reading. The researcher is in the control room/office and after a few minutes presses a button to simulate a partial failure of the lights (i.e. some bulbs fail). The participant re-adjusts the lights to a comfortable level and then continues reading. Vignette ends.

Nine participants took part (in all five vignettes) and in order to provide at least a minimal insight into their background, the qualitative interview data was supplemented by a short questionnaire. It gathered demographic data, such as age, gender, employment and accommodation status, and asked participants to gauge their level of enthusiasm for new technologies by rating themselves out of ten in three areas: personal technologies, work technologies and home technologies. The questionnaire also asked them to tick boxes indicating which technologies they currently owned and which they were aiming to acquire in the next 12 months.

In the second stage of the preliminary evaluation, six of the same participants attended a focus group together in which they were presented with a futuristic scenario based on potential ATRACO capabilities. They discussed various aspects of the scenario in relation to their own visions of future home living.

The interview and focus group data were analysed in terms of themes, resonances and recurrences with the context provided by the questionnaire and demographic data. Standard ethical practice in relation to human research participants was followed throughout.

8.5.4 Preliminary evaluation outcomes and reflections

The initial study found that the majority of participants were positive about the notion of living with some aspects of AmI in the home, although two participants rejected it altogether. Key concerns were:

- Maintaining control over the environment.
- The changeable nature of human moods in relation to an electronic presence.
- The erosion of everyday life skills and domestic cultural norms and rituals.
- The risk of data loss and unauthorized access to the system.

Overall, the study produced valuable data to help shape the final prototype and evaluation approach, including highlighting the importance of ASs and communicating how they are formed and operate as a central concept of ATRACO.

However, reflecting on the study design it was clear that with only minimal and in particular rather scripted exposure to the prototype, participants found it difficult to grasp some of the concepts. It also became clear that embodied in the design of the scenarios and vignettes were many assumptions about the user's acceptance of the vision. The constrained and scripted path of interactions meant that participants found it hard to relate their experience to their own patterns of living, particularly with respect to the learning and adaptation aspects of ATRACO. Also, a lack of any illustration of how they might establish and personalise such a system for their own home gave rise to feelings of being controlled by the technology.

8.6 Design of the final evaluation study

Building on the findings of the preliminary study, a key feature of the final prototype was identified to be that it should support "free-play"; in other words, the unconstrained use of the system within a particular AS. This means the evaluation design can move away from scenarios and vignettes, allowing participants to explore freely the full range of functionality within the sphere according to their own instinctive patterns of thought and action. It is hoped that this will enable them to more closely relate the functionality of ATRACO with their current patterns of activity within the home and therefore overcome feelings of alienation. The prototype will also include a user interface that will allow individual participants to set up and personalise the system according to their own requirements, including for example, their personal choice of music and photos or choice of male or female speech interface etc. Again, it is hoped that this will promote a feeling of connection and ownership with the participants.

The format of the final study will be AS focussed and iterative, with participants returning for up to four separate sessions to 'play' with different ASs or to experience the same AS with multiple users. This is intended to support participants in moving from a superficial to a deeper understanding of ATRACO (and its underlying concepts), and in particular, to allow them over time to experience the 'smart' or learning and anticipatory aspects of the system. Demographic details and metrics relating to participant's disposition towards technology will be collected in the same way as in the earlier study.

8.6.1 Approach to data collection

The move away from the structured 'walkthrough' approach taken in the preliminary evaluation study and allowing the participant uninterrupted 'free-play' with the prototype raises the question of how to gain insight into what moments of understanding, struggle, anxiety, pleasure etc. emerge throughout this much more interactive experience. In this context, we refer to Dervin's Sense-Making [6] which is a methodology drawn from the field of communications and grounded in the paradigm of phenomenology. It provides a framework for investigation based on moments when the making of 'sense' is interrupted, made or remade in communication. It departs from other approaches by theorising humans as being able to reflect on their own experiences in a structured way that emphasises the voice of the participant alongside that of the researcher.

In developing the ontological and epistemological assumptions of Sense-Making, Dervin draws primarily on the work of Richard F. Carter [4, 5] and his assumptions about the discontinuity of reality, 'as well as ideas suggested by Giddens [8]'. In this framework, human experience is seen as being pervaded by 'gaps'. These exist across time, between entities (human, system or otherwise) and between spaces, as Dervin [6] explains:

> This discontinuity condition exists between reality and human sensors, between human sensors and the mind, between mind and tongue, between tongue and message created, between message created and channel [mode of communication], between human at time one and human at time two, between human one at time one and human two at time one, between human and culture, between human and institution, between institution and institution between nation and nation and so on. Discontinuity is an assumed constant of nature generally and the human condition specifically.

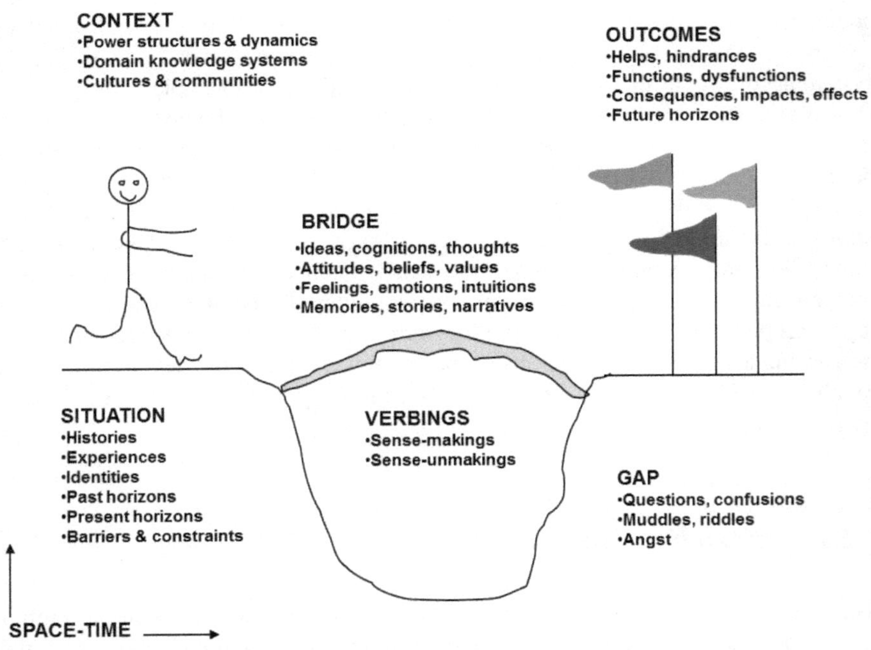

Fig. 8.4 Sense-Making Metaphor (Reproduced with permission from Dervin et al [6]).

As depicted in Fig. 8.4, the human subject in Sense-Making is seen as moving through time-space, possessing an innate need to 'bridge' any gaps that are encountered (here, 'bridging' is the act of communicating, either through internal dialogue or external interaction). The subject is also seen as being 'situated in cultural/historical moments in time-space and that culture, history, and institutions define much of the world within which... [she/he] lives' [6]. At the same time, the actor is assumed to construct her/his own personal sense of her/his relationship to such phenomena drawing on and interpreting her/his existing knowledge. Gaps then, are not rigidly or objectively defined, they are essentially personal moments of struggle, angst or uncertainty; moments where sense cannot immediately be made; moments of reaching out for clues from past experience, from current context or from future expectations, dreams or aspirations.

From a Sense-Making perspective, gathering differing accounts of gaps faced and gaps bridged in relation to the same phenomenon may reveal insight, not into the phenomenon itself, but into the processes, patterns, themes and recursivities relating to the 'experience' of interaction with the phenomena.

Sense-Making, which we deem highly relevant to the UX evaluation requirements of ATRACO, is implemented through a specific approach to interview which as such, in the context of ATRACO we propose to conduct as follows:

- Prior to each interaction session 'training' or talking to the participant about being conscious/mindful of her/his own 'gaps' as they move through the experience.
- The free-play session then takes place and is recorded on video.
- The researcher and the participant then replay the video together. The participant is asked to provide a commentary explaining her/his actions and to stop the video when a gap moment is reached. At each gap moment, the researcher encourages the participant to elaborate in detail, describing the nature of the gap and connecting it to her/his past experience (e.g. existing patterns of living and interaction with personal technologies), future aspirations, attitudes, values etc. The participants are also asked to suggest ideal solutions or preferences in relation to aspects of the system that they do not feel comfortable with.
- The interview concludes with a set of more general questions about the AS concept and the overall experience.

It is envisaged that this approach will result in rich data that include detailed descriptions of the participants' responses as they interact with the system allied to contextual understanding. Subsequent analysis should provide insight into the practical, socially situated usage of each AS, its acceptability and a range of potential user needs that can help inform further refinement of the system and/or its concepts.

Possible drawbacks to the approach may include difficulty in getting participants to think and talk in terms of gap moments. This depends largely on the prior effort of the researcher to communicate the concept effectively. Similarly, the success of the interview is dependent on the skill of the interviewer in allowing participants to elaborate on their personal context while encouraging them to remaining focussed on the task of evaluation.

8.7 The ATRACO contribution

As part of ATRACO, the user experience evaluation has been an essential tool, not simply as a posterior measure of the quality of the designed system and proposed concepts but more importantly, as an active part of the iterative process of developing and shaping the concepts and subsequent prototypes which were implemented to expose, visualize and evaluate ATRACO. As such, while the initially designed scenarios gave structure to the early stages of the project and particular its implementation, the gradual move towards a more 'free' and responsive environment where the user is free to explore her/his augmented environment as part of specific ASs directly guided the implementation and design of the ATRACO components and subsequently their integration towards the final prototype. At the time of writing the integration of this final prototype is in its final stages and it is expected to provide a very rich environment to evaluate the ATRACO concepts, the underlying technologies as well as the user experience .
Finally, it should be noted that the direct involvement of real (lay) users as part of the user evaluation has given the ATRACO project the unique opportunity to ensure the development of concepts and components which are directly useful and acceptable to end-users, a criterion which should facilitate future real life adoption of some of the concepts explored.

8.8 Conclusions

Emerging technologies such as AmI are challenging traditional requirements elicitation and evaluation practices. For example, as discussed earlier, the complexity and personalisation of many AmI systems make it impossible to evaluate every possibility (of interaction, system state,. . .) with the user. This has shifted the emphasis from "optimum performance" as an objective measure, towards the more subjective notion of user experience or user satisfaction. In addition, the advent of Smart Spaces or Living Labs provides a new setting, combing aspects of the controlled clinical environment of the laboratory and the naturalistic but costly and unpredictable field trial.

The development of the ATRACO prototypes and their installation in the iSpace at the University of Essex, UK, has created a valuable opportunity to explore a new user-centred, cross disciplinary approach to evaluation in a simulated home environment. As part of ATRACO, the user experience evaluation has significantly contributed towards guiding the development of the ATRACO concepts (e.g. ASs and their application) and prototypes. It has allowed the project to not only provide advancements in technological, conceptual and visionary terms, but also to validate them by relying on real users, thus ensuring that the specific advancements are relevant to everyday users. This approach in turn is expected to become common practice in AmI research and drive the increasingly prominent real-world adoption of concepts and technologies developed in a research context.

Most importantly, the user experience evaluation studies conducted within ATRACO have highlighted the shortcomings of existing post-implementation, scenario based user evaluations and have clarified the need for a significantly more interactive approach to user requirements elicitations, design and implementation as well as user experience evaluation. As an initial means of addressing this, the final evaluation as part of ATRACO will be based around the use of Dervin's Sense-Making, an approach borrowed from field of Communications which is, adapted to UX , a completely novel approach as far as the authors are aware. In the near future, the final evaluation will be conducted and the authors will present the details in a forthcoming journal article.

In summary, while visionary scenarios have their place in inspiring innovation and rallying resources, it appears the time is now right to look in the other direction and engage users, such as the occupiers of our Victorian semi-detached home, in a process of identifying the next step towards the AmI home. As Aarts [1] suggests "starting from the other side" is quite difficult because it is hard to obtain validated end-user insights that reveal unmet needs leading to successful introduction of new solutions. Therefore we need more insights into the nature of human behaviour.' It is hoped the use of a Sense-Making approach to evaluate ATRACO will contribute to our understanding of how to achieve this.

Finally, compared to the 1890s and the introduction of domestic electricity, the incredible social connectivity of our age may mean that once that steps are identified and the benefits articulated, people could very quickly move towards adoption – and then the house will have new stories to tell.

8.9 Further readings

The reader might be interested in Brenda Dervin and Lois Foreman-Wernet with Eric Lauterbach (Eds.) (2003) Sense-Making Methodology Reader: Selected writings of Brenda Dervin. Creskill NJ: Hampton Press, and Marc Hassenzahl (2010) Experience Design: Technology for All the Right Reasons. San Rafael CA: Morgan and Claypool. Furthermore Malcolm McCullough (2004) Digital Ground: Architecture, Pervasive Computing, and Environmental Knowing. Cambridge MA: MIT Press and Sharlene Nagy Hesse-Biber (in press 2011) The Handbook of Emergent Technologies in Social Research. New York: Oxford University Press provide more in-depth information.

References

1. Aarts, E.: Ambient intelligence: Basic elements and insights. Information Technology **50**, 7–12 (2008)
2. Aarts, E., Marzano, S. (eds.): Cultural Issues in Ambient Intelligence. 101 Publishers, Rotterdam (2003)

3. Alexander, I., Maiden, N. (eds.): Introduction: Scenarios in systems development. John Wiley, Chichester UK (2004)
4. Carter, R.: Discontinuity and communication. In: Proc. Conference on Communication Theory East and West. East-West Centre San Francisco (1980)
5. Carter, R.: What does gap imply? In: Proc. Annual Conference of the International Communication Association. San Francisco (1989)
6. Dervin, B., Foreman-Wernet, L., Lauterbach, E.: Sense-making methodology reader: Selected writings of Brenda Dervin. Hampton Press (2003)
7. Fowler, C., O'Neil, L., van Helvert, J.: Living Laboratories: Social research applications and evaluation. Oxford University Press, New York (2011)
8. Giddens, A.: The Constitution of Society: Outline of the Theory of Structuration. Polity Press, Cambridge UK (1984)
9. Hassenzahl, M.: Experience Design: Technology for all the right reasons. Morgan and Claypool, New York (2010)
10. Hole, L., Williams, O.: Gaining insight into the user experience. Internet draft (2007). URL www.olliewilliams.co.uk/research/paper-2.pdf
11. Institution of Engineering and Technology: Lighting the home. On-line exhibition (2010). URL http://www.theiet.org/about/libarc/archives/exhibition/domestic/lighting.cfm
12. Koskela, T., Vaananen-Vainio-Mattila, K.: Evolution towards smart home environments: empirical evaluation of three user interfaces. Ubiquitous Computing **8**, 234–240 (2004)
13. Law, E.L.C., Roto, V., Hassenzahl, M., Veerman, A., Kort, J.: Understanding, scoping and defining user experience: A survey approach. In: Proc. CHI 2009, pp. 719–728. Boston USA (2009)
14. McCullough, M.: Digital Ground: Architecture, Pervasive Computing and Environmental Knowing. MIT Press, Cambridge MA (2004)
15. Mourouzis, A., Antona, M., Boutsakis, E., Stephanidis, C.: A user-orientation evaluation framework: Assessing accessibility throughout the user experience lifecycle. In: Proc. International Conference on Computers Helping People with Special Needs, pp. 412–428. Linz Austria (2006)
16. Mukherjee, S., Aarts, E., Doyle, T.: Special issue on ambient intelligence. Information System Frontiers **11**, 1–5 (2009)
17. Riva, G.: The Psychology of Ambient Intelligence. IOS Press, Amsterdam (2005)
18. Rodden, T., Benford, S.: The evolution of buildings and the implications for the design of ubiquitous domestic environments. In: Proc. SIGHCI 2003 Conference on Human Factors in Computing Systems, pp. 9–16. Fort Lauderdale Florida (2003)
19. Rogers, Y., Preece, J.: Interaction Design: Beyond Human-Computer Interaction. John Wiley, Chichester UK (2007)
20. Toffler, A.: Future Shock. Bantam Books, New York (1970)
21. Vaananen-Vainio-Mattila, K., Roto, V., Hassenzahl, M.: Towards practical user experience evaluation methods. In: Proc. 5th COST294-MAUSE Open Workshop on Meaningful Measures: Valid Useful user Experience Measurement (VUUM 2008), pp. 9–16. Reykjavik (2008)
22. Wang, Z., Zhou, X., Yu, Z., Wang, H., Ni, H.: Quantitative evaluation of group user experience in smart spaces. Cybernetics and Systems: An International Journal **41**, 105–122 (2010)
23. Wright, S., Steventon, A.: Intelligent Spaces: The vision, the opportunities and the barriers. Springer Verlag, London (2006)

Appendix A
List of Reviewers

Juan Carlos Augusto
University of Ulster, Jordanstown, United Kingdom

Abdelhamid Bouchachia
Alpen-Adria-Universität, Klagenfurt, Austria

Amedeo Cesta
Italian National Research Council, Rome, Italy

Hakan Duman
BT Research and Technology, Ipswich, United Kingdom

Jérôme Euzenat
Institut National de Recherche en Informatique et en Automatique & Laboratoire d'informatique de Grenoble, Grenoble, France

Damianos Gavalas
University of the Aegean, Mytilene, Greece

David K. Hunter
University of Essex, Colchester, United Kingdom

Christophe Jacquet
Supélec, Gif-sur-Yvette, France

Frank Kargl
University of Twente, Enschede, The Netherlands

Christophe Kolski
Université de Valenciennes et du Hainaut-Cambrésis, Valenciennes, France

Andreas Komninos
Glasgow Caledonian University, Glasgow, United Kingdom

Dimitrios Koutsomitropoulos
University of Patras, Patras, Greece

Effie Lai-Chong Law
University of Leicester, Leicester, United Kingdom

Antonio López
University of Oviedo, Gijón, Spain

Elias Manolakos
University of Athens, Athens, Greece

Alessandro Saffiotti
University of Örebro, Örebro, Sweden

Michel Sall
Trialog, Paris, France

Florian Schaub
Ulm University, Ulm, Germany

Vera Stavroulaki
University of Piraeus, Piraeus, Greece

Eran Toch
Carnegie Mellon University, Pittsburgh, USA

Lorenzino Vaccari
European Commission – Joint Research Center, Ispra, Italy

Index

Kolibri 360 GmbH
Ferdinand-Jühlke-Straße 7
99095 Erfurt, Deutschland
produktsicherheit@kolibri360.de

Next Generation Intelligent Environments